PRION BIOLOGY

Research and Advances

PRION BIOLOGY

Research and Advances

Edited by

Vincent Béringue, PhD

Apple Academic Press

TORONTO NEW JERSEY

Apple Academic Press Inc.
3333 Mistwell Crescent
Oakville, ON L6L 0A2
Canada

Apple Academic Press Inc.
9 Spinnaker Way
Waretown, NJ 08758
USA

Printed in the United States of America on acid-free paper

International Standard Book Number-13: 978-1-926895-37-6 (Hardcover)

Library of Congress Control Number: 2012951945

Library and Archives Canada Cataloguing in Publication

Prion biology: research and advances/edited by Vincent Béringue.

Includes bibliographical references and index.
ISBN 978-1-926895-37-6
1. Prions. I. Béringue, Vincent

QR502.P75 2013   579.2'9   C2012-906393-2

About the Editor

Vincent Béringue, PhD

Dr. Vincent Béringue has been working in the prion field for almost twenty years. He obtained his PhD in 1998 from the AgroParisTech Institute in Paris, in the laboratory of late Dr. Dominique Dormont. After an MRC-funded postdoctoral position in the laboratory of Professor John Collinge and Dr. Simon Hawke at Imperial College School of Medicine in London, he joined Dr. Hubert Laude's laboratory at INRA (National Institute for Agricultural Research) in Jouy-en-Josas in 2001, as permanent staff scientist. He is now head of the laboratory in the Molecular Virology Immunology Department at INRA. His primary research interests include the biochemistry, diversity and evolution of animal and human prions.

Contents

List of Contributors

Paul Ajuh
Dundee Cell Products Ltd, James Lindsay Place, Dundee Technopole Dundee, UK

Irina Alexeeva
Medical Research Service, Veterans Affairs Maryland Health Care System, Baltimore, Maryland, USA

Jose M. Arteagoitia
Department of Health and Consumption, Gobierno Vasco, San Sebastian-Donostia Kalea 1, 01010 Vitoria-Gasteiz, Alava, Spain

Begoña Atares
Pathology Service, Hospital de Txagorritxu, José Achótegui s/n, 01009 Vitoria-Gasteiz, Alava, Spain

Sandrine Balzergue
Recherche en Génomique Végétale, UEVE, F-91057 Evry Cedex, France

Ilia V. Baskakov
Center for Biomedical Engineering and Technology, University of Maryland, Baltimore, Maryland, USA

Sylvie L. Benestad
Norwegian Veterinary Institute, P.O. Box 750 Sentrum, 0106 Oslo, Norway

Vincent Beringue
Virologie Immunologie Moléculaires, F-78350 Jouy-en-Josas, France

Miren J. Bilbao
Neurology Service, Hospital de Mendaro, Mendarozabal s/n, 20850 Mendaro, Guipúzcoa, Spain

Eva Birkmann
Institut für Physikalische Biologie, Heinrich-Heine-Universität Düsseldorf, Düsseldorf, Germany

Frederique Bitton
Recherche en Génomique Végétale, UEVE, F-91057 Evry Cedex, France

Alex Bossers
Department of Bacteriology and TSEs, Central Veterinary Institute (CVI) of Wageningen UR, Lelystad, 8200 AB, The Netherlands

Bjørn Bratberg
Norwegian Veterinary Institute, P.O. Box 750 Sentrum, 0106 Oslo, Norway

Bertram Brenig
Institute of Veterinary Medicine, Faculty for Agricultural Sciences, Georg-August University, Burckhardweg 2, 37075 Goettingen, Germany

Sead Chadi
Génétique Animale et Biologie Intégrative, F-78350, Jouy-en-Josas, France

Stefanie Czub
Canadian and OIE Reference Laboratories for BSE, Canadian Food Inspection Agency Lethbridge Laboratory, Lethbridge, Alberta, Canada

Isidro Ferrer
Institut de Neuropatologia, Servei Anatomia Patològica, IDIBELL-Hospital Universitari de Bellvitge, Carrer Feixa Llarga s/n, 08907 Hospitalet de Llobregat, Barcelona, Spain

Vinicius M. Gadotti
Department of Physiology and Pharmacology, Hotchkiss Brain Institute, University of Calgary, Calgary, Canada

Pierluigi Gambetti
Department of Pathology, Case Western Reserve University, Cleveland, Ohio, USA

Carol Man Gao
Ventana Medical Systems, Tucson, Arizona, USA

Joseba M. Garrido
Department of Animal Health, Neiker-Tecnalia, Berreaga 1, 48160 Derio, Bizkaia, Spain

Nuria Gonzalez-Montalban
Center for Biomedical Engineering and Technology, University of Maryland, Baltimore, Maryland, USA

Yoshifumi Iwamaru
Prion Disease Research Center, National Institute of Animal Health, 3-1-5 Kannondai, Tsukuba, Ibaraki 305-0856, Japan

Ramon A. Juste
Department of Animal Health, Neiker-Tecnalia, Berreaga 1, 48160 Derio, Bizkaia, Spain

Kazuo Kasai
Prion Disease Research Center, National Institute of Animal Health, 3-1-5 Kannondai, Tsukuba, Ibaraki 305-0856, Japan

Jae-Il Kim
Department of Physiology and Biophysics, Case Western Reserve University, Cleveland, Ohio, USA

Jan P. M. Langeveld
Department of Bacteriology and TSEs, Central Veterinary Institute (CVI) of Wageningen UR, Lelystad, 8200 AB, The Netherlands

Hubert Laude
Virologie Immunologie Moléculaires, F-78350 Jouy-en-Josas, France

Sandrine Le Guillou
Génétique Animale et Biologie Intégrative, F-78350, Jouy-en-Josas, France

Fabienne Le Provost
Génétique Animale et Biologie Intégrative, F-78350, Jouy-en-Josas, France

Lars Luers
Institut für Physikalische Biologie, Heinrich-Heine-Universität Düsseldorf, Düsseldorf, Germany

Natallia Makarava
Center for Biomedical Engineering and Technology, University of Maryland, Baltimore, Maryland, USA

Marie-Laure Martin-Magniette
Recherche en Génomique Végétale, UEVE, F-91057 Evry Cedex, France

Kentaro Masujin
Prion Disease Research Center, National Institute of Animal Health, 3-1-5 Kannondai, Tsukuba, Ibaraki 305-0856, Japan

Yuichi Matsuura
Prion Disease Research Center, National Institute of Animal Health, 3-1-5 Kannondai, Tsukuba, Ibaraki 305-0856, Japan

Shirou Mohri
Prion Disease Research Center, National Institute of Animal Health, 3-1-5 Kannondai, Tsukuba, Ibaraki 305-0856, Japan

Luitgard Nagel-Steger
Institut für Physikalische Biologie, Heinrich-Heine-Universität Düsseldorf, Düsseldorf, Germany

Hiroyuki Okada
Prion Disease Research Center, National Institute of Animal Health, 3-1-5 Kannondai, Tsukuba, Ibaraki 305-0856, Japan

Valeriy G. Ostapchenko
Center for Biomedical Engineering and Technology, University of Maryland, Baltimore, Maryland, USA

Marian M. de Pancorbo
Department of Zoology and Animal Cellular Biology, Paseo Universidad 7, Universidad del País Vasco, 01006 Vitoria-Gasteiz, Alava, Spain

Giannantonio Panza
Institut für Physikalische Biologie, Heinrich-Heine-Universität Düsseldorf, Düsseldorf, Germany

Bruno Passet
Génétique Animale et Biologie Intégrative, F-78350, Jouy-en-Josas, France

David Peretz
Department of Pathology, University of California, San Diego, La Jolla, California, USA

Coralie Peyre
Unité Expérimentale Animalerie Rongeur, F-78350 Jouy-en-Josas, France

Jan Priem
Department of Bacteriology and TSEs, Central Veterinary Institute (CVI) of Wageningen UR, Lelystad, 8200 AB, The Netherlands

Jean-Pierre Renou
Recherche en Génomique Végétale, UEVE, F-91057 Evry Cedex, France

Detlev Riesner
Institut für Physikalische Biologie, Heinrich-Heine-Universität Düsseldorf, Düsseldorf, Germany

Alan Rigter
Department of Bacteriology and TSEs, Central Veterinary Institute (CVI) of Wageningen UR, Lelystad, 8200 AB, The Netherlands

Ana B. Rodríguez-Martínez
Department of Animal Health, Neiker-Tecnalia, Berreaga 1, 48160 Derio, Bizkaia, Spain

Robert G. Rohwer
Medical Research Service, Veterans Affairs Maryland Health Care System, Baltimore, Maryland, USA

Regina Savatchenk
Center for Biomedical Engineering and Technology, University of Maryland, Baltimore, Maryland, USA

Walter J. Schulz-Schaeffer
Prion and Dementia Research Unit, Department of Neuropathology, University Medical Center, Georg-August University, Robert-Koch Str. 40, 37075 Goettingen, Germany

Yoshihisa Shimizu
Prion Disease Research Center, National Institute of Animal Health, 3-1-5 Kannondai, Tsukuba, Ibaraki 305-0856, Japan

Ludivine Soubigou-Taconnat
Recherche en Génomique Végétale, UEVE, F-91057 Evry Cedex, France

Jan Stohr
Institut für Physikalische Biologie, Heinrich-Heine-Universität Düsseldorf, Düsseldorf, Germany

Krystyna Surewicz
Department of Physiology and Biophysics, Case Western Reserve University, Cleveland, Ohio, USA

Witold K. Surewicz
Department of Physiology and Biophysics, Case Western Reserve University, Cleveland, Ohio, USA

Gaelle Tilly
Génétique Animale et Biologie Intégrative, F-78350, Jouy-en-Josas, France

Drophatie Timmers-Parohi
Pepscan Presto BV, Lelystad, 8243 RC, The Netherlands

Fred G. van Zijderveld
Department of Bacteriology and TSEs, Central Veterinary Institute (CVI) of Wageningen UR, Lelystad, 8200 AB, The Netherlands

Jean-Luc Vilotte
Génétique Animale et Biologie Intégrative, F-78350, Jouy-en-Josas, France

Martha Vilotte
Génétique Animale et Biologie Intégrative, F-78350, Jouy-en-Josas, France

Wibke Wagner
Paul Ehrlich Institute, Paul Ehrlich-Straße 51-59, D-63225 Langen, Germany

Xuemei Wang
Ventana Medical Systems, Tucson, Arizona, USA

Jürgen Weiβ
Institut für klinische Biochemie und Pathobiochemie, Deutsches Diabetes-Zentrum, Düsseldorf, Germany

Wiebke M. Wemheuer
Prion and Dementia Research Unit, Department of Neuropathology, University Medical Center, Georg-August University, Robert-Koch Str. 40, 37075 Goettingen, Germany

Wilhelm E. Wemheuer
Institute of Veterinary Medicine, Faculty for Agricultural Sciences, Georg-August University, Burckhardweg 2, 37075 Goettingen, Germany

Silja Wessler
Paul Ehrlich Institute, Paul Ehrlich-Straße 51-59, D-63225 Langen, Germany

Dieter Wilbold
Institut für Physikalische Biologie, Heinrich-Heine-Universität Düsseldorf, Düsseldorf, Germany

Arne Wrede
Prion and Dementia Research Unit, Department of Neuropathology, University Medical Center, Georg-August University, Robert-Koch Str. 40, 37075 Goettingen, Germany

Ping Wu
Ventana Medical Systems, Tucson, Arizona, USA

Alice Yam
Research & Development, Novartis Vaccines & Diagnostics, Inc., Emeryville, California, USA

Takashi Yokoyama
Prion Disease Research Center, National Institute of Animal Health, 3-1-5 Kannondai, Tsukuba, Ibaraki 305-0856, Japan

Rachel Young
Génétique Animale et Biologie Intégrative, F-78350, Jouy-en-Josas, France

Gerald W. Zamponi
Scientist of the Alberta Heritage Foundation for Medical Research and a Canada Research Chair in Molecular Neurobiology

Juan J. Zarranz
Neurology Service, Hospital de Cruces, Plaza Cruces-gurutzeta 12, 48902 Barakaldo, Bizkaia, Spain

List of Abbreviations

GPI	Glycosylphosphatidylinositol
NBH	Normal hamster brain homogenate
PI-PLC	Phosphatidylinositol-specific phospholipase C
PK	Proteinase K
PMCA	Protein misfolding cyclic amplification
PNGase F	Peptide:N-glycosidase F
PrP	Prion protein
PrP^{C}	Cellular prion protein
PrPres	Proteinase K-resistant form of prion protein
PrP^{Sc}	Disease-associated scrapie isoform of prion protein
rPrP	Recombinant prion protein
rShaPrP	Recombinant full-length Syrian hamster prion protein
SBH	263K scrapie-infected hamster brain homogenate
TSEs	Transmissible spongiform encephalopathies

Introduction

Mammalian prions are infectious pathogens responsible for transmissible spongiform encephalopathies (TSEs), a group of fatal neurodegenerative disorders affecting farmed and wild animals, as well as humans. Scrapie in sheep and goats, bovine spongiform encephalopathy (BSE), chronic wasting diseases in cervids, and human Creutzfeldt-Jakob disease (CJD) are among the most prevalent TSEs worldwide. Long considered in the middle of the last century as "slow viruses" of the central nervous system, prions now stand firmly as a novel class of proteinaceous infectious pathogens. Prions are essentially composed of PrP^{Sc}, a misfolded, aggregation-prone form of the ubiquitously expressed, host-encoded cellular prion protein (PrP^{C}). Due to these unique properties, prions have attracted a strong research interest. Since the "mad cow" crisis, they also have had a profound impact on human and animal health policy and far-reaching implications on the precautionary principle. This book's chapters are aimed at better understanding prion structure and biology.

PrP^{C} is a glycolipid-anchored cell membrane syaloglycoprotein expressed abundantly at the cell surface of neural cells. While evolutionary conserved among mammals, suggesting thus an important role, its physiological function(s) remains poorly understood. PrP^{C} has been potentially involved in many cellular functions, including neuroprotection, response to oxidative stress, cell proliferation, and differentiation, synaptic function, and signal transduction pathways such as those involving response to pain, as shown by Gadotti and Zamponi (chapter 1). One of the complicating factors in elucidating PrP^{C} function is that PrP-knockout mice, cattle, and goat behave normally and show only subtle if any physiological alterations. Chadi et al. (chapter 2) discuss the potential role PrP may have during embryonic mouse development following the marginal differences in the cellular pathways between wild-type and PrP^{C}-ablated mice at zygotic or adult stage, as observed by transcriptomic analyses of 25,000 mouse genes.

Less uncertain is the key role of PrP^{C} in prion diseases. According to the now widely accepted "protein-only" hypothesis, prion propagation is based on the ability of PrP^{Sc} oligomeric seeds to self-promote and -perpetuate PrP^{C} to PrP^{Sc} conformational conversion, though a nucleated polymerization process, leading to PrP^{Sc} deposition in the brain and sometimes in the lymphoid tissue. The PrP^{Sc} seeds would be contained within the contaminated material in iatrogenic TSE. Etiologically, TSE can also occur sporadically or genetically. Human genetic forms of TSE are associated with autosomal dominant mutations in the PrP gene (>30 mutations described), which are assumed to favor the generation of PrP^{Sc} seeds. Sporadic TSE may reflect a rare stochastic event of spontaneous conversion of PrP^{C} into PrP^{Sc}. Whatever the origin of prion diseases, the mechanisms leading to neurodegeneration are not clear: PrP^{Sc} deposits might be directly noxious; a subversion of PrP^{C} neuroprotective functions due to its enrollment in the conversion process might also be deleterious for neurons. A major breakthrough in the demonstration of the protein-only hypothesis has been the possibility to replicate *in vitro* the PrP^{C} into PrP^{Sc} conversion process by a technique called protein misfolding cyclic amplification (PMCA). Gonzalez-Montalban et al. (chapter 3) illustrate how PMCA can powerfully amplify minute amounts of abnormal PrP^{Sc} to render the protein detectable by conventional techniques. Mechanistically the process is still poorly understood and, as illustrated by Kim et al. (chapter 4), it appears yet difficult to substitute the substrate containing extractive PrP^{C} by bacterially produced PrP so as to generate "synthetic" prions. However, it is becoming clear that PMCA will be extensively used in the near future as a surrogate for prion infectivity and for much needed diagnostic purposes. Although PrP^{Sc} is known to be enriched in beta-sheet content, the details of structural rearrangement from PrP^{C} to PrP^{Sc} as well as the critical regions important for conversion are not known. The difficulties in purifying PrP^{Sc} in its native form have prevented so far direct structural analyses by X-ray diffraction or Nuclear Magnetic Resonance. Thus, as illustrated by Yam et al. (chapter 5) and Rigter et al. (chapter 6), only indirect evidence can provide clues on the regions that undergo conformational transition or are important in the conversion process.

Another complicating factor in establishing which structural feature(s) confers a biological activity to PrP is that prions, as conventional pathogens, exhibit strain diversity in the same host species. Phenotypically, prion strains display in the same host distinct heritable traits such as the incubation time and a variety of neuropathological and biochemical features, as illustrated by Wemheuer et al. (chapter 7). In the absence of specific nucleic acids, the strain diversity conundrum is assumed to reflect the existence of stable, structurally distinct PrP^{Sc} conformers, at the level of the tertiary and/or quaternary structure. Due to refined detection techniques and to the (large-scale) testing of human and animals for the presence TSE, the spectrum of prion diseases is in constant evolution. For example, it was thought that the prions responsible for the BSE or "mad cow" epidemic were unique in cattle, until the recent discovery of new uncommon BSE forms, showing distinguishable phenotypes in cattle (chapter 8). It came also as a surprise to discover new sheep scrapie prions although the disease was known since more than 250 years. The newly discovered strain is referred to as "atypical" scrapie because it is very difficult to identify by conventional diagnostic tests based on the detection, in the brain, of the protease-resistant core of PrP^{Sc} (chapter 8). This strain is present worldwide and appears equally or even more prevalent than classical scrapie. Intriguingly, putative forms of prion diseases accumulating protease-sensitive PrP^{Sc} have been recently found in human, as highlighted by Rodriguez-Martinez et al. (chapter 9). The transmissible nature of such forms remains to be demonstrated yet.

Prions can transmit from one species to another. Thus animal prions can therefore represent a risk to humans, as shown by the emergence of the variant form of human Creutzfeldt-Jakob disease due to ingestion of BSE prions-contaminated food. Prions cross-species transmission capacities are limited by a barrier commonly referred to as the "species barrier." Its strength is assumed to depend essentially on structural interactions between host PrP^{C} and the infecting prion strain type(s) or conformation(s). Importantly, interspecies transmission is currently unpredictable. Panza et al. (chapter 10) highlight how to circumvent experimental obstacles to study prion transmission barrier between species.

Although PrP is essential for disease to develop, other proteins or factors are likely to contribute to the conversion process. Cell models propagating prions have proved useful in the identification of such factors, and we can anticipate in the near future that global approaches will further help. Wagner et al. for example (chapter 11) showed that the phosphor-proteome was differentially regulated following prion infection. Such approach may contribute to the identification of novel protein markers, that are much needed to provide a pre-mortem diagnostic test, and for the development of novel therapeutic intervention strategies for these yet incurable diseases.

Together these pieces provide an overview of prion biology and underscore some of the challenges we face if we want to understand how this lively pathogen propagates and evolves in mammals. There is also mounting evidence that studying prion biology has a wider relevance due to similarities in the processes of protein misfolding and aggregation between prion disorders and other neurodegenerative disorders such as Alzheimer's, Parkinson's, and Huntington's diseases.

— Vincent Béringue, PhD

1 Prion Protein's Protection Against Pain

Vinicius M. Gadotti and Gerald W. Zamponi

CONTENTS

1.1 INTRODUCTION

Cellular prion protein (PrP^C) inhibits N-methyl-D-aspartate (NMDA) receptors. Since NMDA receptors play an important role in the transmission of pain signals in the dorsal horn of spinal cord, we thus wanted to determine if PrP^C null mice show a reduced threshold for various pain behaviors.

We compared nociceptive thresholds between wild type and PrP^C null mice in models of inflammatory and neuropathic pain, in the presence and the absence of a NMDA receptor antagonist. Two–three months old male PrP^C null mice exhibited an MK-801 sensitive decrease in the paw withdrawal threshold in response to both mechanical and thermal stimuli. The PrP^C null mice also exhibited significantly longer licking/biting time during both the first and second phases of formalin-induced inflammation of the paw, which was again prevented by treatment of the mice with MK-801, and responded more strongly to glutamate injection into the paw. Compared to wild type animals, PrP^C null mice also exhibited a significantly greater nociceptive response (licking/biting) after intrathecal injection of NMDA. Sciatic nerve ligation resulted in MK-801 sensitive neuropathic pain in wild-type mice, but did not further augment the basal increase in pain behavior observed in the null mice, suggesting that mice lacking PrP^C may already be in a state of tonic central sensitization. Altogether, our data indicate that PrP^C exerts a critical role in modulating nociceptive transmission at the spinal cord level, and fit with the concept of NMDA receptor hyperfunction in the absence of PrP^C.

The dorsal horn of spinal cord is an important site for pain transmission and modulation of incoming nociceptive information arriving from peripheral nociceptors [1, 2]. Glutamate is the key neurotransmitter released by the primary afferent fibers [3, 4] and plays an important role in nociceptor sensitization and in the modulation of allodynia [5]. Glutamate receptors (GluRs) such as NMDA receptors contribute in various ways to pain induction, transmission and control [5-7]. Consequently, NMDA receptor inhibitors exhibit antinociceptive and analgesic effects in rodents [8, 9] as well in humans [10], however, their clinical use for the treatment of pain has been hampered by their CNS side-effects [11, 12]. For this reason, strategies such as src interfering peptides have been proposed as ways to interfere with NMDAR hyper activity in the pain pathway without affecting basal NMDA receptor function [13].

NMDA receptors are regulated by a plethora of cellular signaling pathways that could potentially be targeted for therapeutic intervention [14]. Along these lines, our laboratory has recently shown [15] that NMDA receptor activity in mouse hippocampal neurons is regulated by PrP^C. Specifically, NMDA receptors expressed in mice lacking PrP^C show slowed current decay kinetics, and spontaneous synaptic NMDA currents in pyramidal neurons displayed increased current amplitude [15]. We subsequently showed that PrP^C protects from depressive like behavior by tonically inhibiting NMDA receptor activity [16], thus suggesting that the altered NMDA currents in PrP^C null mice are associated with a clear behavioral phenotype. Given the important role of NMDA receptors in the afferent pain pathway, we hypothesized that absence of PrP^C may give rise to pain hypersensitivity. Here, we show that PrP^C null mice exhibit a decreased nociceptive threshold, both under basal conditions, as well as in models of inflammatory and neuropathic pain. These effects were reversed by treatment of the animals with the NMDA receptor antagonist MK-801, thus implicating NMDA receptor deregulation in the observed pain phenotype.

1.2 MATERIALS AND METHODS

1.2.1 Animals

All experiments were conducted following the protocol approved by the Institutional Animal Care and Use Committee (protocol #M09100) and all efforts were made to minimize animal suffering. Unless stated otherwise, 10-week-old male mice (C57BL/6J wild type and PrP null weighing 25–30 g) were used. Animals were housed at a maximum of five per cage (30 × 20 × 15 cm) with food and water *ad libitum*. They were kept in 12 hr light/dark cycles (lights on at 7:00 a.m.) at a temperature of 23 ± 1°C. All manipulations were carried out between 11.00 am and 3:00 pm. Different cohorts of mice were used for each test and each mouse was used only once. The observer was blind to the experimental conditions in the experiment examining the age dependence of the pain phenotype. Mice with a targeted disruption of the prion gene (*PrP*) of the Zürich 1 strain [48] were obtained from the European Mouse Mutant Archive (EM: 0158; European Mouse Mutant Archive, Rome) and out-bred to generate $PrP^{-/-}$ (PrP-null) littermates used in the experiments. Genotyping was performed by gel electrophoresis of PCR products obtained from genomic DNA that was isolated from tail samples. Primers and PCR parameters were similar to those used previously [48].

1.2.2 Drugs and Treatment

The following drugs were used in the study: *L*-glutamic acid hydrochloride, MK-801, *N*-methyl-*D*-aspartatic acid, formaldehyde (Sigma Chemical Company, St. Louis, MO, USA). All drugs were dissolved in PBS. When drugs were delivered by the intraperitoneal (i.p.) route, a constant volume of 10 ml/kg body weight was injected. When drugs were administered by the intrathecal (i.t.) route, volumes of 5 μl were injected. Appropriate vehicle-treated groups were also assessed simultaneously. The choice of the doses of each drug was based on preliminary experiments in laboratory.

1.2.3 Formalin Test

The formalin test is a widely used model that allows us to evaluate two different types of pain: Neurogenic pain is caused by direct activation of nociceptive nerve terminals, while the inflammatory pain phase is mediated by a combination of peripheral input and spinal cord sensitization [49-51]. Animals received 20 μl of different concentrations of formalin solution (1.25% or 2.5%) made up in PBS injected intraplantarly (i.pl.) in the ventral surface of the right hindpaw. We observed animals individually from 0 to 5 min (neurogenic phase) and 15–30 min (inflammatory phase). Following intraplantarly. injection of formalin, the animals were immediately placed individually in observation chambers and the time spent licking or biting the injected paw was recorded with a chronometer and considered as nociceptive response. In experiments involving MK-801, mice were treated by intrathecal delivery 10 min prior to formalin injection (1.25%) at a dose of 3 nmol.

1.2.4 Intraplantar Glutamate Injection

The procedure used was similar to that described previously [26]. Briefly, a volume of 20 μl of glutamate (3 nmol/paw or 10 nmol/paw prepared in PBS) was injected i.pl. into the ventral surface of the right hindpaw. Animals were observed individually for

15 min following glutamate injection. The amount of time spent licking the injected paw was recorded with a chronometer and was considered as nociceptive response.

1.2.5 Intrathecal NMDA Injection

To directly investigate the role of spinal glutamate receptors in the nociceptive behavior observed for PrP^C knockout mice, we compared the nociceptive behavior of wild-type and PrP^C knockout mice after a single intrathecal injection of NMDA. Animals received an i.t. injection of 5 μl of NMDA solution. Injections were given to non-anaesthetized animals using the method described by Hylden and Wilcox [52]. Briefly, animals were restrained manually and a 30-gauge needle attached to a 25-μl microsyringe was inserted through the skin and between the vertebrae into the subdural space of the L5–L6 spinal segments. NMDA injections (30 pmol or 300 pmol) were given over a period of 5 s. The amount of time that animals spent biting or licking their hind paws, tail, or abdomen was determined with a chronometer and considered as nociceptive response. In experiments involving MK-801, mice were treated by intraperitoneal delivery 30 min prior to NMDA injection at a dose of 30 pmol/site.

1.2.6 Chronic Constriction Injury (CCI)-induced Neuropathy

For neuropathic pain, we used a sciatic nerve injury model according to the method described by Bennett and Xie [53] with minor modifications. Briefly, mice were anaesthetized (isoflurane 5% induction, 2.5% maintenance) and the right sciatic nerve was exposed at the level of the thigh by blunt dissection through the biceps femoris. Proximal to the sciatic nerve trifurcation, about 12 mm of nerve was freed of adhering tissue and three loose ligatures (silk suture 6-0) were loosely tied around it with about 1 mm spacing so that the epineural circulation was preserved. In sham-operated rats, the nerve was exposed but not injured. Mechanical and thermal withdrawal thresholds were determined 3 days after surgery. In another series of experiments, PrP^C null mice received MK-801 (3 pmol/i.t.) and mechanical withdrawal threshold was evaluated 10 min after drug delivery.

1.2.7 Mechanical Withdrawal Threshold

To assess changes in sensation or in the development of mechanical allodynia, sensitivity to tactile stimulation was measured using the DPA (Ugo Basile, Varese, Italy). Animals were placed individually in a small, enclosed testing arena (20 cm × 18.5 cm × 13 cm, length × width × height) with a wire mesh floor for 60 min. The DPA device was positioned beneath the animal, so that the filament was directly under the plantar surface of the foot to be tested. Each paw was tested three times per session. For experiment 1, the same cohort of both wild-type and PrP^C null mice were tested at age 15 months. For neuropathic pain, testing was performed on the ispsilateral (ligated) paw before ligation (day 0) and then on the 3rd day after ligation.

1.2.8 Thermal Withdrawal Threshold

Thermal hyperalgesia was examined by measuring the latency to withdrawal of the hind paws from a focused beam of radiant heat applied to the plantar surface using a Plantar Test apparatus (Ugo Basile). Three trials each for the right hind paws were performed and for each reading, the apparatus was set at a cut-off time of 20 s. As with mechanical pain testing, the same cohorts of either wild-type or PrP^C null mice were used at 15 months. Thermal withdrawal threshold was tested a day after they were used for mechanical testing. For neuropathic pain, testing was performed on the ispsilateral (ligated) paw before ligation (day 0) and then on the 4th day after nerve injury.

1.3 RESULTS

1.3.1 Mechanical and Thermal Withdrawal Threshold of $PrP^{+/+}$ and $PrP^{-/-}$ mice

To determine if PrP^C plays a role in the transmission of pain signals, we have compared nociception in wild type and PrP^C null mice. Paw withdrawal thresholds in response to mechanical and thermal stimuli were measured using the Dynamic Plantar Aesthesiometer and Plantar Test devices, respectively. As shown in Figure 1, a blinded time-course analysis showed that PrP^C null mice exhibit significantly decreased mechanical and thermal withdrawal thresholds when compared to the wild-type group. Specifically, mechanical thresholds were significantly different in 2 month old animals whereas differences in thermal threshold were statistically significant at the age of 3 months. These differences were then maintained up to an age of 5 months, after which the experiment was terminated. To ascertain whether this effect was mediated by spinal NMDA receptor hyperfunction, we intrathecally (i.t.) delivered 3 nmol of the NMDA receptor blocker MK-801 10 min prior to assessing mechanical withdrawal threshold. As shown in Figure 1(C), MK-801 reversed the decreased mechanical withdrawal threshold of PrP^C null mice. Two-way ANOVA revealed a significant difference of genotype [$F(2, 57) = 5.6$ $P< 0.05$] and genotype X-treatment interaction [$F(1, 43 = 5.5$ $P < 0.05)$]. Altogether, these data indicate that the PrP^C inhibits nociceptive signaling through an NMDA receptor dependent mechanism.

A

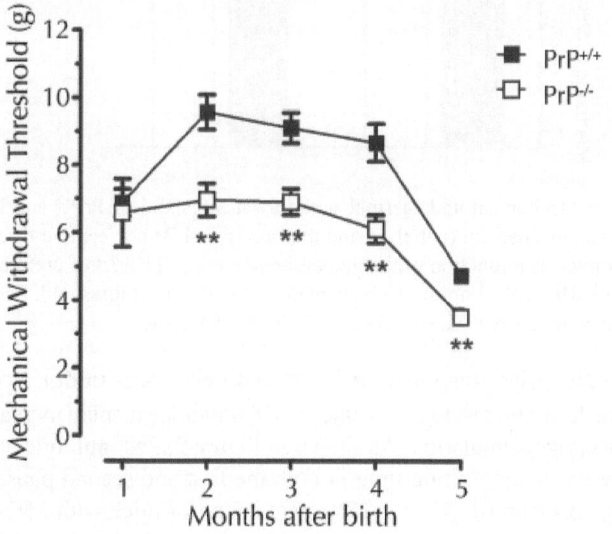

FIGURE 1 *(Continued)*

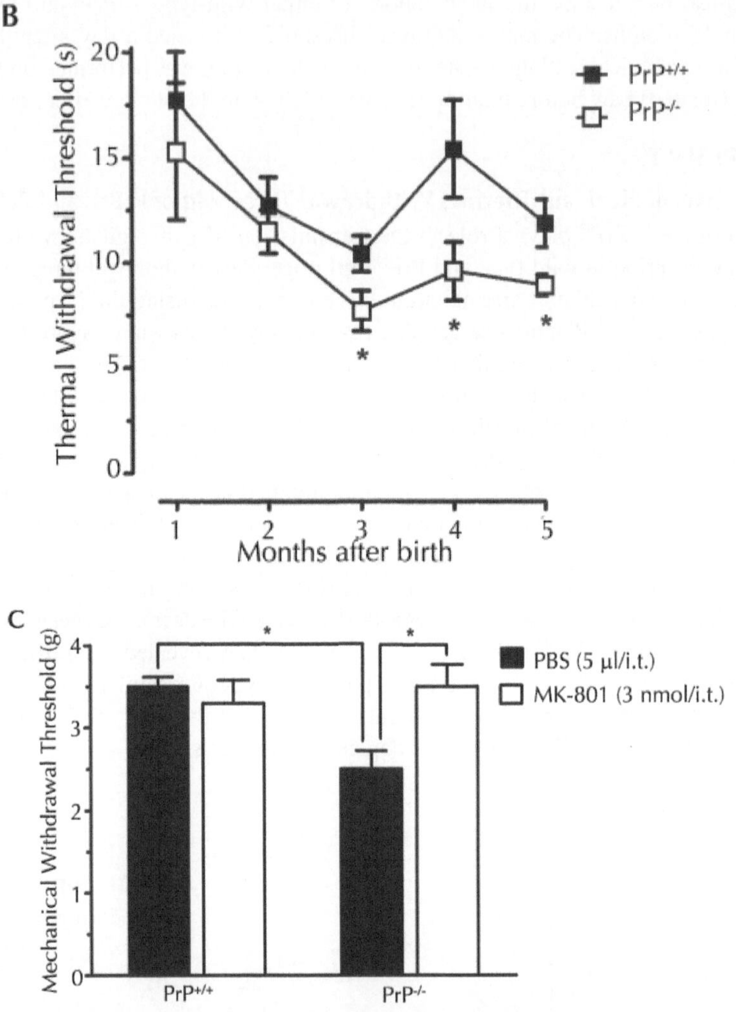

FIGURE 1 Mechanical and thermal withdrawal threshold of $PrP^{+/+}$ and $PrP^{-/-}$ mice. The time course of basal mechanical (panel A) and thermal (panel B) nociceptive threshold of wild type or PrP^{C} null mice as a function of the age of the animal. (C) Effect of pretreatment of 3 months old mice with MK-801 (3 nmol/i.t.) on mechanical withdrawal threshold. Each point or column represents the mean ± S.E.M. (n = 5 - 6). $^{*}P< 0.05$, $^{**}P< 0.01$.

1.3.2 Nociceptive Response of $PrP^{+/+}$ and $PrP^{-/-}$ Mice under Acute Stimulation

We used the formalin test to determine if PrP^{C} null mice display increased sensitivity to acute nociceptive stimulation. As shown in Figure 2, PrP^{C} null mice exhibited significantly elevated licking/biting time in both the first and second phases of nociception induced by formalin (0.7% or 1.25%). Treatment of mice with MK-801 (3 nmol/i.t., 10 min prior to testing) resulted in a significant reduction of the nociceptive behavior of $PrP^{C-/-}$ null mice for both the first and second phases of formalin-induced nociception. Two-way ANOVA revealed a significant difference of genotype [F(2, 31) = 3.4, P< 0.05] and genotype X-treatment interaction [F(6, 52) = 3.4, P< 0.05)] for the earlier (first) phase, and for the second phase (genotype [F(11, 80) = 4.3, P < 0.05] and genotype X-treatment interaction [F(16, 35) = 3.4, P< 0.001)]).

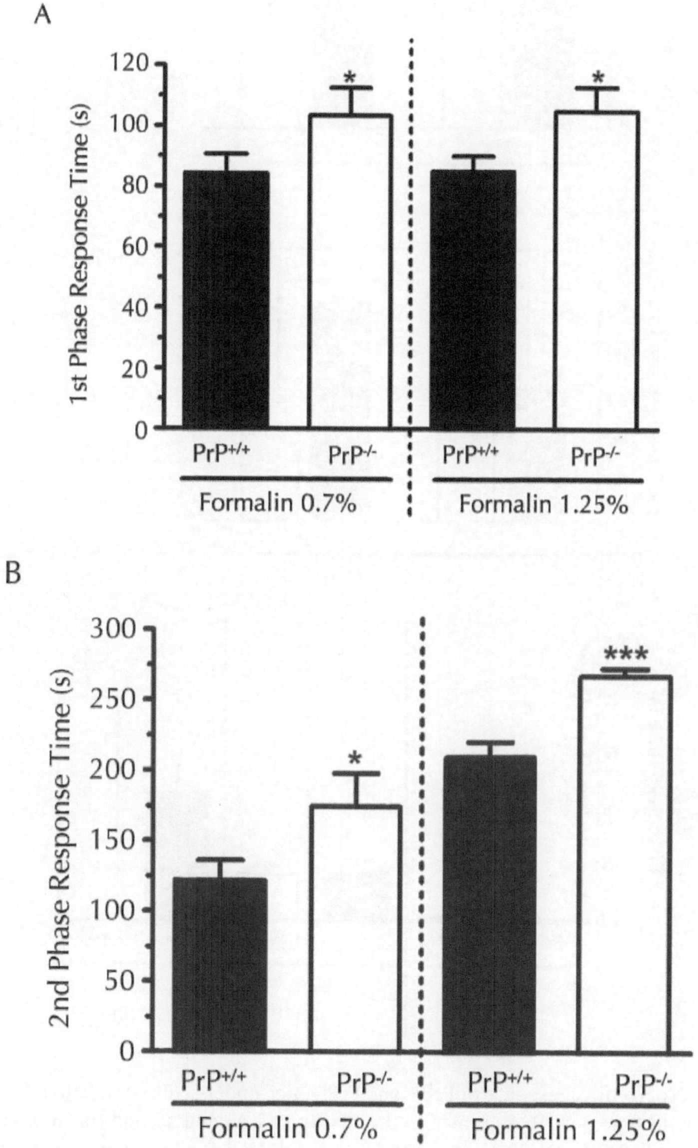

FIGURE 2 *(Continued)*

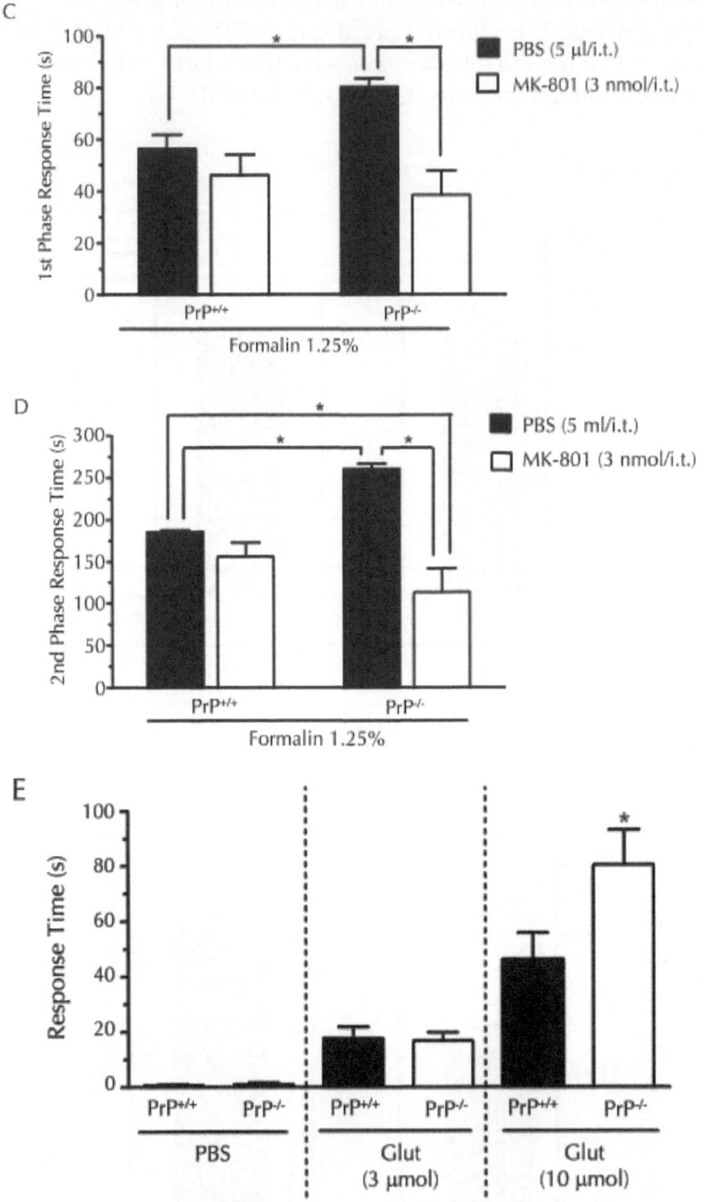

FIGURE 2 Nociceptive response of $PrP^{+/+}$ and $PrP^{-/-}$ mice under acute stimulation. Nociceptive response of wild type or PrP^{C} null mice in the first (panel A) and second (panel B) phases of formalin-induced (0.7% or 1.25%) nociception. (C) and (D) Effect of pretreatment of animals with MK-801 (3 nmol/i.t.) for the first (panel C) and second (panel D) phases of the formalin response. (E) Nociceptive responses of WT and null mice following intraplantar injection of glutamate. Each column represents the mean ± S.E.M. (n = 69). $^{*}P < 0.05$, $^{***}P < 0.001$.

We also directly injected glutamate (3 μmol or 10 μmol) into the paws of wild type and null mice, and determined the time that the animals spent licking and biting over a 15-min time course. As shown in Figure 2(E), the higher dose of glutamate resulted in a significantly greater increase in response time compared to wild type animals.

1.3.3 Nociceptive Response of $PrP^{+/+}$ and $PrP^{-/-}$ Mice in Response to NMDA Treatment

To further investigate the involvement of spinal NMDA receptors in the decreased nociceptive threshold of PrP^C null mice we directly activated these receptors *via* i.t. NMDA injection. As shown in Figure 3A, $PrP^{C-/-}$ mice exhibited a significantly higher nociceptive response (licking/biting) induced by different concentrations of intrathecally delivered NMDA. Interestingly, intrathecal injection of the lower dose of NMDA (30 pmol), which did not appear to affect wild-type mice, increased licking/biting time in $PrP^{C-/-}$ null mice. These data suggest that hyper activity of spinal NMDA receptors of PrP^C null mice may account for the decreased nociceptive sensitivity observed for those animals. As expected, treatment of animals with MK-801 (0.005 mg/kg, i.p., 30 min prior) prevented the effects of NMDA.

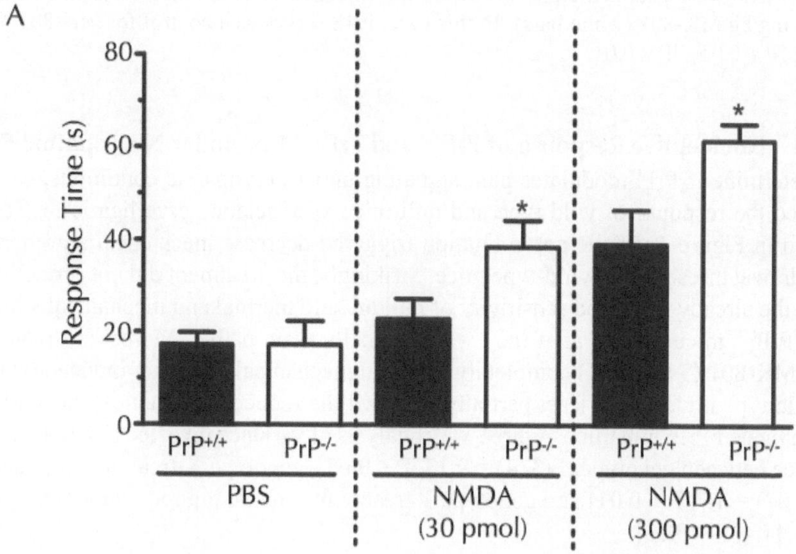

FIGURE 3 *(Continued)*

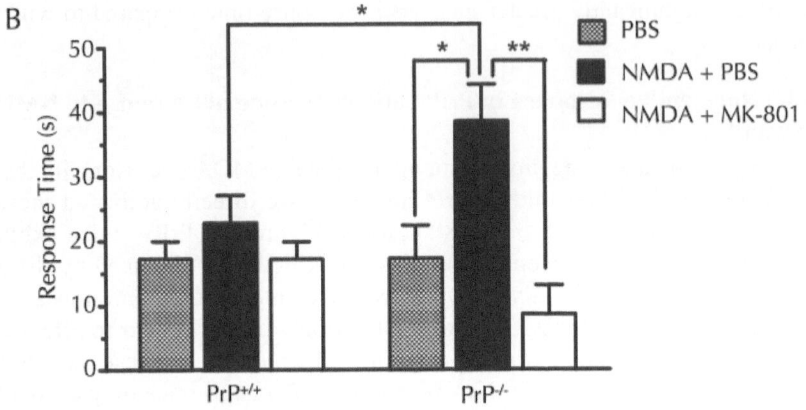

FIGURE 3 Nociceptive response of $PrP^{+/+}$ and $PrP^{-/-}$ mice in response to NMDA treatment. (A) Nociceptive response of wild type or PrP^C null mice following intrathecal injection of NMDA (30 pmol/ i.t. or 300 pmol/i.t.). Each column represents the mean + S.E.M. (n = 5 - 6). $^*P < 0.05$. (B) Effect of MK-801 on the pain behavior induced by intrathecal injection of NMDA (30 pmol). Each column represents the mean ± S.E.M. Control data (hatched bars) were obtained following i.t. injection of 5 μl PBS (i.e., the same route of delivery as for NMDA). The NMDA data were obtained either following i.p. injection of 10 ml/kg PBS (black bars) or 0.005 mg/kg MK-801 (white bars). In this case, PBS serves as a control for MK-801. (n = 6 - 9). $^*P < 0.05$, $^{**}P < 0.01$.

1.3.4 Nociceptive Response of $PrP^{+/+}$ and $PrP^{-/-}$ Mice under Neuropathic Pain

To determine if PrP^C modulates pain signaling under neuropathic conditions, we examined the response of wild type and null mice after sciatic nerve ligation CCI. As shown in Figure 4, sciatic nerve ligation triggered decrease mechanical and thermal withdrawal thresholds in wild-type mice. Strikingly, this treatment did not further augment the already increased sensitivity of null mice to thermal and mechanical stimuli, as if $PrP^{C-/-}$ mice behave as if they were tonically neuropathic. Moreover, treatment with MK-801 (3 nmol/i.t.) completely reversed mechanical allodynia induced by CCI in wild type mice and at least partially reversed the reduced mechanical threshold in nerve ligated PrP^C null mice. A three-way analysis of variance revealed a statistical difference between genotype [$F(3, 4) = 9.13$, $P < 0.05$], genotype X-treatment interaction [$F(1, 07) = 4.7$, $P < 0.01$] and genotype × treatment × nerve injured interactions [$F(1, 75) = 11.4$, $P < 0.05$].

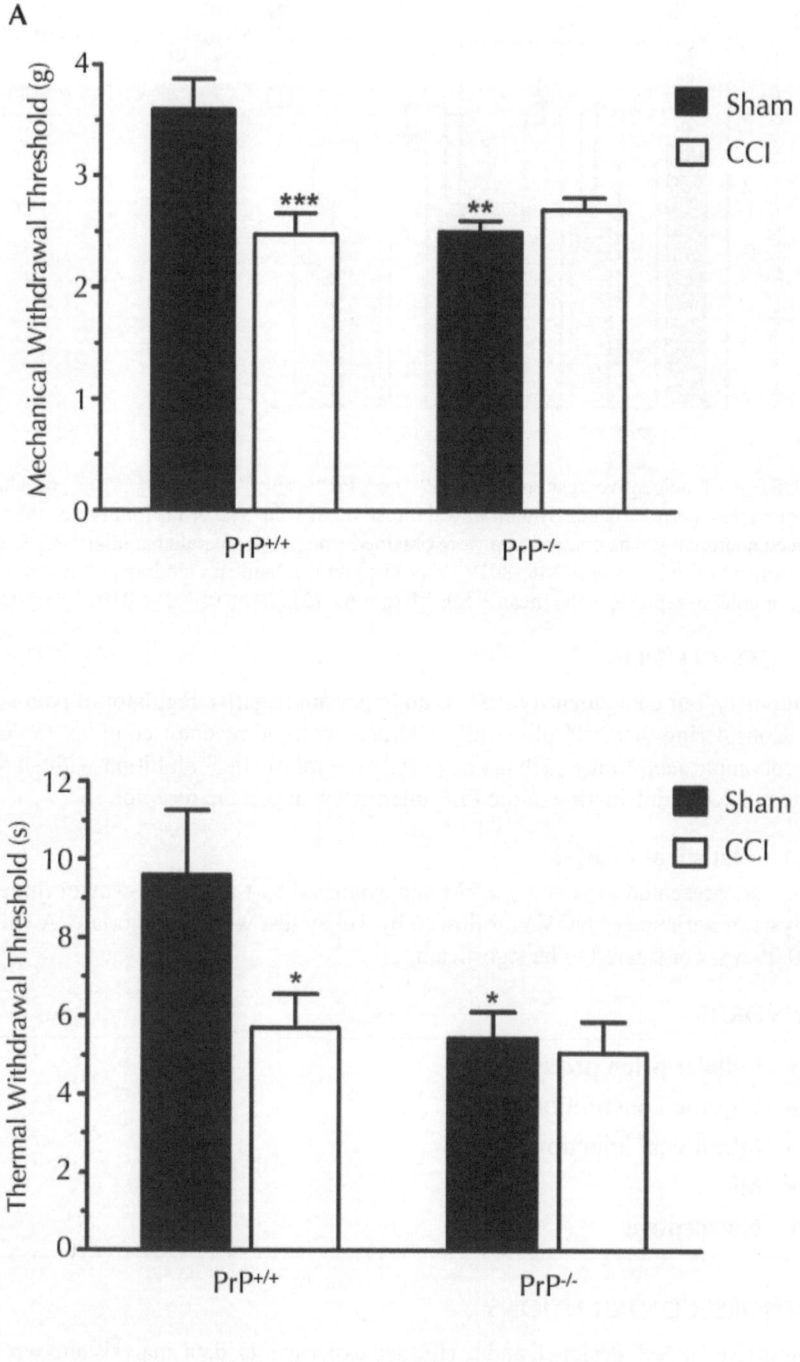

FIGURE 4 *(Continued)*

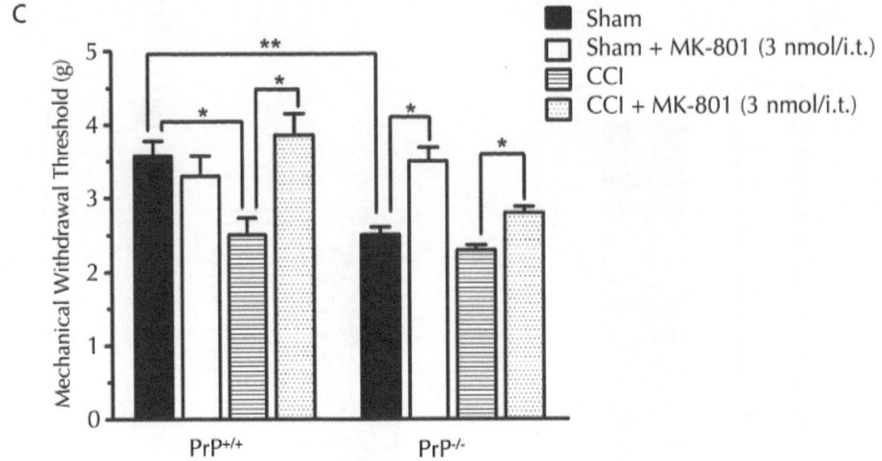

FIGURE 4 Nociceptive response of $PrP^{+/+}$ and $PrP^{-/-}$ mice under neuropathic pain conditions. Mechanical (panel A) and thermal (panel B) withdrawal threshold of wild type or PrP^C null mice after CCI-induced neuropathy. The control data were obtained from sham-operated animals. (C) Effect of pretreatment of animals with MK-801 (3 nmol/i.t.) on mechanical withdrawal threshold. Each point or column represents the mean ± S.E.M. (n = 6 - 12). $^*P < 0.05$, $^{**}P < 0.01$, $^{***}P < 0.001$.

1.4 CONCLUSION

In summary, our data indentify PrP^C as an important negative regulator of pain signaling. Considering that PrP^C physically interacts with the receptor complex to depress current amplitude, it may perhaps be possible to mimic this inhibition with small organic molecules interacting at the PrP^C interaction site on the receptor.

1.4.1 Statistical Analysis

Data were presented as means ± SEM and evaluated by t-tests, two-way or three-way analysis of variance (ANOVA) followed by Tukey test when appropriate. A value of $P < 0.05$ was considered to be significant.

KEYWORDS

- **Cellular prion protein**
- **Chronic constriction injury**
- **Intrathecal injection**
- **Mice**
- **Nociceptive**

AUTHORS' CONTRIBUTIONS

Vinicius M. Gadotti designed and performed experiments, data analysis and wrote the article. Gerald W. Zamponi supervised the research project and edited the manuscript. The authors read and approved the final manuscript.

ACKNOWLEDGMENT

This work was supported by an operating grant to Gerald W. Zamponi from the Prio-Net Canada. Gerald W. Zamponi is a Scientist of the Alberta Heritage Foundation for Medical Research (AHFMR) and a Canada Research Chair in Molecular Neurobiology. Vinicius M. Gadotti is supported by an AHFMR Fellowship and by a Fellowship from the Hotchkiss Brain Institute (HBI). We thank Dr. Clint Doering for genotyping and breeding paradigms, Dr. Frank R. Jirik for providing the outbred PrP null mouse line, and Stephan Bonfield for the blind analysis for experiment 1.

REFERENCES

1. Yaksh, T. L. Regulation of spinal nociceptive processing: where we went when we wandered onto the path marked by the gate. *Pain*, **6**, 149–152 (1999).
2. Hill, R. G. Molecular basis for the perception of pain. *Neuroscientist*, **7**, 282–292 (2001).
3. Jackson, D. L., Graff, C. B., Richardson, J. D., and Hargreaves, K. M. Glutamate participates in the peripheral modulation of thermal hyperalgesia in rats. *Eur. J. Pharmacol.*, **284**, 321–325 (1995).
4. Larsson, M. Ionotropic glutamate receptors in spinal nociceptive processing. *Mol. Neurobiol.*, **40**, 260–288 (2009).
5. Minami, T., Matsumura, S., Okuda-Ashitaka, E., Shimamoto, K., Sakimura, K., Mishina, M., Mori, H., and Ito, S. Characterization of the glutamatergic system for induction and maintenance of allodynia. *Brain Res.*, **895**, 178–185 (2001).
6. Carlton, S. M. and Coggeshall, R. E. Inflammation-induced changes in peripheral glutamate receptor populations. *Brain Res.*, **820**, 63–70 (1999).
7. Nakanishi, O., Ishikawa, T., and Imamura, Y. Modulation of formalin-evoked hyperalgesia by intrathecal N-type Ca channel and protein kinase C inhibitor in the rat. *Cell. Mol. Neurobiol.*, **19**, 191–197 (1999).
8. Berrino, L., Oliva, P., Massimo, F., Aurilio, C., Maione, S., Grella, A., and Rossi, F. Antinociceptive effect in mice of intraperitoneal N-methyl-D-aspartate receptor antagonists in the formalin test. *Eur. J. Pain*, **7**, 131–137 (2003).
9. Villetti, G., Bergamaschi, M., Bassani, F., Bolzoni, P. T., Maiorino, M., Pietra, C., Rondelli, I., Chamiot-Clerc, P., Simonato, M., and Barbieri, M. Antinociceptive activity of the N-methyl-D-aspartate receptor antagonist N-(2-Indanyl)-glycinamide hydrochloride (CHF3381) in experimental models of inflammatory and neuropathic pain. *J. Pharmacol. Exp. Ther.*, **306**, 804–814 (2003).
10. Wiech, K., Kiefer, R. T., Töpfner, S., Preissl, H., Braun, C., Unertl, K., Flor, H., and Birbaumer, N. A placebo-controlled randomized crossover trial of the N-methyl-D-aspartic acid receptor antagonist, memantine, in patients with chronic phantom limb pain. *Anesth. Analg.*, **98**, 408–413 (2004).
11. Chizh, B. A. Novel approaches to targeting glutamate receptors for the treatment of chronic pain. *Amino Acids*, **23**, 169–176 (2002).
12. Millan, M. J. The induction of pain. An integrative review. *Prog. Neurobiol.*, **57**, 1–164 (1999).
13. Yu, X. M., Askalan, R., Keil, G. J., and Salter, M. W. NMDA channel regulation by channel-associated protein tyrosine kinase Src. *Science*, **275**, 674–678 (1997).
14. Gladding, C. M. and Raymond, L. A. Mechanisms underlying NMDA receptor synaptic/extra-synaptic distribution and function. *Mol. Cell. Neurosci.*, in press.
15. Khosravani, H., Zhang, Y., Tsutsui, S., Hameed, S., Altier, C., Hamid, J., Chen, L., Villemaire, M., Ali, Z., Jirik, F. R., and Zamponi, G. W. Prion protein attenuates excitotoxicity by inhibiting NMDA receptors. *J. Cell Biol.*, **181**, 551–565 (2008).
16. Gadotti, V. M., Bonfield, S. P., and Zamponi, G. W. Depressive-like behavior of mice lacking cellular prion protein. *Behav. Brain Res.*, in press.

17. Latremoliere, A. and Woolf, C. J. Central sensitization. A generator of pain hypersensitivity by central neural plasticity. *J. Pain*, **10**, 895–926 (2009).
18. Woolf, C. J. and Salter, M. W. Neuronal plasticity. Increasing the gain in pain.*Science*, **288**, 1765–1769 (2000).
19. Herrero, J. F., Laird, J. M., and López-García, J. A. Wind-up of spinal cord neurones and pain sensation. Much ado about something? *Prog. Neurobiol.*, **61**, 169–203 (2000).
20. Casals-Díaz, L., Vivó, M., and Navarro, X. Nociceptive responses and spinal plastic changes of afferent C-fibers in three neuropathic pain models induced by sciatic nerve injury in the rat. *Exp. Neurol.*, **217**, 84–95 (2009).
21. Yoshimura, M. and Yonehara, N. Alteration in sensitivity of ionotropic glutamate receptors and tachykinin receptors in spinal cord contribute to development and maintenance of nerve injury-evoked neuropathic pain. *Neurosci. Res.*, **56**, 21–28 (2006).
22. Zhou, Q., Price, D. D., Callam, C. S., Woodruff, M. A., and Verne, G. N. Effects of the N-methyl-D-aspartate receptor on temporal summation of second pain (wind-up) in irritable bowel syndrome. *J. Pain*, **12**, 297–303 (2011).
23. Paszcuk, A. F., Gadotti, V. M., Tibola, D., Quintão, N. L., Rodrigues, A. L., Calixto, J. B., and Santos, A. R. Anti-hypernociceptive properties of agmatine in persistent inflammatory and neuropathic models of pain in mice. *Brain Res.*, **23**, 124–133 (2007).
24. Khosravani, H., Zhang, Y., and Zamponi, G. W. Cellular prion protein null mice display normal AMPA receptor mediated long term depression. *Prion*, **2**, 48–50 (2008).
25. Davidson, E. M., Coggeshall, R. E., andCarlton, S. M. Peripheral NMDA and non-NMDA glutamate receptors contribute to nociceptive behaviors in the rat formalin test. *Neuroreport* **8**, 941–946 (1997).
26. Beirith, A., Santos, A. R. S., and Calixto, J. B. Mechanisms underlying the nociception and paw oedema caused by injection of glutamate into the mouse paw. *Brain Res.*, **924**, 219–228 (2002).
27. Beirith, A., Santos, A. R. S., and Calixto, J. B. The role of neuropeptides and capsaicin sensitive fibres in glutamate-induced nociception and paw oedema in mice. *Brain Res.*, **969**, 110–116 (2003).
28. Ji, R. R. and Rupp, F. Phosphorylation of transcription factor CREB in rat spinal cord after formalin-induced hyperalgesia. relationship to c-fos induction. *J. Neurosci.*, **17**, 1776–1785 (1997).
29. Santos, A. R. S., Gadotti, V. M., Oliveira, G. L., Tibola, D., Paszcuk, A. F., Neto, A., Spindola, H. M., Souza, M. M., Rodrigues, A. L. S., and Calixto, J. B. Mechanisms involved in the antinociception caused by agmatine in mice. *Neuropharmacology*, **48**, 1021–1034 (2005).
30. Davis, A. M. and Inturrisi, C. E. Attenuation of hyperalgesia by LY235959, a competitive N-methyl-D-aspartate receptor antagonist. *Brain Res.*, **894**, 150–153 (2001).
31. Li, T. T., Ren, W. H., Xiao, X., Nan, J., Cheng, L. Z., Zhang, X. H., Zhao, Z. Q., and Zhang, Y. Q. NMDA NR2A and NR2B receptors in the rostral anterior cingulate cortex contribute to pain-related aversion in male rats. *Pain*, **146**, 183–193 (2009).
32. Quartaroli, M., Carignani, C., Dal Forno, G., Mugnaini, M., Ugolini, A., Arban, R., Bettelini, L., Maraia, G., Belardetti, F., Reggiani, A., Trist, D. G., Ratti, E., Di Fabio, R., and Corsi, M. Potent antihyperalgesic activity without tolerance produced by glycine site antagonist of N-methyl-D-aspartate receptor GV196771A. *J. Pharmacol. Exp. Ther.*, **290**, 158–169 (1999).
33. Svendsen, F., Tjølsen, A., Rygh, L. J., and Hole, K. Expression of long-term potentiation in single wide dynamic range neurons in the rat is sensitive to blockade of glutamate receptors. *Neurosci. Lett.*, **259**, 25–28 (1999).
34. Suzuki, R., Matthews, E. A. and Dickenson, A. H. Comparison of the effects of MK-801, ketamine and memantine on responses of spinal dorsal horn neurones in a rat model of mononeuropathy. *Pain*, **91**, 101–109 (2001).
35. Kalliomäki, J., Granmo, M. and Schouenborg, J. Spinal NMDA-receptor dependent amplification of nociceptive transmission to rat primary somatosensory cortex (SI). *Pain*, **104**, 195–200 (2003).

36. Meotti, F. C., Carqueja, C. L., Gadotti, V. M., Tasca, C. I., Walz, R., and Santos, A. R. Involvement of cellular prion protein in the nociceptive response in mice. *Brain Res.*, **1151**, 84–90 (2007).
37. Martins, V. R., Beraldo, F. H., Hajj, G. N., Lopes, M. H., Lee, K. S., Prado, M. M., and Linden, R. Prion protein. Orchestrating neurotrophic activities. *Curr. Issues Mol. Biol.*, **12**, 63–86 (2010).
38. Zamponi, G. W. and Stys, P. K. Role of prions in neuroprotection and neurodegeneration. a mechanism involving glutamate receptors? *Prion*, **3**, 187–189 (2009).
39. Lapergue, B., Demeret, S., Denys, V., Laplanche, J. L., Galanaud, D., Verny, M., Sazdovitch, V., Baulac, M., Haïk, S., Hauw, J. J., Bolgert, F., Brandel, J. P., and Navarro, V. Sporadic Creutzfeldt-Jakob disease mimicking nonconvulsive status epilepticus.*Neurology*, **74**, 1995–1999 (2010).
40. Ratté, S., Vreugdenhil, M., Boult, J. K. Patel, A., Asante, E. A., Collinge, J., and Jefferys, J. G. Threshold for epileptiform activity is elevated in prion knockout mice.*Neuroscience*, **179**, 56–61 (2011).
41. Prusiner, S. B. Prion diseases and the BSE crisis. *Science*, **278**, 245–251 (1997).
42. Prusiner, S. B. Prions. *Proc. Natl. Acad. Sci. U. S. A.*, **95**, 13363–13383 (1998).
43. Colby, D. W., Prusiner, and S. B. Prions. *Cold Spring Harb. Perspect. Biol.*, (2011).
44. Muller, W. E., Ushijima, H., Schroder, H. C., Forrest, J. M., Schatton, W. F., Rytik, P. G., and Heffner-Lauc, M. Cytoprotective effect of NMDA receptor antagonists on prion protein (PrionSc)-induced toxicity in rat cortical cell cultures. *Eur. J. Pharmacol.*, **246**, 261–267 (1993).
45. Spencer, M. D., Knight, R. S., and Will, R. G. First hundred cases of variant Creutzfeldt-Jakob disease. Retrospective case note review of early psychiatric and neurological features. *BMJ*, **324**, 1479–1482 (2002).
46. Macleod, M. A., Stewart, G. E., Zeidler, M., Will, R., and Knight, R. Sensory features of variant Creutzfeldt-Jakob disease. *J. Neurol.*, **249**, 706–711 (2002).
47. Reichman, O., Tselis, A., Kupsky, W. J., and Sobel, J. D. Onset of vulvodynia in a woman ultimately diagnosed with Creutzfeldt-Jakob disease. *Obstet. Gynecol.*, **115**, 423–425 (2010).
48. Bueler, H., Fischer, M., Lang, Y., Bluethmann, H., Lipp, H. P., DeArmond, S. J., Prusiner, S. B., Aguet, M., and Weissmann, C. Normal development and behavior of mice lacking the neuronal cell-surface PrP protein. *Nature*, **356**, 577–582 (1992).
49. Dubuisson, D. and Dennis, S. G. The formalin test. A quantitative study of the analgesic effects of morphine, meperidine, and brain stem stimulation in rats and cats. *Pain*, **4**, 161–74 (1977).
50. Hunskaar, S. and Hole, K. The formalin test in mice. dissociation between inflammatory and non-inflammatory pain. *Pain*, **30**, 103–114 (1987).
51. Tjølsen, A. and Hole, K. Animal models of analgesia. In *The Pharmacology of Pain. Volume 130.* Berlin, Springer-Verlag, pp. 1–20 (2001).
52. Hylden, J. L. and Wilcox, G. L. Intrathecal morphine in mice. A new technique. *Eur. J. Pharmacol.*, **67**, 313–316 (1980).
53. Bennett, G. J. and Xie, Y. K. A peripheral mononeuropathy in rat that produces disorders of pain sensation like those seen in man. *Pain*, **33**, 87–107 (1988).

2 Prion Protein-Encoding Genes

Sead Chadi, Rachel Young, Sandrine Le Guillou, Gaëlle Tilly, Frédérique Bitton, Marie-Laure Martin-Magniette, Ludivine Soubigou-Taconnat, Sandrine Balzergue, Marthe Vilotte, Coralie Peyre, Bruno Passet, Vincent Béringue, Jean-Pierre Renou, Fabienne Le Provost, Hubert Laude, and Jean-Luc Vilotte

CONTENTS

2.1 INTRODUCTION

The physiological function of the prion protein remains largely elusive while its key role in prion infection has been expansively documented. To potentially assess this conundrum, we performed a comparative transcriptomic analysis of the brain of wild-type mice with that of transgenic mice invalidated at this locus either at the zygotic or at the adult stages.

Only subtle transcriptomic differences resulting from the Prion Protein (*Prnp)* knockout could be evidenced, beside *Prnp* itself, in the analyzed adult brains following microarray analysis of 24,109 mouse genes and Quantive Polymerase Chain Reaction (QPCR) assessment of some of the putatively marginally modulated loci. When performed at the adult stage, neuronal *Prnp* disruption appeared to sequentially induce a response to an oxidative stress and a remodeling of the nervous system. However, these events involved only a limited number of genes, expression levels of which were only slightly modified and not always confirmed by Real-time Quantive Polymerase Chain Reaction (RT-qPCR). If not, the qPCR obtained data suggested even less pronounced differences.

These results suggest that the physiological function of PrP is redundant at the adult stage or important for only a small subset of the brain cell population under classical breeding conditions. Following its early reported embryonic developmental regulation, this lack of response could also imply that PrP has a more detrimental role during mouse embryogenesis and that potential transient compensatory mechanisms have to be searched for at the time this locus becomes transcriptionally activated.

The pivotal role that the prion protein (PrP) plays in transmissible spongiform encephalopathies (TSE) is now well established ([1, 2] for recent reviews). The conversion of this host-encoded protein to an abnormal, partially proteinase K resistant, isoform is a hallmark of most TSEs and PrP is the only known constituent of mammalian prions [3]. The *Prnp* gene that encodes for PrP, is expressed in a broad range of vertebrate tissues but most abundantly in the central nervous system [4].

Although PrP is evolutionary conserved, suggesting that it has an important role, its physiological function remains unclear even though its implication in neuroprotection, response to oxidative stress, cell proliferation and differentiation, synaptic function and signal transduction has been proposed [5, 6]. Its temporal regulation led also to suspect an implication of this protein in early embryogenesis [7-9] but *Prnp*-knockout mice [10, 11], cattle [12] and goat [13] were obtained with no drastic developmental phenotype and only subtle alterations of their circadian rhythm, hippocampus function and of their behavior. A similar observation was made when this gene was invalidated in adult neurons [14, 15]. To explain these data, it was hypothesized that another host-encoded protein is able to compensate for the lack of PrP [16]. However, this protein has not yet been identified.

Transcriptomic analysis has emerged as a powerful tool to decipher cellular pathways that are modified following a gene expression alteration as it does not pre-require the need of restricting hypothesis. Such approaches have been conducted to analyse the mechanisms underlying prion replication and neurotoxicity (see [17-24] for recent examples). The obtained results appeared however inconsistent and closely related to the cell type and/or strain and animal model used, leading to difficulties in identifying the metabolic pathways involved.

Fewer studies have used a similar approach to try to understand the biological function of the PrP protein in immortalized non-neuronal cells [25-27]. The obtained results appear again to correlate with the cell line used as experimental model and no shared pathway has emerged from the comparison of these different experiments. In parallel, proteomic studies have been conducted either using two cell lines [28] or transgenic knockout mice [29]. While different sets of proteins were found to be affected by the PrP expression level in cells according to their origin, no significant difference was detected in the brain proteome of the analyzed 129/Sv-C57/Bl6 transgenic mice, bearing in mind that variations occurring for low abundant proteins might not have been detected [29].

In the present study, we report the comparisons of the whole brain transcriptomes of PrP knockout or wild type mice, both on an FVB/N genetic background, and of that of mice invalidated for the *Prnp* locus in adult neurons.

2.2 METHODS

2.2.1 Mouse Brain Material and DNA or RNA Extraction

Mouse brains from five 6 weeks old female FVB/N and FVB/N *Prnp*$^{-/-}$ animals [10, 30] were collected and frozen in liquid nitrogen immediately after decapitation. Homozygous Tg37 mice were crossed with heterozygous NFH-Cre (Cre 22) mice and the genotype of the resultsing pups determined by PCR analysis of their tail-extracted genomic DNA as previously described [14]. Sets of two males and two females of Tg37$^{+/-}$ NFH-Cre$^{+/-}$ or of Tg37$^{+/-}$ NFH-Cre$^{-/-}$ genotype, respectively, and of either 10 or 14 weeks old were obtained or their brains collected and frozen in liquid nitrogen immediately after decapitation. All animal manipulations were done according to the recommendations of the French Commission de Génie Génétique. RNA extractions for the microarray were made using the RNeasy Lipid Tissue Midi kit (Qiagen cat no. 75842). Each brain sample was treated independently. RNA concentration was calculated by electro-spectrophotometry and the RNA integrity checked with the Agilent Bioanalyser (Waldbroom, Germany). Pools were obtained by mixing equal amounts of total RNA from each individual sample.

2.2.2 Microarray Analysis

Microarray analysis was carried out at the Unité de Recherche en Génomique Végétale (URGV, Evry, France) using the mouse 25K array [38] containing 24,109 mouse gene-specific oligonucleotides. Amplified RNAs were produced from 2 µg of total RNA from each pool with the "Amino Allyl Message Amp aRNA amplification kit" (Ambion). 5 gm of amplified RNAs were reversed transcribed with SuperScript II Reverse Transcriptase kit (Invitrogen) in the presence of cy3-dUTP or cy5-dUTP for each slide as previously described [39]. Hybridizations, array scanning and image analyses were performed as previously described [40], using a GenePix 4200A scanner and GenePix Pro 3.0 (Axon Instruments).

The statistical analysis was based on two dye-swap. For each array, the raw data comprised the logarithm of median feature pixel intensity at wavelengths 635 nm (red) and 532 nm (green). No background was subtracted. In the following description, log-ratio refers to the differential expression between the two tissues analysed: either log2 (red/green) or log2 (green/red), according to the experimental design. An array-by-array normalization was performed to remove systematic biases. First, features that were considered by the experimenter to be badly formed (e.g. because of dust) were excluded (flagged) 100 in the GENEPIX software. Then we performed a global intensity-dependent normalization using the Loess procedure [41] to correct the dye bias. Finally, on each block the log-ratio median was subtracted from each value of the log-ratio of the block to correct a print-tip effect.

To determine differentially expressed genes, we performed a paired t-test. We assumed that the variance of the log-ratios was the same for all genes, by calculating the average of the gene-specific variance. In order to assess this assumption, we excluded spots with a variance too small or too large. Raw P-values were adjusted by the Bonferroni method, which controls the family-wise error rate [42]. A gene is declared differentially expressed if its adjusted P-value is lower than 0.05. The statistical analysis was performed by using the package R anapuce

2.2.3 QPCR Analysis

Three micrograms of purified RNA was reverse transcribed with SuperScript II Reverse Transcriptase kit (Invitrogen) according to the manufacturer's protocol. Quantitative real-time PCR analysis was performed using ABI PRISM 7000 Sequence Detection System (Applied Biosystems) and SYBR Green (Applied Biosystems). Primers used are listed in Table 6. These primers were designed over exon-exon borders if possible. If not, a RT-control was added in the experiment to control for the absence of DNA contaminant. Normalization was done using the β-actin housekeeping gene. The temperature cycle used comprised 45 cycles at 95°C for 15 s and 60°C for 1 min. A dissociation curve followed, this was comprised of 95°C for 15 s, 60°C for 1 min, and 95°C for 10 s. Each sample was analyzed in triplicate and data analyzed using the Delta-Delta C_t method.

2.2.4 Scg5 Promoter Analysis

Genomic DNA was extracted from tail biopsies as previously described [43]. The *Scg5* proximal promoter was amplified by PCR using the set of primers 5′-CCAGGAATCTCCTAAGATCCTGG-3′ and 5′-GACATCCTCTAGATTTTAGAATTACC-3′ [33]. The amplified DNA fragment was gel purified and sequenced [44].

2.3 DISCUSSION AND RESULTS

2.3.1 Comparative Transcriptional Analysis of FVB/N versus FVB/N *Prnp*$^{-/-}$ Mouse Brains

A search for differentially expressed genes was done by comparison of the expression profiles of FVB/N versus FVB/N *Prnp*$^{-/-}$ [10, 30] mouse brains. To this aim, RNA samples were prepared by pooling RNAs from five brains of 6-week old mice of each genotype. After statistical analysis, two genes were found to be differentially expressed, including the *Prnp* one (Table 1). The relatively low log ratio observed for the variation in *Prnp* expression is explained by the fact that the knockout experiment was done in such a way that the gene remains expressed although at a lower level, around 2- to 3-fold, as observed by Northern blotting (data not shown), but that the resulting mRNA does not encode for PrP anymore [10].

TABLE 1 Candidate genes resulting from microarray studies comparing FVB/N *Prnp*$^{-/-}$ versus FVB/N mice.

Genes	FVB/N *Prnp*$^{-/-}$ Microarray (Log2 ratio)	FVB/N *Prnp*$^{-/-}$ QPCR (ΔΔ CT)	Protein
Downregulated			
Prnp	−1.63	−15	Prion protein or PrP^c
Scg5	−1.37	−2.13	Neuroendocrine secretory protein 7B2
Upregulated			
None			

Differentially expressed genes detected by the microarray analysis are listed with the calculated differential ratio (log2) alongside the observed delta-delta CT resulting from the QPCR experiment. The comparative C_t method is known as the $2^{[delta][delta]Ct}$ method, where delta delta C_t = [delta]$C_{t,\ sample}$ −[delta]$C_{t,\ reference}$.

The QPCR was applied to confirm the suspected differential expression of the detected genes using the same pools of brain RNAs. Both *Prnp* and *Scg5* differential expression were confirmed (Table 1). The higher relative fold-change observed by QPCR for the *Prnp* gene compared to that detected in the micro-array experiment is related to the location of the used primers in exon 2 and in exon 3 of the PrP-encoding gene, respectively, which will not amplify the retro-transcribed RNA expressed by the invalidated locus.

The *Scg5* encodes the 7B2 neuroendocrine secretory protein, a specific chaperone for the proprotein convertase 2 [31]. Invalidation of this gene leads to a hypersecretion of cortocitropin that induces early lethality. This protein function was of interest in regards with PrP since hypercorticism is a phenotype associated with scrapie in ewes [32]. However, a search in the mouse genome database for the chromosomal localization of the *Prnp* and *Scg5* loci revealed that these two genes are physically linked and only 11 cM apart. It has been reported that the level of expression of 7B2, at least in the pancreas, differs between mouse strains and is related to a genetic polymorphism that occurs within its proximal promoter [33].

Since the *Prnp* knockout was done on 129/Sv ES cells, we hypothesized that the *Scg5* gene could still be of a 129/Sv genetic origin in FVB/N *Prnp*$^{-/-}$ mice while it is of FVB/N genetic origin in wild-type mice. We amplified by PCR and sequenced the –200/-60 *Scg5* promoter region starting from genomic DNA of three FVB/N, 129/Sv and FVB/N *Prnp*$^{-/-}$ mice, respectively. A single nucleotide polymorphism (G/T) could be detected at position -97 that discriminates between the FVB/N and the 129/Sv or FVB/N *Prnp*$^{-/-}$ genotypes (Table 2). These results thus indicate that in the FVB/N

Prnp$^{-/-}$ mice, the *Scg5* locus remains of 129/Sv genetic origin. It is worth mentioning that the detected single nucleotide polymorphism abolishes a putative AML 1A transcription factor binding site, TGGGGT, in the FVB/N *Scg5* promoter possibly explaining the different levels of expression observed. Altogether, these data suggest that the brain *Scg5* differential expression between the FVB/N and FVB/N *Prnp*$^{-/-}$ mice results from a different genetic origin of this locus.

TABLE 2 Single nucleotide polymorphism observed within the *Scg5* proximal promoter region.

–111	**-87**
FVB/N	..CAGGGCTTAAGTGC**G**GGGGTAGGAAA
FVB/N *Prnp*$^{-/-}$	..CAGGGCTTAAGTGCTGGGGTAGGAAA
129	..CAGGGCTTAAGTGC **T**GGGGTAGGAAA

The sequences were obtained from three independent mice of each genotype. The sequences are numbered backward starting from the reported distal-most transcription initiation site [33]. The observed single nucleotide polymorphism is indicated in bold-faced type.

The observed poor transcriptional alteration in the brain of mouse depleted for PrP could suggest that these animals adapted to this genetic environment during embryogenesis. If such, invalidation of the PrP-encoding gene at an adult stage might induce detectable transient modification of the genome transcriptomic regulation. Using already validated conditional knockout transgenic lines [14], we analyzed the potential impact of an adult neuronal PrP depletion on the overall brain transcriptome.

2.3.2 Incidence of a Post-Natal Neuronal PrP Depletion on the Brain Transcriptome

This experiment was performed using Tg37 mice, transgenic mice expressing physiological levels of mouse PrP from a transgene composed of a the floxed coding sequences inserted within the hamster-based CosShaTet expression vector, crossed with NFH-Cre transgenic mice [14]. Both transgenic mice were under a mouse *Prnp*-knockout genetic background. The brain RNA pools consisted of littermates of 2 males and 2 females of either 10 or 14 weeks old for each Tg37$^{+/-}$ NFH-Cre$^{-/-}$ or Tg37$^{+/-}$ NFH-Cre$^{+/-}$ genotypes. These two ages were chosen since activation of the NFH promoter results in ablation of PrP in neurons after 9 weeks [14]. After statistical analysis, 11 and 47 genes were found to be differentially expressed at 10 (Table 3) or 14 (Table 4) weeks, respectively. At 10 weeks, 3 genes were over-expressed and 8 under-expressed in *Prnp* depleted mice as compared with the NFH-Cre-negative control animals. At 14 weeks, 21 and 26 genes were over-expressed or under-expressed, respectively. Because the oligonucleotide that recognized *Prnp* in the microarray is located within the 3′ UTR of the gene, a region poorly conserved, it was not expected to detect expression of the Tg37 transgene that encompasses the hamster *Prnp* 3′ UTR sequence and indeed differential expression of this gene was not revealed. The absence of the *Scg5* locus within the detected differentially expressed genes further supports the above hypothesis that explains its detection in the previous experiment by a physical link between the *Scg5* and the *Prnp* loci rather than by a functional one.

TABLE 3 Differentially expressed genes detected at 10 weeks by microarray studies comparing Tg37$^{+/-}$ NFH-Cre$^{-/-}$ and Tg37$^{+/-}$ NFH-Cre$^{+/-}$ brain tissues.

Microarray ID	*Locus name*	*Ratio (log2)*	*Bonf Pval*	**Top functions**
301955	Atp13al	0, 81	2, 90E−09	*Response to oxidative stress*
217673	A630023P12Rik	−0, 62	3, 16E−04	*Unknown*
250792	*Txnl2*	−0, 60	8, 65E−04	*Cardiovascular system development, Neuronal*
Differentiation				
262559	*Zfp819*	−0, 58	2, 89E−03	*Embryonic Development*
203516	Sepx1	−0, 56	7, 29E−03	Genetic Disorder
284620	Slc8al	−0, 54	1, 74E−02	*Response to oxidative stress*
235326	*Kcnj5*	−0, 54	2, 31E−02	Cell Death, Neurological Disease, Nervous System
Development and Function				
202068	1810030J14Rik	−0, 53	3, 99E−02	*Cancer, Cell death*
268919	A130070M06	−0, 52	4, 95E−02	*Ribosome release*
253726	*Synaptotagmin11*	0, 52	4, 38E−02	*Synaptic vesicle trafficking, Nervous system function,*
Response to oxidative Stress				
247145	0610038F07Rik	0, 58	3, 43E−03	*Mitochondrial function*

Differentially expressed genes detected by the microarray analysis are listed with the calculated differential ratios (log2) and the Bonferroni p values (Bonf Pval). The top functions were deduced either using the Ingenuity pathways analysis software or by looking at the expression pattern and putative functions attributed to those genes (italized annotations). Italic names genes potentially involved in cellular development and differentiation. Bold-faced type names genes potentially involved in cell death and disorders, including response to oxidative stress. Italic and bold-faced type names genes potentially involved in both sets of functions.

TABLE 4 Differentially expressed genes detected at 14 weeks by microarray studies comparing $Tg37^{+/-}$ $NFH\text{-}Cre^{-/-}$ and $Tg37^{+/-}$ $NFH\text{-}Cre^{+/-}$ brain tissues.

Microarray ID	Locus name	*Ratio (log2)*	*Bonf Pval*	Top functions
275404	*EST1* (*Genebank* AV451297.1)	–2, 62	0, 00E+00	Unknown (Embryonic Development)
197253	*Ifitm3*	–1, 41	0, 00E+00	DNA replication, Nervous System
Development and Function				
237827	AY036118	–1, 19	0, 00E+00	*Eukaryotic polypeptide chain releasing factor*
245680	4931406E20Rik	–0, 92	0, 00E+00	*Unknown*
202068	1810030J14Rik	–0, 89	0, 00E+00	*Unknown (Cancer, Cell death)*
196280	*Prelid2*	–0, 81	1, 40E–10	Cardiovascular Disease, Cellular
Development, Embryonic Development				
211028	4930428E07Rik	–0, 79	6, 77E–10	*Unknown (Reproductive System)*
272796	6430604K15Rik	–0, 76	7, 67E–09	*Unknown (Zinc Finger protein)*
312533	BM229693	–0, 71	2, 99E–07	*Unknown (Embryonic Development)*
231366	Tmem98	–0, 69	7, 27E-–07	*Unknown (Transmembrane protein)*
242062	*Sdccag3*	–0, 65	1, 11E–05	Cancer, Cardiovascular System
Development and Function, Reproductive				
System Disease				
226542	*Ifng*	–0, 62	6, 35E–05	Cardiovascular Disease, Cellular
Development, Embryonic Development				
Cell Death, Neurological Disease, Nervous				
System				
281732	*Papss2*	–0, 60	1, 62E–04	Development and Disease Cell Death,

TABLE 4 *(Continued)*

Microarray ID	Locus name	*Ratio (log2)*	*Bonf Pval*	Top functions
Neurological Disease, Nervous System				
202885	*Ramp1*	–0, 58	7, 58E–04	Development and Disease
213956	9930022N03Rik	–0, 58	8, 07E–04	*Unknown (expressed in dendtritic cells)*
253726	*Synaptotagmin11*	–0, 56	1, 77E–03	*Synaptic vesicle trafficking, Nervous system function, Response to oxidative stress*
218976	4833414E09Rik	–0, 55	3, 39E–03	*Unknown (expressed in skin and neonate*
head)				
273728	*Zbtb33*	–0, 55	3, 81E–03	DNA replication, Nervous System
Development and Function				
277491	*Adam24*	–0, 54	5, 33E–03	Reproductive System Development and
Function				
300948	4930579C12Rik	–0, 54	5, 41E–03	*Unknown (Reproductive system)*
207253	C330013F16Rik	–0, 54	7, 11E–03	*Unknown*
279418	A530088H08Rik	–0, 53	8, 21E–03	*Unknown*
287029	*Fryl*	–0, 53	9, 99E–03	Cardiovascular Disease, Cellular
Development, Embryonic Development				
287258	Grin1	–0, 52	1, 26E–02	Carbohydrate metabolism, Lipid
metabolism, Small molecule Biochemistry				
312507	9030411K21Rik	–0, 51	2, 50E–02	*Unknown (Embryonic Development)*
192336	BC043118	–0, 50	3, 79E–02	*Unknown (Nervous System Development)*
217372	Sec1	0, 50	3, 28E–02	*Synaptic transmission and general*

TABLE 4 *(Continued)*

Microarray ID	Locus name	*Ratio (log2)*	*Bonf Pval*	Top functions
secretion				
241944	2810471M01Rik	0, 51	2, 15E–02	*Unknown*
214574	Cabp1	0, 52	1, 61E–02	*Calcium transport, Response to oxidative*
stress				
248843	T2bp	0, 52	1, 32E–02	*Cell Death*
240393	*Nfe2*	0, 53	1, 00E–02	DNA replication, Nervous System
Development and Function				
305580	*Abcc12*	0, 53	9, 52E–03	Cardiovascular Disease, Cellular
Development, Embryonic Development				
253103	*Adck4*	0, 53	9, 28E–03	Cardiovascular Disease,
Cellular Development,				
Embryonic Development				
235246	Cyp2d26	0, 54	5, 56E–03	*Detoxification, Clearance of drugs*
200884	*Sall3*	0, 54	5, 56E–03	Cancer, Cell growth and proliferation,
Respiratory Disease				
235326	*Kcnj5*	0, 54	4, 30E–03	Cell Death, Neurological Disease, Nervous
System Development and Disease				
187962	1700055C04Rik	0, 56	2, 11E–03	*Unknown (Reproductive system)*
310508	*Arfgef2*	0, 56	1, 79E–03	Cardiovascular Disease, Cellular
Development, Embryonic Development				
271541	Dusp4	0, 56	1, 56E–03	Carbohydrate metabolism, Lipid

TABLE 4 *(Continued)*

Microarray ID	Locus name	*Ratio (log2)*	*Bonf Pval*	Top functions
metabolism, Small molecule Biochemistry				
189538	*Hist2h3c1*	0, 57	1, 29E–03	DNA replication, Nervous System
Development and Function				
192597	*Grit*	0, 58	6, 36E–04	*Neural Development*
273128	*Nrbp2*	0, 59	2, 86E–04	Embryonic mouse brain development,
Neuronal differentiation				
259059	Ralb	0, 62	6, 20E–05	Carbohydrate metabolism, Lipid
metabolism, Small molecule Biochemistry				
194060	Defb13	0, 65	1, 47E–05	*Host's innate defense*
197262	*Sprr2g*	0, 67	3, 88E–06	Cell Death, Neurological Disease, Nervous
System Development and Disease				
278391	1110038D17Rik	0, 79	8, 61E–10	*Unknown (Embryonic Development)*
284995	GeneBank A530045M11, AI604229, AA174363	0, 81	1, 62E–10	Unknown

Differentially expressed genes detected by the microarray analysis are listed with the calculated differential ratios (log2) and the Bonferroni p values (Bonf Pval). The top functions were deduced either using the Ingenuity pathways analysis software or by looking at the expression pattern and putative functions attributed to those genes (italized annotations). Italic names genes potentially involved in cellular development and differentiation. Bold-faced type name gene potentially involved in cell death and disorders, including response to oxidative stress. Italic and bold-faced type names genes potentially involved in both sets of functions.

The QPCR was applied to confirm the microarray results for the genes suspected differentially expressed with the observed highest log2 ratios, using the same pools of brain RNAs and to assess the Cre-induced *Prnp* invalidation (Table 5). This latter point was confirmed at both 10 and 14 weeks with a highly significant knockdown of the Tg37-transgene expression in the brain of Tg37-NFH-Cre transgenic mice. The down-regulation observed is less important than that detected in FVB/N *Prnp*$^{-/-}$ mice which is an expected result since the Cre deletion is limited to the neurons, due to the tissue-specificity of the NFH promoter. The slight difference observed between 10 and 14 weeks might suggest that at 10 weeks, the deletion process is not fully complete. The percentage of brain cells that have a deleted Tg37 transgene following Cre activation was previously estimated to be around 29–37% [14]. Our data suggest that these cells are among those that express the transgene the most.

TABLE 5 The QPCR analysis of the expression of candidate genes resulting from microarray studies comparing Tg37xNFH-Cre versus Tg37 mice.

Genes	Tg37xNFH-Cre	Tg37xNFH-Cre	
	Microarray	QPCR	Protein
	(log2 ratio)	(ΔΔ CT)	
10 weeks			
Glrx3	–0.6	0.34	glutaredoxin 3
Atp13a1	0.81	–0.08	ATPase type 13A1
Prnp Tg37	ND	-5.62	Prion protein
14 weeks			
AV451297.1	–2.62	–0.02	Hypothetical protein
Ifitm3	–1.41	–0.12	Interferon-induced transmembrane
protein 3			
Erf1	–1.19	–1.12	Eukaryote class I release factor
Bace1	–0.48	–0.02	Beta-site APP-cleaving enzyme 1
BB217622.2	0.81	0.64	Unknown
Riken D17	0.79	0.31	Unknown
Fgf2	0.48	0.47	Fibroblast growth factor 2
Prnp Tg37	ND	-6.43	Prion protein

Differentially expressed genes detected by the microarray analysis are listed with the calculated differential ratio (log2) alongside the observed delta-delta CT resulting from the QPCR experiment. The comparative C_t method is known as the $2^{-[delta][delta]Ct}$ method, where delta delta C_t = $[delta]C_{t,\ sample}$ - $[delta]C_{t,\ reference}$. Age of the analyzed mice is mentioned.

The microarray data were confirmed for 5 out of 9 analyzed genes (Table 5). Among the non-confirmed genes is the aforementioned AV451297.1 putative gene. This gene, located in mouse chromosome 17 and/or mouse chromosome 6, encodes for a hypothetical protein, and its transcription was only reported as an EST in ES cells. The non-confirmation of the differential expression of this gene was surprising since its estimated log ratio was relatively high. Blast alignment of the microarray oligonucleotide corresponding to this gene allowed us to design primers that recognized a family of mouse ESTs that encompass this sequence (Table 6).

However, this oligonucleotide aligns with various regions of the mouse genome (data not shown), located on several chromosomes, and we therefore cannot exclude that a transcript, originating from one of these regions, that will not be amplified by our set of primers is responsible for the observed differential expression. Although showing a down-regulated expression in Tg37$^{+/-}$ NFH-Cre$^{+/-}$ mice in both the microarray analysis and the QPCR experiment, the ratio observed by QPCR was relatively lower than could be expected for the *ifitm3* gene. The primers used for the QPCR were chosen in order not to amplify the other *ifitm* gene family mRNAs (see Table 6 for the QPCR primer sequences). However, they share some homology with the ankyrin repeat domain 12 (data not shown) which might interfere with the obtained results. The other non-confirmed genes correspond to differentially expressed genes showing very low log2 ratios on the microarray results, between –0.8 and +0.8. Overall the qPCR obtained data for these genes strengthen the relative transcriptomic stability of the *Prnp* knockout brain. The microarray and QPCR data were consistent for the *Erf1* transcriptional deregulation (Table 5). We further analyzed the expression level of this gene in the brain of *Prnp*-knockout mice expressing or not the NFH-Cre transgene. No difference was observed (data not shown), demonstrating that the Cre expression does not significantly influence the expression profile of this locus and thus that its observed deregulation in our experiment results from the *Prnp* invalidation. Although we cannot formally exclude that some of the other deregulated genes listed in Tables 3 and 4 results from the Cre expression, this data strongly suggest that the neuronal PrP depletion is responsible for the observed transcriptional modifications.

TABLE 6 List of the used Oligonucleotides.

FVB/N *PrP*$^{-/-}$ mice	
Name	**SEQUENCE (5′–3′)**
Prnp 5′	CAACCGAGCTGAAGCATTCTG
Prnp 3′	CGACATCAGTCCACATAGTC
scg5 5′	CCTTTATGAGAAAATGAAGGG
scg5 3′	GGACAGATTTCTTTGCCACA
Tg37xNFH-Cre mice	
Name	SEQUENCE (5′–3′)
Ifitm3 5′	TCAGCATCCTGATGGTTGTT
Ifitm3 3′	TGTTACACCTGCGTGTAGGG

TABLE 6 *(Continued)*

FVB/N *PrP*$^{-/-}$ mice	
Name	**SEQUENCE (5′–3′)**
AV451297.1 5′	CCCGAAGCGTTTACTTTGAA
AV451297.1 3′	CCCTCTTAATCATGGCCTCA
Erfl 5′	TCGCTCCACCAACTAAGAAC
Erfl 3′	AAACACGGGAAACCTCACC
Prnp Tg37 5′	GAAGGAGTCCCAGGCCTATT
Prnp Tg37 3′	GCAGGAATGAGACACCACCT
Glrx3 5′	CATAAGCATGGTGTCCAAGG
Glrx3 3′	TGCCTTCTCTGCTTCGTAGA
Riken D17 5′	AAGCCTTCATAGCGAGTGGA
Riken D17 3′	TTCCAGACAAGTGGACCTGA
Bace1 5′	TCGACCACTCGCTATACACG
Bace1 3′	CTCCTTGCAGTCCATCTTGAG
Fgf2 B 5′	AGCGGCTCTACTGCAAGAAC
Fgf2 B 3′	GCCGTCCATCTTCCTTCATA
Atp13a1 5′	CGTGACAAGGGTGAAGATGG
Atp13a1 3′	ATAGTAAGAGAAGGCATTCC
BB217622.2 5′	CCAGTTCCGTCAAAGTACCC
BB217622.2 3′	CATGCAGATCTTCAGGTCCA
β-actin 5′	TGTTACCAACTGGGACGACA
β-actin 3′	GGGGTGTTGAAGGTCTCAAA

The sequences of the oligonucleotides used in the QPCR experiments are listed; including those of the housekeeping gene that was used in the three described analyses. The sets of primers for the Erf1, Glrx3, BB217622.2 and the AV451297.1 loci were designed within a single exon. All the other sets were designed over exon-exon borders.

The log ratios observed for the other detected differentially expressed genes were relatively low. However, it has to be kept in mind that this invalidation only involved neuronal cells, and probably not all of them [14], and thus that transcriptomic modifications occurring within this cell population will be diluted by the heterogeneous cell composition of the analyzed adult brain tissues. It could also suggest that the biological relevance of the observed variation is doubtful. Only 3 genes were found to be differentially expressed at both stages, *Kcnj5*, *1810030J14Rik* and *Synaptotagmin 11* (Tables 3 and 4), of which only *1810030J14Rik* was found to behave similarly between these two time-points. This apparent discrepancy could be explained when the function of the differentially express genes was further analysed, either using the Ingenuity pathways analysis or by looking at the expression pattern and putative functions attributed to those genes (Tables 3 and 4). At 10 weeks, the detected genes appear to reflect a cellular response to an oxidative stress (Table 3), which is in phase with putative physiological functions attributed to PrP [5, 6]. Some of the detected genes also suggest that at that stage, the PrP depletion might induce damaged synaptic trafficking and cell death, two cellular pathways into which PrP is also suspected to have a role (Table 3 and [5, 6]). At 14 weeks, the differentially expressed genes are rather evocative of a remodeling of the nervous system (Table 4).

Most of the identified genes are indeed involved in cellular development and neuronal differentiation. The stage-specific modulation of the *Kcnj5* and *Synaptotagmin II* are in agreement with this proposed scenario. So although a few genes are found to be differentially expressed with low detected log ratios, the functions of these genes appear relevant and consistent.

Overall, our data suggest that invalidation of the *Prnp* gene does not induce gross modification of the adult mouse brain transcriptome. When this event happens a few days before the analysis is performed, we cannot however exclude that the few moderately differentially expressed genes that are then detected indicate a physiological PrP role in adult neuronal homeostasis, synaptic transmission, survival and differentiation. Several hypotheses might explain this unexpected low responsiveness to the invalidation of such an evolutionary conserved protein at least in mammals. An explanation might be that the cellular response to the lack of PrP does not involve transcriptomic alteration but modifications of post-transcriptomic regulations. This latter suggestion is attractive in regards with the recently published miRNA specific signature observed in mouse-scrapie affected brains [34]. However, it is in contradiction with the lack of detectable modification of the brain proteome of *Prnp*$^{-/-}$ mice [29], which could rather suggest that the variations observed in the miRNA profile is a consequence of the scrapie infection rather than of a PrP loss of function. It is also possible that the overall brain lack of response is due to the fact that the invalidation of the *Prnp* gene only affects a small subset of the brain cellular population and is therefore not detectable in our present transcriptomic analysis or in the proteomic experiment of Crecelius *et al.* [29]. Indeed, if as suggested PrP positively regulates neural precursor proliferation in adult [35], the effect of its invalidation might be difficult to assess without prior purification of this cell type. PrP might also be essential for brain response to specific stressful physiological conditions and that the physiological role of this gene was therefore not challenged in the classical presently used breeding conditions. Another attractive explanation would be that PrP has a key function only during early embryogenesis, as its developmental regulation [7-9] and recently published experimental data involving gene knockdowns [36, 37] suggest. Following this early developmental stage, the physiological role of PrP might be less crucial and/or redundant under normal physiological conditions. If so, it would be important to repeat transcriptomic and proteomic analyses at earlier embryonic stages, at the time *Prnp* is turned on or under specific breeding conditions.

2.4 CONCLUSION

This chapter documents the lack of drastic brain transcriptomic modification following the *Prnp* invalidation either at the zygotic stage or in adult neuronal cells of the brain tissue. It is consistent with the recently reported proteomic stability of the brain of such PrP-knockout mice [29] and questions some of the obtained results using *in vitro* cell cultures [25-27]. It might suggest that either this gene knockdown affects the animal physiology at a different developmental stage than the one studied here or that it has to be analyzed in certain particular environmental conditions and/or in more specific cell types.

KEYWORDS

- **Prion protein**
- **Quantitative real-time PCR**
- **Scrapie**

AUTHORS' CONTRIBUTIONS

Sead Chadi , Rachel Young , and Gaëlle Tilly performed the RNA purifications and labeling and microarray hybridizations, Rachel Young, Sandrine Le Guillou, and Gaëlle Tilly performed the QPCR experiments. Marthe Vilotte, Bruno Passet, and Coralie Peyre bred and obtained the transgenic mice and Bruno Passet.; Gaëlle Tilly and Rachel Young collected the tissue samples used. Frédérique Bitton, Ludivine Soubigou-Taconnat, Sandrine Balzergue, Marie-Laure Martin-Magniette and Jean-Pierre Renou supervised the microarray experiment and performed statistical analysis of the results. Vincent Béringue, Fabienne Le Provost, , Hubert Laude, and Jean-Luc Vilotte designed and supervised the overall experiment and prepared the manuscript. All authors read and approved the final manuscript.

ACKNOWLEDGMENT

We are most grateful to John Collinge and colleagues for providing the Tg37 and NFH-Cre transgenic mice and for agreeing to include the data obtained with them in this article and to S. Prusiner for providing the FVB/N *Prnp*$^{-/-}$ mice. RY is a post-doctorant supported by the French Ministry of Research and the ANR-09-BLAN-OO15-01.

REFERENCES

1. Aguzzi, A. and Heikenwalder, M. Polymenidou M. Insights into prion strains and neurotoxicity. *Nat. Rev. Mol. Cell. Biol.***8**, 552–561 (2007).
2. Kovacs, G. G. and Budka, H. Molecular pathology of human prion diseases. *Int. J. Mol. Sci.* **10**, 976–999 (2009).
3. Prusiner, S. B. Novel proteinaceous infectious particles cause scrapie. *Science* **216**, 136–144 (1982).

4. Bendheim, P. E., Brown, H. R., Rudelli, R. D., Scala, L. J., Goller, N. L., Wen, G. Y., Kascsak, R. J., Cashman, N. R., and Bolton, D. C. Nearly ubiquitous tissue distribution of the scrapie agent precursor protein. *Neurology* **42**, 149–156 (1992).
5. Linden, R., Martins, V. R., Prado, M. A., Cammarota, M., Izquierdo, I., and Brentani, R. R. Physiology of the prion protein. *Physiol. Rev.* **88**, 673–728 (2008).
6. Martins, V. R., Beraldo, F. H., Hajj, G. N., Lopes, M. H., Lee, K. S., Prado, M. M., and Linden, R. Prion protein. Orchestrating neurotrophic activities. *Curr. Issues Mol. Biol.* **12**, 63–86 (2009).
7. Manson, J., West, J. D., Thomson, V., McBride, P., Kaufman, M. H., and Hope, J. The prion protein gene. A role in mouse embryogenesis? *Development* **115**, 117–122 (1992).
8. Miele, G., Alejo Blanco, A. R., Baybutt, H., Horvat, S., Manson, J., and Clinton, M. Embryonic activation and developmental expression of the murine prion protein gene. *Gene Expr.* **11**, 1–12 (2003).
9. Tremblay, P., Bouzamondo-Bernstein, E., Heinrich, C., Prusiner, S. B., and DeArmond, S. J. Developmental expression of PrP in the post-implantation embryo. *Brain Res.* **1139**, 60–67 (2007).
10. Bueler, H., Fischer, M., Lang, Y., Bluethmann, H., Lipp, H. P., DeArmond S. J., Prusiner, S. B., Aguet, M., and Weissmann, C. Normal development and behavior of mice lacking the neuronal cell-surface PrP protein. *Nature* **356**, 577–582 (1992).
11. Manson, J. C., Clarke, A. R., Hooper, M. L., Aitchison, L., McConnell, and I. Hope, J. 129/Ola mice carrying a null mutation in PrP that abolishes mRNA production are developmentally normal. *Mol. Neurobiol.* **8**, 121–127 (1994).
12. Richt, J. A., Kasinathan, P., Hamir, A. N., Castilla, J., Sathiyaseelan, T., Vargas, F., Sathiyaseelan, J., Wu, H., Matsushita, H., Koster, J., et al. Production of cattle lacking prion protein. *Nat. Biotechnol.* **25**, 132–138 (2007).
13. Yu, G., Chen, J., Xu, Y., Zhu, C., Yu, H., Liu, S., Sha, H., Chen, J. Xu, X. Wu, Y., et al. Generation of goats lacking prion protein. *Mol. Reprod. Dev.* **76**, 3 (2009).
14. Mallucci, G. R., Ratte, S., Asante, E. A., Linehan, J., Gowland, I., Jefferys, J. G., and Collinge, J. Post-natal knockout of prion protein alters hippocampal CA1 properties, but does not result in neurodegeneration. *EMBO J.* **21**, 202–210 (2002).
15. White, M. D., Farmer, M., Mirabile, I., Brandner, S., Collinge, J., and Mallucci, G. R. Single treatment with RNAi against prion protein rescues early neuronal dysfunction and prolongs survival in mice with prion disease. *Proc. Natl. Acad. Sci. U. S. A.* **105**, 10238–10243 (2008).
16. Shmerling, D., Hegyi, I., Fischer, M., Blattler, T., Brandner, S., Gotz, J., Rulicke, T., Flechsig, E., Cozzio, A., von Mering, C., et al. Expression of amino-terminally truncated PrP in the mouse leading to ataxia and specific cerebellar lesions. *Cell* **93**, 203–214 (1998).
17. Xiang, W., Windl, O., Wunsch, G., Dugas, M., Kohlmann, A., Dierkes, N., Westner, I. M., and Kretzschmar, H. A. Identification of differentially expressed genes in scrapie-infected mouse brains by using global gene expression technology. *J. Virol.* **78**, 11051–11060 (2004).
18. Brown, A. R., Rebus, S., McKimmie, C. S., Robertson, K., Williams, A., and Fazakerley, J. K. Gene expression profiling of the preclinical scrapie-infected hippocampus. *Biochem. Biophys. Res. Commun.* **334**, 86–95 (2005).
19. Skinner, P. J., Abbassi, H., Chesebro, B., Race, R. E., Reilly, C., and Haase, A. T. Gene expression alterations in brains of mice infected with three strains of scrapie. *BMC Genomics* **7**, 114 (2006).
20. Xiang, W., Hummel, M., Mitteregger, G., Pace, C., Windl, O., Mansmann, U., and Kretzschmar, H. A. Transcriptome analysis reveals altered cholesterol metabolism during the neurodegeneration in mouse scrapie model. *J. Neurochem.* **102**, 834–847 (2007).
21. Sorensen, G., Medina, S., Parchaliuk, D., Phillipson, C., Robertson, C., and Booth, A. Comprehensive transcriptional profiling of prion infection in mouse models reveals networks of responsive genes. *BMC Genomics* **9**, 114 (2008).
22. Julius, C., Hutter, G., Wagner, U., Seeger, H., Kana, V., Kranich, J., Klohn, P. C., Weissmann, C., Miele, G., and Aguzzi, A. Transcriptional stability of cultured cells upon prion infection. *J. Mol. Biol.* **375**, 1222–1233 (2008).

23. Miele, G., Seeger, H., Marino, D., Eberhard, R., Heikenwalder, M., Stoeck, K., Basagni, M., Knight, R., Green, A., Chianini, F., et al. Urinary alpha1-antichymotrypsin. A biomarker of prion infection. *PLoS One* **3**, e3870 (2008).
24. Hwang, D., Lee, I. Y., Yoo, H., Gehlenborg, N., Cho, J. H., Petritis, B., Baxter, D., Pitstick, R., Young, R., Spicer, D., et al. A systems approach to prion disease. *Mol. Syst. Biol.* **5**, 252 (2009).
25. Liang, J., Luo, G., Ning, X., Shi, Y., Zhai, H., Sun, S., Jin, H., Liu, Z., Zhang, F., Lu, Y., et al. Differential expression of calcium-related genes in gastric cancer cells transfected with cellular prion protein. *Biochem. Cell Biol.* **85**, 375–383 (2007).
26. Satoh, J., and Yamamura, T. Gene expression profile following stable expression of the cellular prion protein. *Cell. Mol. Neurobiol.* **24**, 793–814 (2004).
27. Satoh, J., Kuroda, Y., and Katamine, S. Gene expression profile in prion protein-deficient fibroblasts in culture. *Am. J. Pathol.* **157**, 59–68 (2000).
28. Ramljak, S., Asif, A. R., Armstrong, V. W., Wrede, A., Groschup, M. H., Buschmann, A., Schulz-Schaeffer, W., Bodemer, W., and Zerr, I. Physiological role of the cellular prion protein (PrP^c). Protein profiling study in two cell culture systems. *J. Proteome Res.* **7**, 2681–2695 (2008).
29. Crecelius, A. C., Helmstetter, D., Strangmann, J., Mitteregger, G., Frohlich, T., Arnold, G. J., and Kretzschmar, H. A. The brain proteome profile is highly conserved between Prnp-/- and Prnp+/+ mice. *Neuroreport* **19**, 1027–1031 (2008).
30. Giri, R. K., Young, R., Pitstick, R., DeArmond, S. J., Prusiner, S. B., and Carlson, G. A. Prion infection of mouse neurospheres. *Proc. Natl. Acad. Sci. U. S. A.* **103**, 3875–3880 (2006).
31. Mbikay, M., Seidah, N. G., and Chretien, M. Neuroendocrine secretory protein 7B2. structure, expression and functions. *Biochem. J.* **357**, 329–342 (2001).
32. Gayrard, V., Picard-Hagen, N., Grino, M., Sauze, N., Grandjean, C., Galea, J., Andreoletti, O., Schelcher, F., and Toutain, P. L. Major hypercorticism is an endocrine feature of ewes with naturally occurring scrapie. *Endocrinology* **141**, 988–994 (2000).
33. Schmidt, G., Sirois, F., Anini, Y., Kauri, L. M., Gyamera-Acheampong, C., Fleck, E., Scott, F. W., Chretien, M., and Mbikay, M. Differences of pancreatic expression of 7B2 between C57BL/6J and C3H/HeJ mice and genetic polymorphisms at its locus (Sgne1). *Diabetes* **55**, 452–459 (2006).
34. Saba, R., Goodman, C. D., Huzarewich, R. L., Robertson, C., and Booth, S. A. A miRNA signature of prion induced neurodegeneration. *PLoS One* **3**, e3652 (2008).
35. Steele, A. D., Emsley, J. G., Ozdinler, P. H., Lindquist, S., and Macklis, J. D. Prion protein (PrP^c) positively regulates neural precursor proliferation during developmental and adult mammalian neurogenesis. *Proc. Natl. Acad. Sci. U. S. A.* **103**, 3416–3421 (2006).
36. Malaga-Trillo, E., Solis, G. P., Schrock, Y., Geiss, C., Luncz, L., Thomanetz, V., and Stuermer, C. A. Regulation of embryonic cell adhesion by the prion protein. *PLoS Biol.* **7**, e55 (2009).
37. Young, R., Passet, B., Vilotte, M., Cribiu, E. P., Beringue, V., Le Provost, F., Laude, H., and Vilotte, J. L. The prion or the related Shadoo protein is required for early mouse embryogenesis. *FEBS Lett.* **583**, 3296–3300 (2009).
38. Le Brigand, K., Russell, R., Moreilhon, C., Rouillard, J. M., Jost, B., Amiot, F., Magnone, V., Bole-Feysot, C., Rostagno, P., Virolle, V., et al. An open-access long oligonucleotide microarray resource for analysis of the human and mouse transcriptomes. *Nucleic Acids Res.* **34**, e87 (2006).
39. Lurin, C., Andres, C., Aubourg, S., Bellaoui, M., Bitton, F., Bruyere, C., Caboche, M., Debast, C., Gualberto, J., Hoffmann, B., et al. Genome-wide analysis of Arabidopsis pentatricopeptide repeat proteins reveals their essential role in organelle biogenesis. *Plant Cell* **16**, 2089–2103 (2004).
40. Minic, Z., Jamet, E., San-Clemente, H., Pelletier, S., Renou, J. P., Rihouey, C., Okinyo, D. P., Proux, C., Lerouge, P., and Jouanin, L. Transcriptomic analysis of Arabidopsis developing stems. A close-up on cell wall genes. *BMC Plant Biol.* **9**, 6 (2009).
41. Yang, Y. H. Dudoit, S. Luu, P. Lin, D. M. Peng, V. Ngai, J. and Speed, T. P. Normalization for cDNA microarray data. A robust composite method addressing single and multiple slide systematic variation. *Nucleic Acids Res.* **30**, e15 (2002).

42. Ge, X. Tsutsumi, S. Aburatani, H. and Iwata, S. Reducing false positives in molecular pattern recognition. *Genome Inform* **14**, 34–43 (2003).
43. Vilotte, J. L. Soulier, S. Stinnakre, M. G. Massoud, M. and Mercier, J. C. Efficient tissue-specific expression of bovine alpha-lactalbumin in transgenic mice. *Eur. J. Biochem.* **186**, 43–48 (1989).
44. Sanger, F. Air, G. M. Barrell, B. G. Brown, N. L. Coulson, A. R. Fiddes, C. A. Hutchison, C. A. Slocombe, P. M. and Smith, M. Nucleotide sequence of bacteriophage phi X174 DNA. *Nature* **265**, 687–695 (1977).

3 Protein Misfolding Cyclic Amplification

Nuria Gonzalez-Montalban, Natallia Makarava, Valeriy G. Ostapchenko, Regina Savtchenk, Irina Alexeeva, Robert G. Rohwer, and Ilia V. Baskakov

CONTENTS

3.1 INTRODUCTION

Protein misfolding cyclic amplification (PMCA) provides faithful replication of mammalian prions *in vitro* and has numerous applications in prion research. However, the low efficiency of conversion of PrP^C into PrP^{Sc} in PMCA limits the applicability of

PMCA for many uses including structural studies of infectious prions. It also implies that only a small sub-fraction of PrP^C may be available for conversion. Here we show that the yield, rate, and robustness of prion conversion and the sensitivity of prion detection are significantly improved by a simple modification of the PMCA format. Conducting PMCA reactions in the presence of Teflon beads (PMCAb) increased the conversion of PrP^C into PrP^{Sc} from 10% to up to 100%. In PMCAb, a single 24-hr round consistently amplified PrP^{Sc} by 600–700-fold. Furthermore, the sensitivity of prion detection in one round (24 hr) increased by two–three orders of magnitude. Using serial PMCAb, a 10^{12}-fold dilution of scrapie brain material could be amplified to the level detectible by Western blotting in three rounds (72 hr). The improvements in amplification efficiency were observed for the commonly used hamster 263K strain and for the synthetic strain SSLOW that otherwise amplifies poorly in PMCA. The increase in the amplification efficiency did not come at the expense of prion replication specificity. The current study demonstrates that poor conversion efficiencies observed previously have not been due to the scarcity of a sub-fraction of PrP^C susceptible to conversion nor due to limited concentrations of essential cellular cofactors required for conversion. The new PMCAb format offers immediate practical benefits and opens new avenues for developing fast ultrasensitive assays and for producing abundant quantities of PrP^{Sc} *in vitro*.

The PMCA provides faithful replication of mammalian prions *in vitro*. While PMCA has become an important tool in prion research, its application is limited because of low yield, poor efficiency and, sometimes, stochastic behavior. The current study introduces a new PMCA format that dramatically improves the efficiency, yield, and robustness of prion conversion *in vitro* and reduces the time of the reaction. These improvements have numerous implications. The method opens new opportunities for improving prion detection and for generating large amounts of PrP^{Sc} *in vitro*. Furthermore, the results demonstrate that *in vitro* conversion is not limited by lack of convertible PrP^C nor by concentrations of cellular cofactors required for prion conversion.

The PMCA provides faithful amplification of mammalian prions *in vitro* and, since its introduction in 2001 [1], has become an important tool in prion research. To date, PMCA provides the most sensitive approach for detecting miniscule amounts of prion infectivity [2-5], including detection of prions in blood or peripheral tissues at preclinical stages of the disease [6-8]. In recent studies, PMCA was employed for generating infectious prions (PrP^{Sc}) *in vitro de novo* in crude brain homogenate [9], and for producing infections prions from the cellular prion protein (PrP^C) purified from normal mammalian brains [10] and recombinant PrP (rPrP) produced in *E. coli* [11]. Furthermore, PMCA has been used for identifying cofactors that are involved in prion replication [12-15] and assessing the impact of glycosylation on replication of prion strains [16]. PMCA has also been utilized for assessing the prion transmission barrier [17, 18], prion interference [5] and adaptation to new hosts [19].

PMCA reactions consist of two alternating steps incubation and sonication. Sonication fragments PrP^{Sc} particles or fibrils into smaller pieces, a process that that is believed to result in the multiplication of active centers of PrP^{Sc} growth. During the incubation step, small PrP^{Sc} particles grow by recruiting and converting PrP^C molecules into PrP^{Sc}.

While the discovery of PMCA has provided new opportunities for exploring the prion replication mechanism, the low yield of PrP^{Sc} has limited its utility for structural studies. Furthermore, the efficiency of amplification in PMCA varies dramatically depending on minor variations in experimental parameters, including those that are difficult to control, such as the age of the sonicator's horn and individual patterns of horn corrosion. Previous strategies for improving the efficiency of PMCA focused on increasing the number of cycles within a single PMCA round [2] or increasing the substrate concentration by using a normal brain homogenate (NBH) from transgenic mice overexpressing PrP^C [20, 21]. Here we describe a new PMCA format that employs beads (referred to as PMCAb). Supplementing the reaction with beads resulted in remarkable improvements in the yield, rate and robustness of prion conversion, as well as in the sensitivity of prion detection. This simple modification of the PMCA format enables fast and efficient production of high quantities of PrP^{Sc}. This result also shows that the low yield observed previously has not been due to a lack of PrP^C susceptible to conversion, nor it has been limited by cellular cofactors.

3.2 METHODS

3.2.1 Protein Misfolding Cyclic Amplification

Healthy hamsters were euthanized and immediately perfused with PBS, pH 7.4, supplemented with 5 mM EDTA. Brains were dissected, and 10% brain homogenate (w/v) was prepared using ice-cold conversion buffer and glass/Teflon tissue grinders cooled on ice and attached to a constant torque homogenizer (Heidolph RZR2020). The brains were ground at low speed until homogeneous, then five additional strokes completed the homogenization. The composition of conversion buffer was as previously described [29]. Ca^{2+}-free and Mg^{2+}-free PBS, pH 7.5, supplemented with 0.15 M NaCl, 1.0% Triton and 1 tablet of complete protease inhibitors cocktail (Roche, Cat. # 1836145) per 50 ml of conversion buffer. The resulting 10% normal brain homogenate in conversion buffer was used as the substrate in PMCA reactions. To prepare seeds, 10% scrapie brain homogenates in PBS were serially diluted 10 to 10^{14} fold, as indicated, in the conversion buffer and 10μl of the dilution were used to seed 90μl of NBH in PMCA. Samples in 0.2ml thin wall PCR tubes (Fisher, Cat. # 14230205) were placed in a rack fixed inside Misonix S-3000 or S-4000 microplate horn, filled with 300ml water. Two coils of rubber tubing attached to a circulating water bath were installed for maintaining 37°C inside the sonicator chamber. The standard sonication program consisted of 30s sonication pulses delivered at 50% to 70% efficiency applied every 30min during a 24hr period. For PMCA with beads, small (1.58mm diameter) or large (2.38mm diameter) Teflon beads (McMaster-Carr, Los Angeles, CA) were placed into the 0.2 ml tubes first using tweezers, then NBH and seeds were added. The following beads from Small Parts (www.smallparts.com) were tested in Figure 1. PTFE Ball Grade II (Teflon); Stainless Steel 440C Ball Grade 24; Neoprene Ball; Nylon Ball; EPDM Ball; Nitrile Rubber Ball; Stainless Steel 302 Ball Grade 100; Acetal Ball Grade I. The diameter of all beads was 2.38 mm except of Stainless Steel 440C Ball, which was 2 mm in diameter. The following low binding beads showed no effects on efficiency in PMCA. Silica Beads Low Binding 800 or 400 μm diameter, and

Zirconium Beads Low Binding 200 or 100 μm diameter (all from OPS Diagnostics LLS, Lebanon, NJ).

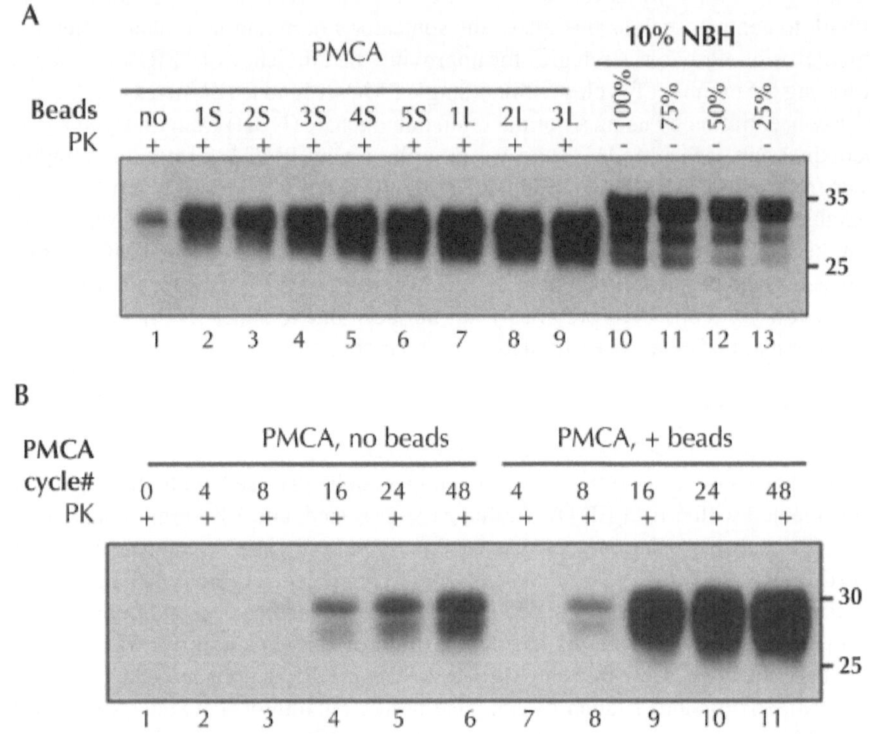

FIGURE 1 Effect of bead material on efficiency of amplification.

In experience, the amplification efficiency in PMCA depended strongly on the position of the tube within a microplate horn, that is distance of a tube from horn's surface to the tube and its center; and the age of the sonicator's horn. In the current studies, several Misonix sonicators were used, all equipped with horns less than one year old. The tubes were placed only in positions between 1.5 cm and 5 cm from the horn's center. Nevertheless, we experienced substantial variations in amplification efficiency in standard PMCA (no beads), which appear due to differences in the age of horns, individual patterns of horn corrosion or differences in the horizontal coordinates of tubes. In the presence of beads, the amplification was much more robust and showed only minor variations.

3.2.2 Bioassay and Estimation of the Incubation Times to Disease

Weanling golden Syrian hamsters were inoculated intracerebrally with 50 μl each using the following inocula. animals of group 1 were inoculated with 263K brain homogenate diluted 10^4-fold relative to whole brain in PBS, 1% BSA. For groups 2 and

3, 10 μl of 10% 263K scrapie brain homogenate were mixed with 90 μl of PBS and subjected to a sonication procedure equivalent to a single PMCA round (48 sonication cycles) in the absence of NBH. Sonication was performed either without beads (for group 2) or with three large beads (for group 3). Then, the sonication products were diluted 100-fold into PBS, 1% BSA to obtain final dilution of 10^{-4} relative to whole 263K brain for inoculation. For groups four and five, PMCA reactions were seeded with 10^{-4}-diluted 263K brain material, then six serial PMCA rounds were conducted in the absence of beads (for group 4) or presence of three large beads (for group 5) using 1:10 dilutions between rounds. After the sixth round of serial PMCA, the amplification products were diluted an additional 10-fold into PBS, 1% BSA to obtain final 10^{10}-fold dilutions of the initial 263K brain material prior to inoculation.

In hamsters inoculated with the 263K scrapie strain, the asymptomatic period of infection lasts 60 to 160 days followed by a stereotypic clinical progression leading to death two to three weeks later. Individual symptoms, such as wobbling gait and head bobbing, are readily recognized but their onset is subtle and subject to large inter (and even intra) observer variability. Incubation time determinations are greatly improved by an empirical determination of endpoint [23]. The adult body weight of asymptomatic or uninfected hamsters is stable or increases slowly during adulthood but drops precipitously during symptomatic disease. Hamsters showing clear signs of early scrapie were individually caged and weighed daily. The weight history was plotted against time with a reference endpoint line marking 20% of the maximum weight registered. Animals were euthanized when their body weights dropped below 20% of maximum body weight and the incubation endpoint was taken as the time intercept of the 20% line.

3.2.3 Proteinase K assay

To analyse production of PK-resistant PrP material in PMCA, 15 μl of each sample were supplemented with 2.5 μl SDS and 2.5 μl PK, to a final concentration of SDS and PK of 0.25% and 50 μg/ml respectively, followed by incubation at 37°C for 1 hr. The digestion was terminated by addition of SDS-sample buffer and boiling for 10 min. Samples were loaded onto NuPAGE 12% BisTris gels, transferred to PVDF membrane, and stained with 3F4 or D18 antibody for detecting hamster or mouse PrPs, respectively.

To analyse scrapie brain homogenates, an aliquot of 10% brain homogenate was mixed with an equal volume of 4% sarcosyl in PBS, supplemented with 50 mM Tris, pH 7.5, and digested with 20 μg/ml PK for 30 min at 37°C with 1000 rpm shaking (Eppendorf thermomixer). The reaction was stopped by adding 2 mM PMSF and SDS sample buffer. Samples were boiled for 10 min and loaded onto NuPAGE 12% BisTris gels. After transfer to PVDF membrane, PrP was detected with 3F4 antibody.

3.2.4 Quantification of PrP^{Sc} by Dot Blotting

To obtain calibration curves for calculating of PMCA fold amplification, 10% brain homogenate from 263K animals was sonicated for 1 min and serially diluted into 10% NBH sonicated for 30 s. PMCA samples as well as 263K dilutions were digested with 50 μg/ml PK for 1 hr at 37°C. The reaction was stopped by addition of 2 mM PMSF.

All samples were diluted 10-fold in PBS, and analysed using a 96-well immunoassay similar to those previously published [30]. Procedure employed the Bio-Dot microfiltration system (Bio-Rad, Hercules, and CA) used according to the instruction manual. 50 μl of diluted samples were loaded into each well and allowed to bind to a 0.45 μm Trans-Blot nitrocellulose membrane (Bio-Rad, Hercules, and CA). Following two washes with PBS, the membrane was removed, incubated for 30 min in 6 M GdnHCl to enable PrP denaturation, washed, and probed with 3F4 antibody according to the standard immunoblotting procedure. Chemiluminescent signal from the membrane was collected with a Typhoon 9200 Variable Mode Imager (Amersham Biosciences, Piscataway, NJ) and quantified with ImageQuant software (Amersham Biosciences).

3.2.5 Atomic Force Microscopy Imaging of rPrP-Fibrils

Full-length hamster rPrP (residues 23–231, no tags) was expressed and purified as previously described [27, 28]. To prepare fibrils, the fibrillation reactions were conducted in 2 M GdnHCl, 50 mM MES, pH 6.0 at 37°C at slow agitation (~60 rpm) and rPrP concentration of 0.25 mg/ml [31]. 0.2 ml PCR tubes containing 100 μl solution of rPrP fibrils (10 μg/ml) dialyzed into 5 mM sodium acetate, pH 5.0, were placed in a rack fixed on the top of Misonix S-4000 microplate horn filled with 300 ml water and sonicated in the absence or presence of five small Teflon beads at 200 W ultrasound power for 30 s. Then, 5 μl of each sample was placed on a freshly cleaved piece of mica, incubated for 5 min, washed gently with Milli Q water, dried on air, and analysed using a Pico LE AFM system (Agilent Technologies, Chandler, AZ) equipped with a PPP-NCH probe (Nanosensors, Switzerland) and operated in tapping mode. Topography/amplitude images (square size of 2 to 10 μm, 512×512 pixels) were obtained at 1 line/s scanning speed. Tip diameter was calibrated by obtaining images of 5 nm gold particles (BBinternational, UK) under the same scanning conditions. Images were analysed with PicoScan software supplied with the instrument. Particle height was calculated directly from the topography profiles, while width and length were measured at the half-height and corrected for the tip diameter. Dimensions of 130–150 particles for each group from three independent experiments were measured.

3.3 DISCUSSION

The current studies demonstrated that the yield and the rate of prion conversion in PMCA can be substantially improved by including beads. Remarkably, substantial improvements in the amplification efficiency and robustness did not come at the cost of prion replication specificity. While beads were found to increase the amplification rate of two hamster strains in hamster NBHs, no detectible amplification of these strains were observed in mouse NBHs within three rounds. This shows that the species specificity was preserved (Figure 2). Furthermore, beads were found to help counteract the negative effect of rPrP on amplification. It is tempting to speculate that the PMCAb format will help to improve the sensitivity of prion detection in body fluids such as blood or urine that might contain inhibitory compounds. Considering substantial enhancement in amplification yield, efficiency and robustness, PMCAb is a promising new platform for developing sensitive and rapid tests for prions, and producing PrP^{Sc} *in vitro* for structural studies.

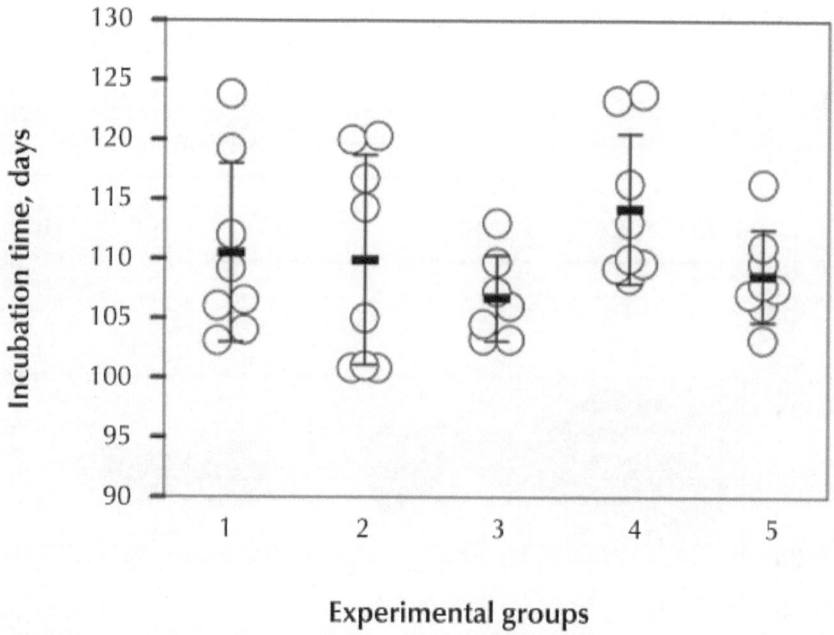

FIGURE 2 The PMCA with beads preserves species barrier.

The effect of beads on prion amplification can be explained by several mechanisms. Using amyloid fibrils produced from rPrP, we showed that sonication in the presence of beads effectively fragmented rPrP fibrils into pieces that were substantially smaller than those observed in the absence of beads (Figure 3). This result suggests that beads might improve the efficiency of PrP^{Sc} fragmentation. Consistent with this mechanism, beads were found to enhance significantly the amplification efficiency of SSLOW PrP^{Sc}, a strain that is deposited in the form of large plaques [25]. Sonication may not only fragment PrP^{Sc} particles but could also irreversibly damage or denature PrP^{Sc} and/or PrP^{C}. We observed that during sonication, the beads rose from the bottom of the tubes and vibrated in the reaction mixtures. Perhaps, the presence of beads helps to redistribute the cavitation energy of bubbles into the much "softer" energy of mechanical vibration, making the conditions for breaking PrP^{Sc} particles more optimal.

In addition to more efficient fragmentation of PrP^{Sc} particles, the effect of beads could be attributed to a breakage of cellular debris and an increase in the accessibility of PrP^{C} and/or cellular cofactors essential for conversion. Considering that different strains or strains from different species might utilize a variety of cellular cofactors of different chemical natures [13], optimizing PMCA amplification might require a different bead material for some strains. Nevertheless, it is currently not known whether any of the proposed mechanisms provides an actual physical explanation for the effect of beads, which at this time should be considered empirical. While the mechanism of bead-induced effect remains to be elucidated in future studies, the PMCAb format offers immediate practical benefits.

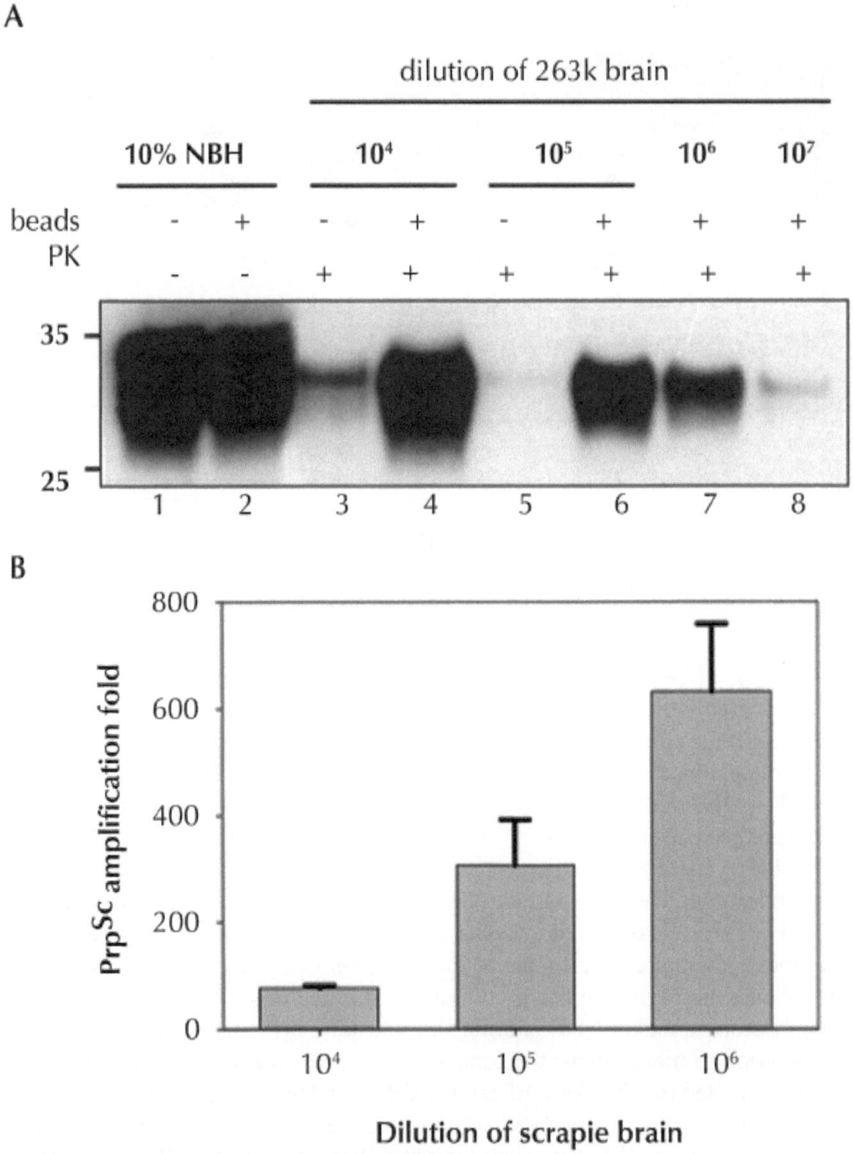

FIGURE 3 The AFM imaging of rPrP fibril fragmentation.

In previous studies, only a small subfraction of PrP^C could be converted into PrP^{Sc} in PMCA, which raised concerns that only a fraction of PrP^C is susceptible to conversion. In an attempt to improve the conversion yield, an increase in the number of

PMCA cycles [2] or the application of NBH from transgenic mice with high expression of endogenous PrP^{C} was employed [20, 21]. However, it is unclear whether these approaches can favorably change the balance between productive conversion and the competing reactions, which might include spontaneous oxidative modification of PrP^{C} [27], the self-cleavage of PrP^{C} [28] and unproductive misfolding. Increasing the time of a PMCA round or the concentration of a substrate is likely to impact both productive and unproductive pathways. The current work shows that an alternative approach that relies on a simple technical modification in the reaction format could be much more rewarding than biochemical approaches. An increase in the conversion yield suggested that beads selectively accelerate the rate of productive conversion of PrP^{C} into PrP^{Sc} without affecting competing reactions. Remarkably, a substantial fraction if not 100% of PrP^{C} could be converted into PrP^{Sc} in PMCAb (Figure 4). This result argues that the amplification yield is not limited to a small subfraction of PrP^{C} susceptible to conversion or by cellular cofactors involved in the conversion reactions.

The most beneficial effect of beads on amplification was observed at high speed dilutions, that is . at high PrP^{C}/PrP^{Sc} ratios when the supply of PrP^{C} was unlimited (Figure 6). In this case, beads improved the sensitivity of detection by at least two or three orders of magnitude. When seeded with high concentrations (10^3 or 10^4-dilution of scrapie brain material), the differences in amplification yield between PMCA and PMCAb was approximately 10-fold (Figure 4). A 10-fold difference was also consistent with the difference in mean incubation time observed between the two groups after inoculation (Figure 5). However, this difference must be considered tentative due to the low statistical significance of this measurement. Regardless, the bioassay confirmed that PMCAb amplifies prion infectivity at least equivalently to PMCA.

In experience, prion amplification in PMCA is very sensitive to technical settings such as the precise position of a tube within the microplate horn, that is the distance of a tube from horn's surface and its center; the age of the sonicator's horn; the tube's shape. Furthermore, aging of sonicatior's horn and individual patterns of horn erosion with age cause time- and position dependent variations in sonication power. As a result, it is difficult to obtain consistent amplification of PrP^{Sc} in experiments performed in different sonicators or even using the same sonicator as it ages. For instance, the differences in the yield of PrP^{Sc} amplification seen in lanes B6 and A1 in Figure 4 were attributed to the aging of the sonicator's horn, as both of these experiments were performed using the same sonicator but at a slightly different age. In experience, Teflon beads significantly improve the robustness of PMCA making prion amplification less sensitive to technical variations, which are difficult to control. The new format should help to establish a PMCA-based approach for assays of prion infectivity.

3.4 RESULTS

3.4.1 Beads Significantly Improve the Yield and Rate of PMCA Reactions

In the past, we found that beads with diameters of 1.59 mm (referred to as small or S) or 2.38 mm (referred to as large or L) significantly accelerated the formation of amyloid fibrils of rPrP *in vitro* [22]. Here, we tested whether beads have any effects on the rate of prion amplification in PMCA. In standard PMCA (sonication for 30 s every 30 min, 48 cycles total, no beads), the typical yield of conversion of PrP^{C} into PrP^{Sc}

was approximately 10% as judged by Western blotting. This amplification yield was consistent with previous studies on amplification of the 263K strain. In the presence of beads, however, the conversion yield improved significantly and approached 100% when three large or five small beads were used.

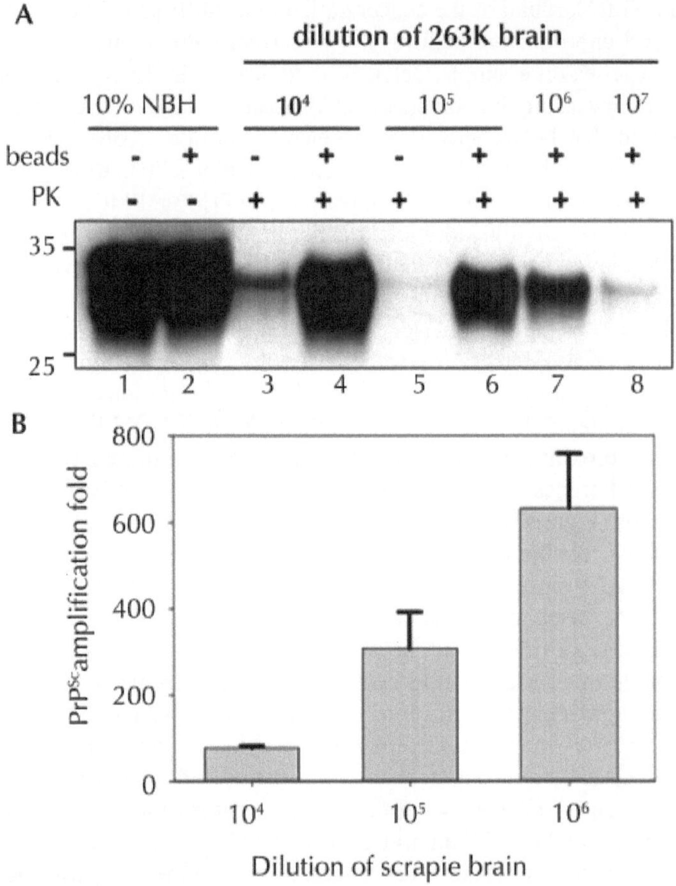

FIGURE 4 Beads improve the yield and rate of PrP^{Sc} conversion.

The kinetics of PrP^{Sc} amplification monitored by Western blotting revealed that in the absence of beads the newly generated PrP^{Sc} was detected by the 16th cycle, whereas in the presence of beads it already was seen by the 8th cycle. Furthermore, in the presence of beads the reaction reached a plateau in only 24 cycles and produced a much higher yield. These results illustrated that beads with diameters of 1.59 or 2.38 mm improved both the yield and the rate of 263K conversion. When beads of submillimeter diameter (800, 400, 200 or 100 μm, see Methods) were used instead, no noticeable increase in PrP^{Sc} amplification was observed.

3.4.2 PMCAb Amplifies Prion Infectivity

To test whether the products of PMCAb were infectious, the reactions were seeded with 10^4-diluted 263K brain material and subjected to amplification in the presence or absence of beads for 6 rounds of 48 cycles each. The products of each round were diluted 10-fold into fresh NBH for the subsequent round. The PMCA products from the final round were then diluted an additional 10-fold prior to inoculation of 50 μl per animal. The final dilution of the initial 263K brain material was 10^{10}-fold. In laboratory the concentration of 263K scrapie in the brains of hamsters in the late stages of symptomatic disease is consistently between 1 and 2×10^{10} infectious $dose_{50}$/g of brain [23]. In the absence of amplification, a 10^{10} dilution of 263K brain would contain 1 ID_{50}/ml giving a probability of infection of 0.05 from a 50 μl inoculation [23]. Nevertheless, all animals inoculated with PMCA products formed in the presence or absence of beads developed clinical disease with the mean value of endpoint 108.6 ± 3.9 or 114.2 ± 6.3 days post inoculation, respectively (Figure 5, groups 4 and 5, respectively). Incubation time endpoints were determined empirically as described in the methods and [23]. The reference group inoculated with 10^4-diluted 263K brain reached the endpoint by 110.5 ± 7.5 days (Figure 5, group 1). Bioassays of two 263K brain homogenates sonicated for 48 PMCA cycles (1 round) in the absence of a substrate revealed that sonication of PrP^{Sc} per se did not notably change its infectivity level regardless of whether beads were present or absent during the sonication cycles (Figure 5, groups 2 and 3, respectively). Similar amounts of PrP^{Sc} were found in the brains from all animal groups.

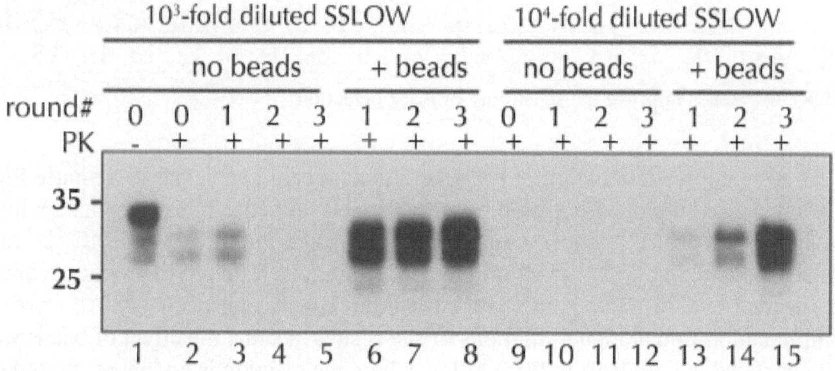

FIGURE 5 Bioassay of PMCA products.

The bioassay experiment confirmed that prion infectivity is amplified in PMCAb. Without a titration experiment, it is difficult to establish accurate infectivity titers of PMCA or PMCAb products. Nevertheless, considering that group 5 gave the same incubation times as group 1, even though the amplification products were diluted an additional 10-fold prior to inoculation, the infectivity dose of PMCAb products appeared to be 10-fold higher than the dose in 10^4-diluted 263K brain material.

The sensitivity of PrP^{Sc} detection is improved in PMCAb. To test whether the application of beads improves the detection limit, serially diluted 263K brain homogenate was used to seed the PMCA reactions that consisted of 48 cycles. In the absence of beads, seeding with 10^3-fold and with 10^4-fold diluted scrapie brains gave sufficient amplification of PrP^{Sc} to be detected by Western blotting. In the presence of beads, however, the reactions seeded with 10^6-fold diluted 263K brains showed consistent, reproducible amplification for subsequent detection by Western blotting (Figure 6). Frequently, sufficient amplification of PrP^{Sc} for detection by Western blotting was observed in the reactions with beads seeded with 10^7-fold diluted scrapie brains. Therefore, within 48-cycle PMCA, beads improved the sensitivity of detection by at least 2 or 3 orders of magnitude. To rule out the possibility of the PrP^{Sc} formation *de novo*, 32 unseeded reactions were conducted, each of which consistent of three rounds of serial PMCAb (sPMCAb). None of them showed PK-resistant material on Western blotting.

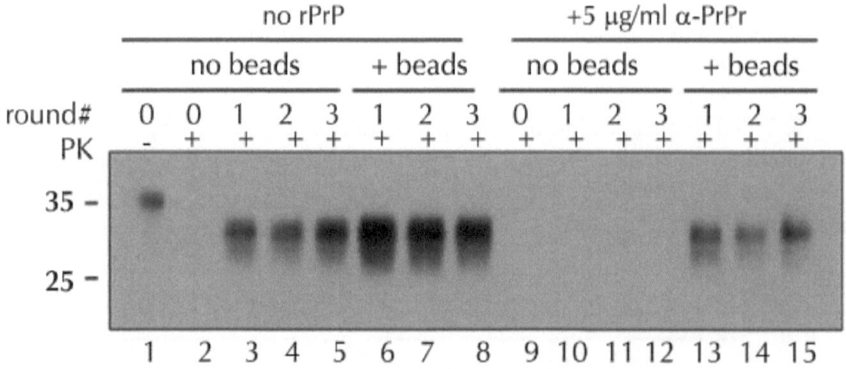

FIGURE 6 Beads improve the sensitivity of PrP^{Sc} detection.

To estimate quantitatively the PrP^{Sc} amplification fold achieved in a single PMCAb round, we employed dot blotting as it provides a better linear response within a broader range of PrP^{Sc} concentrations than the Western blotting. The PMCAb reactions seeded with 10^4-, 10^5-, or 10^6-diluted 263K brain were found to produce reliable amplification by ~75-, 300-, and 635-fold within 48 cycles, respectively. An increase in amplification fold at higher dilutions of seeds suggests that the effect of beads was most beneficial at high PrP^C to PrP^{Sc} ratios, where the reaction is no longer limited by the concentration of a substrate and/or cofactors.

To estimate the PrP^{Sc} amplification fold using an alternative approach, sPMCA reactions consisted of three rounds were performed with the dilution factors between the rounds ranging from 1:10 to 1:1000. In the absence of beads, we observed a gradual decrease in the signal intensity as a function of PMCA round at dilutions of 1:20 indicating that the amplification fold in each round was slightly lower than 20. In the presence of beads, however, the signal was stable at 1:100 dilutions but decayed at

1:1000 dilutions, suggesting that the amplification fold in each round was higher than 100 but less than 1000. This experiment confirmed that up to several hundred fold amplification could be achieved in one PMCAb round consisted of 48 cycles, if the reaction is not limited by substrate and cofactors.

3.4.3 Detection of Minute Amounts of PrP^{Sc}

In previous studies, sPMCA of serially diluted 263K brain homogenate was used to determine the last dilution that still contained PrP^{Sc} particles [2]. Three out of four reactions seeded with 10^{12}-diluted 263K brain material were found to be positive, while five to seven sPMCA rounds, each consisting of 144 cycles, were required to amplify 10^{12}-diluted 263K to levels detectible by Western blotting [2]. To test the effectiveness of PMCAb in amplifying minute quantities of PrP^{Sc}, 263 K brain homogenate was serially diluted up to 10^{14}-fold and then amplified in sPMCAb, where each round consisted of 48 cycles. 10^{12}-diluted 263K brain material was detected in four out of eight reactions in the third round. An increase in number of rounds to six did not increase the percentile of positive reactions seeded with 10^{12}-diluted 263K brain nor did it reveal any positive signals in reactions seeded with 10^{14}-diluted 263K brain. 10^{10}-diluted 263K brains showed a positive signal in all independent reactions. Non-seeded reactions or reactions seeded with NBH from old animals showed no positive signals in PMCAb. These results are consistent with the previous studies where brain material diluted 10^{12}-fold detected PrP^{Sc} and showed stochastic behavior [2] consistent with a limiting dilution of the signal [24]. In the current experiments, PMCAb achieved the same level of sensitivity as PMCA in $1/7^{th}$ of the time and with no evidence of spontaneous conversion from NHB substrate.

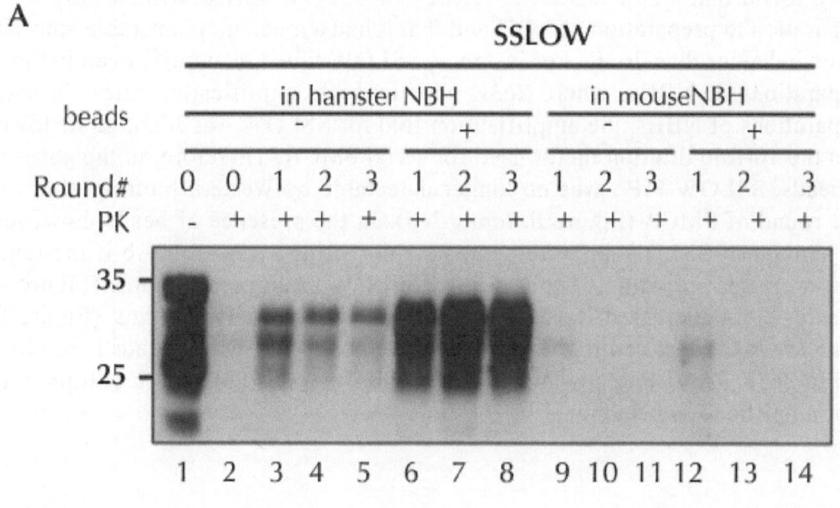

FIGURE 7 *(Continued)*

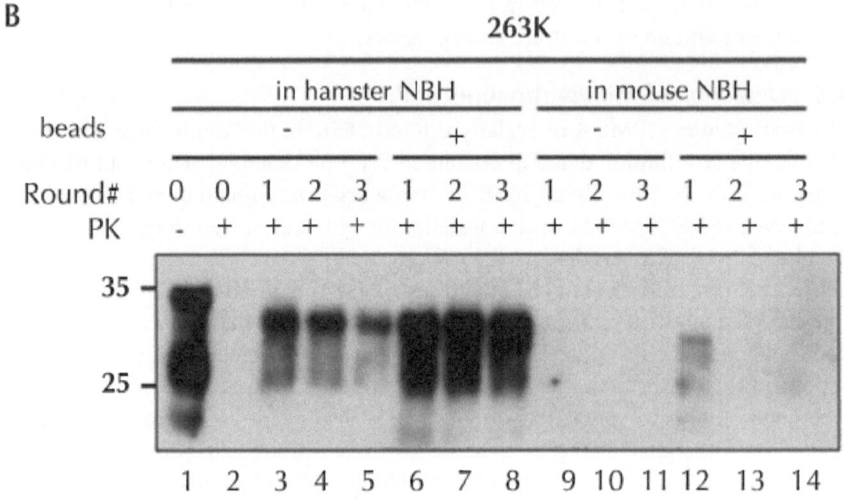

FIGURE 7 Amplification of minute amounts of PrP^{Sc} in PMCAb beads improve the amplification rate of a synthetic prion strain.

To test whether the positive effect of beads on prion amplification was limited to 263K, we used a synthetic prion strain, SSLOW, which was previously found to have a very peculiar amplification behavior in PMCA [25]. Previously we found that amplification efficiency of SSLOW varied significantly from preparation to preparation of NBH and that it had a much more unstable amplification behavior than 263K. For instance, SSLOW failed to amplify even in those preparations of NBHs, where 263K showed high amplification rates. In such preparations of NBHs, the amplification fold for SSLOW was found to be lower than the 10-fold dilution factor used for serial PMCA. Therefore, in the absence of beads, SSLOW PrP^{Sc} was no longer detectable by Western blotting after the first round of PMCA (Figure 8, *lanes 3–5*). In the presence of beads, however, the amount of SSLOW PrP^{Sc} remained stable during serial PMCAb if the reactions were seeded with 10^3-fold diluted SSLOW brain homogenates (Figure 8, *lanes 6–8*), or increased if 10^4-fold dilutions were used for seeding (Figure 8, *lanes 13–15*). These results illustrate that the positive effect of beads is not limited to 263K and that beads improved the robustness of PMCA for a strain with poor amplification behavior.

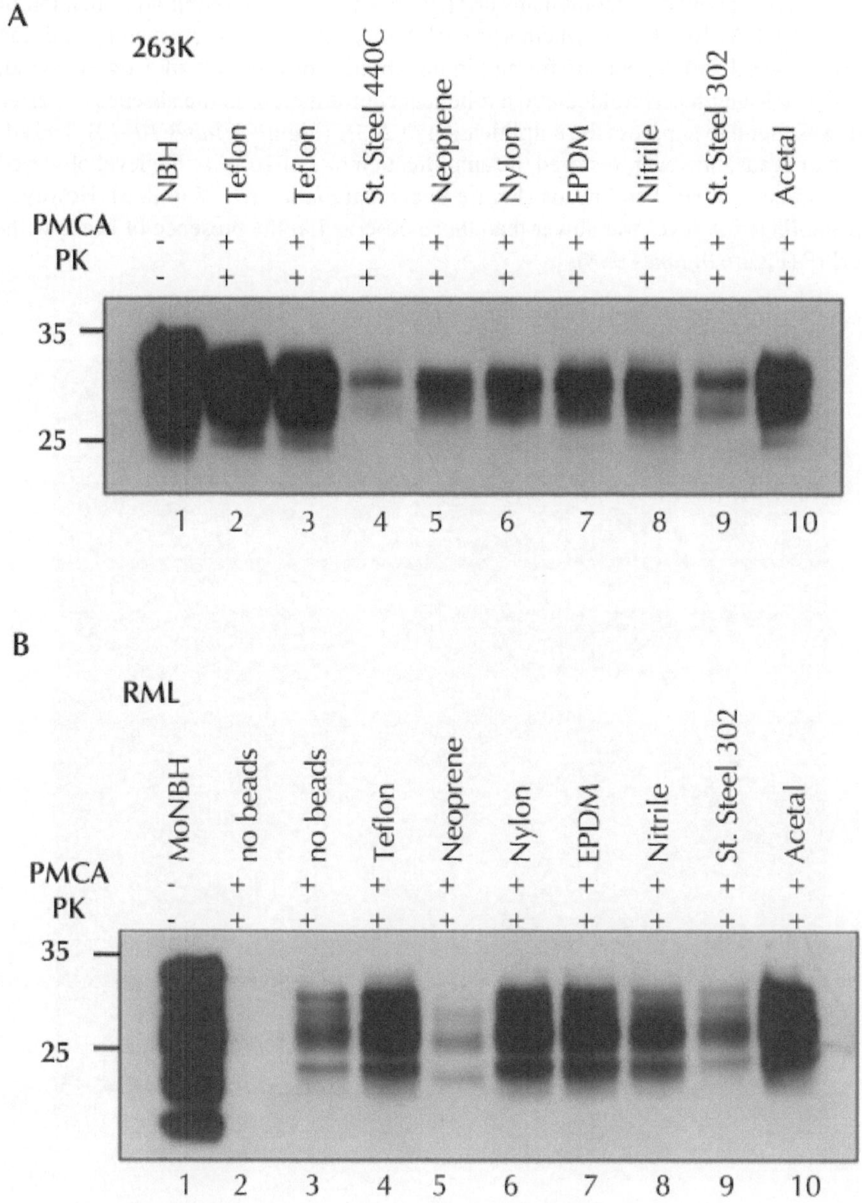

FIGURE 8 Beads improve the amplification efficiency of SSLOW.

In previous studies, recombinant PrP (rPrP) was found to inhibit amplification of PrP^{Sc} in PMCA [16]. To test whether the inhibitory effect can be rescued by addition of beads, serial PMCA was performed in the absence or presence of 5 μg/ml Syrian hamster full-length rPrP folded into a α-helical conformation. In the absence of beads, rPrP was found to suppress the amplification of 263K (Figure 9, *lanes 10–12*). The addition of beads, however, restored the amplification rate of 263K to the level observed in the absence of rPrP and beads (Figure 9, compare *lanes 13–15* to *3–5*). However, this amplification level was lower than those observed in the presence of beads without rPrP (Figure 9, *lanes 6–8*).

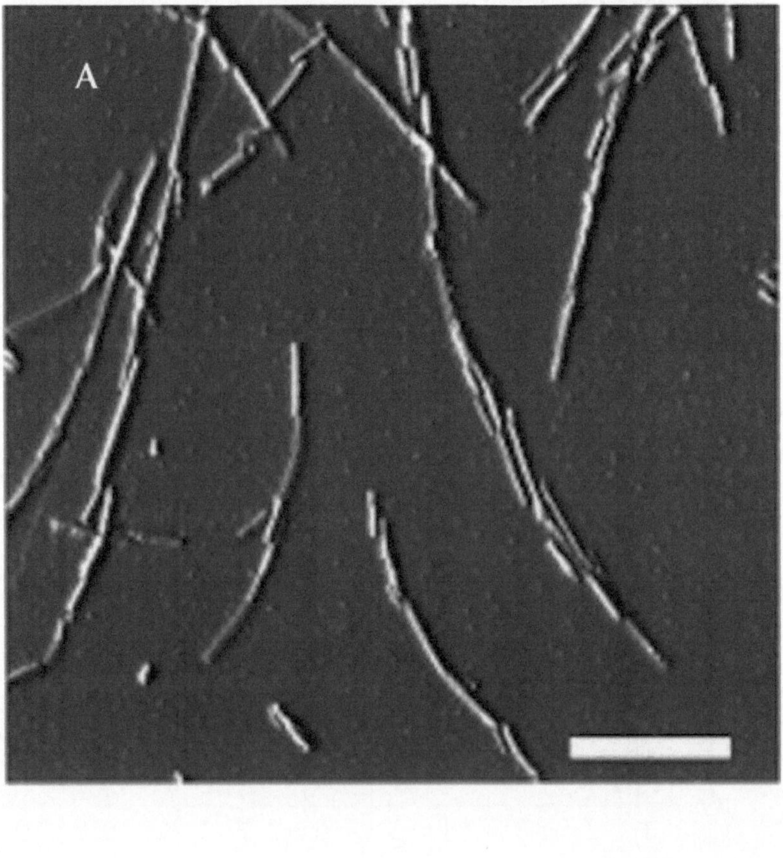

FIGURE 9 *(Continued)*

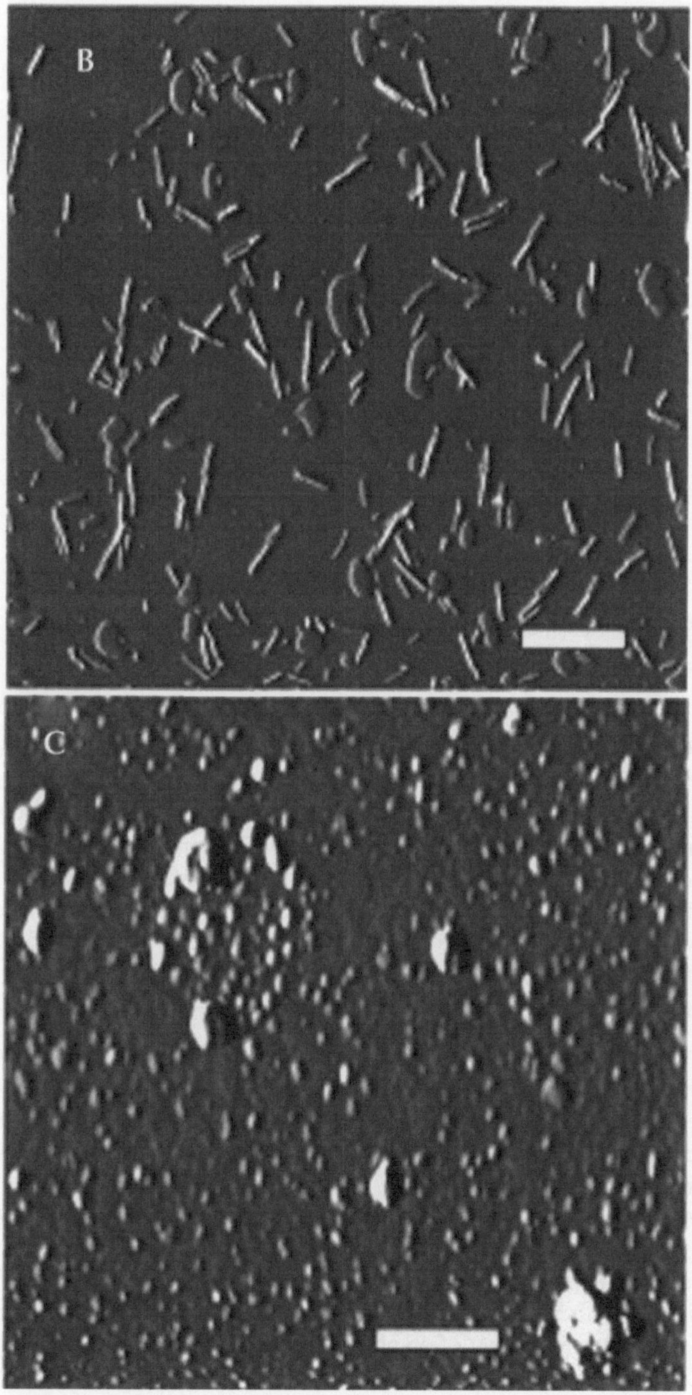

FIGURE 9 *(Continued)*

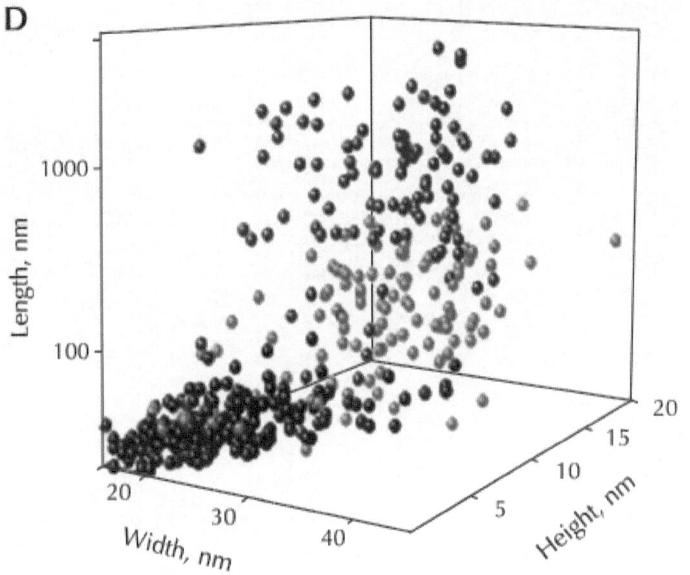

FIGURE 9 Beads counteract the negative effect of rPrP on PrP^{Sc} amplification. Species-specificity is preserved in PMCAb.

3.4.4 Effect of Bead Material on Efficiency of Amplification

To test whether efficiency of amplification depends on the bead material, beads made from eight different materials including Teflon beads purchased from two companies were used for amplification of 10^5-fold diluted 263K or 10^4-fold diluted RML. Beads made from Teflon and acetal showed the best amplification efficiency for both strains. Nylon and EPDM beads showed very good performance in amplifying RML, but were less efficient for 263K. Notably, the ranking orders in amplification efficiency for different materials appeared to be strain- or species-dependent. The detailed relationship between the bead material and their efficiencies to amplify different scrapie strains or strains from different species will be explored in future studies.

Sonication with beads fragments amyloid fibrils of rPrP into small pieces

To gain insight into the effect of beads on prion amplification, we tested whether beads affect the fragmentation efficiency of PrP aggregates during sonication. Amyloid fibrils produced from rPrP were sonicated in the presence or absence of beads, and the size of fibrillar fragments was analysed using atomic force microscopy (AFM) imaging. Consistent with studies [26], sonication was found to break fibrils into smaller fragments. Sonication in the presence of beads, however, reduced the size of fibrillar fragments even more producing smaller particles. In fact, AFM imaging revealed that after sonication with beads, the fibrillar fragments appeared as small oligomers.

To test whether efficiency of amplification depends on the bead material, beads made from eight different materials including Teflon beads purchased from two companies were used for amplification of 10^5-fold diluted 263K or 10^4-fold diluted RML. Beads made from Teflon and acetal showed the best amplification efficiency for both strains. Nylon and EPDM beads showed very good performance in amplifying RML, but were less efficient for 263K. Notably, the ranking orders in amplification efficiency for different materials appeared to be strain or species-dependent. The detailed relationship between the bead material and their efficiencies to amplify different scrapie strains or strains from different species will be explored in future studies.

Sonication with beads fragments amyloid fibrils of rPrP into small pieces

To gain insight into the effect of beads on prion amplification, we tested whether beads affect the fragmentation efficiency of PrP aggregates during sonication. Amyloid fibrils produced from rPrP were sonicated in the presence or absence of beads, and the size of fibrillar fragments was analyzed using atomic force microscopy (AFM) imaging. Consistent with studies [26], sonication was found to break fibrils into smaller fragments. Sonication in the presence of beads, however, reduced the size of fibrillar fragments even more producing smaller particles. In fact, AFM imaging revealed that after sonication with beads, the fibrillar fragments appeared as small oligomers.

3.5 CONCLUSION

Prion amplification in PMCA was previously shown to exhibit species specificity that faithfully reflects the transmission barrier observed in animals [17, 18]. Considering that beads were found to improve significantly the amplification efficiency. Consistent with our results, beads improved the conversion yield for both strains when they were amplified in Syrian hamster NBH. However, when 263K or SSLOW were diluted with mouse NBH no detectible amplification was observed for at least three serial PMCA rounds in the presence or absence of beads. A control experiment revealed that mouse (Rocky Mountain Laboratory) strain could be amplified in mouse NBH. Therefore, the lack of detectible amplification of hamster strains in serial PMCA in mouse NBH confirmed that the presence of beads does not eliminate the species barrier. Taken together, these results illustrate that significant improvements in amplification efficiency do not come at the expense of amplification specificity.

KEYWORDS

- **Normal brain homogenate**
- **Protein misfolding cyclic amplification**
- **Scrapie**
- **Sonication**
- **Western blotting**

ACKNOWLDEGMENT

This work was supported by NIH grant NS045585 to Ilia V. Baskakov Baltimore Research and Education Foundation and the Beatriu de Pinos Fellowship to Nuria

Gonzalez-Montalban with the support of the Commission for Universities and Research of the Department of Innovation, Universities and Enterprise of the Government of Catalonia. The funders had no role in study design, data collection and analysis, decision to publish, or preparation of the manuscript.

ETHICS STATEMENT

This study was carried out in strict accordance with the recommendations in the Guide for the Care and Use of Laboratory Animals of the National Institutes of Health. The protocol was approved by the Institutional Animal Care and Use Committee of the University of Maryland, Baltimore (Assurance Number A32000-01, Permit Number. 0309001).

REFERENCES

1. Saborio, G. P., Permanne, B., and Soto C. Sensitive detection of pathological prion protein by cyclic amplification of protein misfolding. *Nature* **411**, 810–813 (2001).
2. Saa, P., Castilla, J., and Soto C. Ultra-efficient replication of infectious prions by automated protein misfolding cyclic amplification. *J. Biol. Chem.* **281**, 35245–35252 (2006).
3. Gonzalez-Romero, D., Barria, M. A., Leon, P., Morales, R., and Soto, C. Detection of infectious prions in urine. *FEBS Lett.* **582**, 3161–3166 (2008).
4. Murayama, Y., Yoshioka, M., Okada, H., Takata, M., Tokashi, Y., et al. Urinary excretion and blood level of prions in scrapie-infected hamsters. *J. Gen. Virol.* **88**, 2890–2898 (2007).
5. Shikiya, R. A., Ayers, J. I., Schutt, C. R., Kincaid, A. E., and Bartz, J. C. Coinfecting prion strains compete for a limiting cellular resource. *J. Virol.* **84**, 5706–5714 (2010).
6. Saa, P., Castilla, J., and Soto, C. Presymptomatic detection of prions in blood. *Science* **313**, 92–94 (2006).
7. Tattum, M. H. Jones, S., Pal, S., Collinge, J., and Jackson, G. S. Discrimination between prion-infected and normal blood samples by protein misfolding cyclic amplification. *Transfusion* **50**, 996–1002 (2010).
8. Haley, N. J., Mathiason, C. K., Zabel, M. D., Telling, G. C., and Hoover, E. A. Detection of sub-clinical CWD infection in conventional test-negative deer long after oral exposure to urine and feces from CWD+ deer. *PLoS One* **4**, e7990 (2009).
9. Barria, M. A., Mukherjee, A., Gonzalez-Romero D., Morales, R., and Soto, C. De Novo generation of infectious prions in vitro produces a new disease phenotype. *PLoS Pathog.* **5**, e1000421(2009).
10. Deleault, N. R., Harris, B. T., Rees, J. R., and Supattapone, S. Formation of native prions from minimal components in vitro. *Proc. Acad. Natl. Sci. U. S. A.* **104**, 9741–9746 (2007).
11. Wang, F., Wang, X., Yuan, C-G., Ma, J. Generating a prion bacterially expressed recombinant prion protein. *Science* **327**, 1132–1135 (2010).
12. Deleault, N. R., Geoghegan, J. C., Nishina, K., Kascsak, R., Williamson, R. A., et al. Protease-resistant prion protein amplification reconstituted with partially purified substrates and synthetic polyanions. *J. Biol. Chem.* **280**, 26873–26879 (2005).
13. Deleault, N. R., Kascsak, R., Geoghegan, J. C., and Supattapone, S. Species-dependent differences in cofactor utilization for formation of the protease-resistant prion protein in vitro. *Biochemistry* **49**, 3928–3934 (2010).
14. Deleault, N. R., Lucassen, R. W., and Supattapone, S. RNA molecules stimulate prion protein conversion. *Nature* **425**, 717–720 (2003).
15. Mays C. E., and Ryou, C. Plasminogen stimulates propagation of protease-resistant prion protein in vitro. *FASEB J.* (2010), in press.

16. Nishina, K., Deleault, N. R., Mahal, S., Baskakov, I., Luhrs, T., et al. The stoichiometry of host PrP^{C} glycoforms modulates the efficiency of PrP^{Sc} formation in vitro. *Biochemistry* **45**, 14129–14139 (2006).
17. Castilla, J., Gonzalez-Romero, D., Saa, P., Morales, R., De Castro, J. et al. Crossing the species barrier by PrP^{Sc} replication in vitro generates unique infectious prions. *Cell* **134**, 757–768 (2008).
18. Green, K. M., Castilla, J., Seward, T. S. Napier, D. L. Jewell, J. E. et al., Accelerated high fidelity prion amplification within and across prion species barriers. *PLoS Pathog.* **4**, e1000139 (2008).
19. Meyerett, C., Michel, B., Pulford, B., Sparker, T. R., Nichols, T. A., et al. In vitro strain adaptation of CWD prions by serial protein misfolding cyclic amplification. *Virology* **382**, 267–276 (2008).
20. Mays, C. E., Titlow, W., Seward, T., Telling, G. C., and Ryou, C. Enhancement of protein misfolding cyclic amplification by using concentrated cellular prion protein source. *Biochem. Biophys. Res. Commun.* **388**, 306–310 (2009).
21. Kurt, T. D., Perrott, M. R., Wilusz, C. J., Wilusz, J., Supattapone S., et al. Efficient in vitro amplification of chronic wasting disease PrP-res. *J Virol* **81**:9605–9608(2007).
22. Bocharova, O. V., Breydo, L., Parfenov, A. S., Salnikov, V. V., and Baskakov, I. V. In vitro conversion of full length mammalian prion protein produces amyloid form with physical property of PrP^{Sc}. *J. Mol. Biol.* **346**, 645–659 (2005).
23. Gregori, L., Lambert, B.C., Gurgel P. V., Gheorghiu, L., Edwardson, P., et al. Reduction of transmissible spongiform encephalopathy infectivity from human red blood cells with prion protein affinity ligands. *Transfusion* **46**, 1152–1161 (2006).
24. Gregori, L., McCombie, N., Palmer, D., Birch, P., Sowemimo-Coker, S. O., et al. Effectiveness of leucoreduction for removal of infectivity of transmissible spongiform encephalopathies from blood. *Lancet* **364**, 529–531 (2004).
25. Makarava, N., Kovacs, G. G., Bocharova, O. V., Savtchenko, R., Alexeeva, I., et al. Recombinant prion protein induces a new transmissible prion disease in wild type animals. *Acta Neuropathol.* **119**, 177–187 (2010).
26. Sun, Y., Makarava, N., Lee, C. I., Laksanalamai, P., Robb, F. T., et al. Conformational stability of PrP amyloid firbils controls their smallest possible fragment size. *J. Mol. Biol.* **376**, 1155–1167 (2008).
27. Breydo, L., Bocharova, O. V., Makarava, N., Salnikov, V. V., Anderson, M., et al. Methionine oxidation interferes with conversion of the prion protein into the fibrillar proteinase K-resistant conformation. *Biochemistry* **44**, 15534–15543 (2005).
28. Ostapchenko, V. G., Makarava, N., Savtchenko, R., and Baskakov, I. V. The polybasic N-terminal region of the prion protein controls the physical properties of both the cellular and fibrillar forms of PrP. *J. Mol. Biol.* **383**, 1210–1224 (2008).
29. Castilla, J., Saa, P., Hetz, C., and Soto, C. In vitro generation of infectious scrapie prions. *Cell* **121**, 195–206 (2005).
30. Kramer, M. L. and Bartz, J. C. Rapid, high-throughput detection of PrP^{Sc} by 96-well immunoassay. *Prion* **3**, 44–48 (2009).
31. Makarava, N. and Baskakov, I. V. The same primary structure of the prion protein yields two distinct self-propagating states. *J. Biol. Chem.* **283**, 15988–15996 (2008).

4 Amplification of the Scrapie Isoform of Prion Protein

Jae-Il Kim, Krystyna Surewicz, Pierluigi Gambetti, and Witold K. Surewicz

CONTENTS

4.1 INTRODUCTION

Transmissible spongiform encephalopathies are associated with an autocatalytic conversion of normal prion protein (PrP^C); to a protease-resistant form, PrPres. This autocatalytic reaction can be reproduced *in vitro* using a procedure called protein misfolding cyclic amplification (PMCA). Here we show that, unlike brain-derived PrP^C.; bacterially expressed recombinant prion protein (rPrP) is a poor substrate for PrPres amplification in a standard protein misfolding cyclic amplification (PMCA) reaction. The differences between PrP^C and rPrP appear to be due to the lack of the glycophosphatidylinositol (GPI) anchor in the recombinant protein. These findings shed a new light on prion protein conversion process and have important implications for the efforts to generate synthetic prions for structural and biophysical studies.

Prion diseases, or transmissible spongiform encephalopathies (TSEs), are infectious neurodegenerative disorders that affect many mammalian species and include scrapie in sheep, bovine spongiform encephalopathy in cattle, chronic wasting disease in elk and deer, and Creutzfeldt-Jakob disease (CJD) in humans [1-3]. These diseases are associated with conformational conversion of the cellular prion protein, PrP^{C}; to a misfolded form, PrP^{Sc}. In contrast to PrP^{C}; which is α-helical and sensitive to proteolytic digestion, PrP^{Sc} is rich in β-sheet structure and shows resistance to proteolytic enzymes, with the proteinase K (PK)-resistant core corresponding the C-terminal ~140 residues [1-3]. An increasing body of evidence indicates that the infectious TSE agent is devoid of replicating nucleic acids, consisting mainly—if not solely—of PrP^{Sc} [1-3]. This highly unusual pathogen is believed to be self-propagate by an autocatalytic mechanism involving binding to PrP^{C} and templating its conformational conversion to the PrP^{Sc} state.

Many efforts have been made to recapitulate prion protein conversion and prion propagation in cell-free systems. While early studies have shown that PrP^{C} can be converted to PK-resistant form, PrPres, simply by incubation with PrP^{Sc} from TSE-affected animals [4], the yields of this reaction were very low and no infectivity could be attributed to the newly converted material [5]. An important recent advance was the development of a procedure called PMCA, which, using successive rounds of incubation and sonication, is able to replicate and indefinitely amplify the PrPres conformer employing PrP^{C} present in brain homogenate as a substrate [6]. Remarkably, the newly generated PrPres was shown to cause TSE disease in experimental animals [7], suggesting that it faithfully replicates the structure of brain PrP^{Sc}. Furthermore, PMCA-based reactions could reproduce the phenomena of species barriers and prion strains [8, 9]. Infectious prions could also be produced by PMCA employing purified PrP^{C} as a substrate (though only in the presence of co-purified lipids and synthetic poly(A) RNA molecules) [10], providing strong support to the protein-only hypothesis of prion diseases.

Despite growing importance of PMCA technology in prion research, the molecular basis of these conversion reactions remains unclear. Here we report data pointing to an important role of the (GPI) anchor as a modulator of PrPres amplification *in vitro* by PMCA.

4.2 MATERIALS AND METHODS

4.2.1 Reagents

Phosphatidylinositol-specific phospholipase C (PI-PLC) and PK were purchased from Sigma Aldrich, and peptide:N-glycosidase F (PNGase F) was from New England BioLabs. Mouse monoclonal anti-PrP antibodies 3F4 (recognizing epitope 109–112) and 3F10 (recognizing epitope 135–150) were kindly provided by Drs. Richard Kascsak and Yong-Sun Kim, respectively. Recombinant full-length Syrian hamster prion protein (rShaPrP) was expressed in *Escherichia coli* and purified as described previously [11].

4.2.2 Preparation of Brain Homogenates

Ten percent (w/v) homogenates of normal Syrian golden hamster and PrP-knockout (FVB/$Prnp^{0/0}$) mice brains were prepared in PMCA buffer (PBS containing 1% Triton X-100, 0.15 M NaCl, 4 mM EDTA.; and the Complete protease inhibitor cocktail [Roche]) as described previously [7]. Samples containing delipidated PrP^{C} were prepared by incubating normal hamster brain homogenate (NBH) with PI-PLC (1 unit/ml) for 2 hr at 37°C with shaking. In control experiment, NBH was subjected to the same treatment using PI-PLC inactivated by boiling for 1 hr. For enzymatic deglycosylation, samples were treated with PNGase F according to the manufacturer's instruction. Hamster brains infected with 263K scrapie strain were kindly provided by Dr. Richard Carp.

4.2.3 PMCA Procedures and Immunodetection of PrP

An aliquot (1 µl) of scrapie brain homogenate (SBH) was mixed with 100 µl of NBH or NBH pretreated with active PI-PLC or heat-inactivated PI-PLC.; and the samples were subjected to 48 cycles of PMCA as described previously [7]. For serial PMCA; an aliquot of the product of previous PMCA reaction was diluted ten times into fresh NBH.; and the mixture was again subjected to 48 cycles of PMCA. For PMCA reactions with the rPrP, rShaPrP substrate at various concentrations (1–25 µg/ml) was added to PMCA buffer containing SBH (1/100 dilution) and the samples were subjected to 24, 48, or 80 cycles of PMCA. When indicated, 10% (w/v) PrP-knockout brain homogenate was included in the reaction mixture. For competition experiments, rShaPrP (2.5–25 µg/ml) was added to PMCA mixtures containing SBH seed and NBH (1:100 ratio), and the samples were subjected to 24 cycles of PMCA.

Western blot analysis of PrP using anti-PrP antibodies (3F4 [1:10, 000] or 3F10 [1:5, 000]) was performed as described previously [12]. For the analysis of PrPres amplification by PMCA; samples were treated with PK (50 µg/ml, 37°C.; 1 hr) prior to Sodium dodecyl sulfate polyacrylamide gel electrophoresis (SDS-PAGE). For slot blotting analysis, samples in SDS-PAGE sample buffer were diluted 25-fold with Tris-buffered saline (TBS.; pH 7.5) and applied to nitrocellulose membrane using Bio-Dot SF microfiltration apparatus (Bio-Rad). The membrane was then treated with 3 M guanidine hydrochloride (HCl) (10 min, room temperature), rinsed extensively with TBS.; and probed with anti-PrP 3F10 antibody (which shows greater sensitivity in probing slot blots than 3F4 antibody). Although it has been suggested that negatively charged nylon membrane may be more useful for slot blotting of delipidated PrP^C [13], using our protocol we found nitrocellulose membrane to be equally suitable.

4.3 DISCUSSION

The PMCA, a procedure that allows essentially indefinite amplification of PrPres and TSE infectivity *in vitro* using PrP^C present in normal brain homogenate as a substrate, has emerged as a powerful tool in prion research [7-9]. The molecular mechanism of this amplification reaction remains, however, largely unexplored. Furthermore, there is major interest in expanding this approach to the rPrP , both from the perspective of prion diagnostics as well as fundamental studies of the $PrP^C \rightarrow PrP^{Sc}$ conversion mechanism [19].

Here we report that, under the conditions of the standard PMCA protocol, bacterially expressed rPrP is a very poor substrate for amplification of PrPres from scrapie-infected hamsters. Furthermore, we found that, when added to NBH.; rPrP acts as an efficient inhibitor of the PMCA conversion of brain PrP^C. Building on these observations, we tested the role of PrP^C GPI anchor, finding that removal of this anchor by

PI-PLC treatment renders PrP^C incompetent to act as a substrate for the amplification of PrPres by the standard PMCA protocol.

The finding that native GPI anchor is important for the amplification of PrPres conformation *in vitro* by the PMCA protocol is unexpected, especially since previous studies have shown that both GPI-free PrP^C derived from mammalian cell culture [4, 20] as well as bacterially expressed rPrP [21] can be converted to PrPres in a "discontinuous" cell-free conversion assay developed by Caughey and coworkers [4, 22]. However, the findings of those previous cell-free conversion studies and our present PMCA experiments are not necessarily contradictory as—despite an apparent similarity—the assays used in these two types of studies probe distinct phenomena. Thus, in PMCA experiments a minute quantity of PrP^{Sc} is used as a seed, and essentially an infinite number of substrate PrP^C molecules are converted to PrPres conformation that appears to faithfully replicate that of the PrP^{Sc} template, with the newly generated material being infectious [7]. This implies that, upon binding to the surface of PrP^{Sc} seed, each molecule of the substrate protein acquires catalytic properties of the seed, acting as a new template for the conversion of additional substrate molecules. By contrast, the reactions in discontinuous cell-free conversion assays are characterized by very low yields and appear to lack any autocatalytic properties, with the number of PrP^C molecules converted to PrPres being substantially lower compared to input PrP^{Sc} [4, 22]. Given these substoichiometric yields, it is likely that in these reactions only single substrate molecule can be recruited and converted to PrPres state by each catalytic site of PrP^{Sc} oligomers, effectively "capping" these sites and preventing the conversion of additional substrates. While PrP molecules converted in this reaction acquire seed-like PK-resistance, their three-dimensional structure might be not identical to that of PrP^{Sc}. This would explain both the lack of autocatalytic properties of this reaction as well as an apparent lack of infectivity of the newly converted material [5]. The capping mechanism described earlier may also be operational in our PMCA experiments using rPrP as a substrate, accounting for the observed inhibitory effect of the latter protein on the formation of PrPres from brain-derived PrP^C.

It should be noted that a modified PMCA protocol has been recently developed in which the original PMCA buffer (containing Triton X-100 as a sole detergent) was supplemented with an acidic detergent, SDS [19]. Using this modified PMCA method, it is possible to convert, in an autocatalytic fashion, bacterially expressed rPrP to fibrils apparently displaying PrP^{Sc}-like PK-digestion pattern (with a 16–17 kDa PK-resistant fragment similar to that observed for nonglycosylated PrP^{Sc}). However, this scrapie-like PK-resistance persists only if PK-digestion is performed in the presence of specific detergents used in the PMCA buffer (especially SDS), whereas upon removal of the detergents from already converted material the size of the longest PK-resistant fragment is reduced to ~12 kDa [12]. Thus, while this modified PMCA protocol allows efficient, PrP^{Sc}-seeded conversion of rPrP.; the major structure propagating in this reaction does not fully match that of the PrP^{Sc} seed.

The present findings in the cell-free environment seem to echo previous observations that the absence of the GPI moiety in PrP^C diminishes propagation of PrP^{Sc} in cell culture models of prion infectivity [23, 24]. However, the GPI anchor is not obligatory for prion replication *in vivo*. This is clearly indicated by the recent study showing that transgenic mice expressing anchorless PrP were susceptible to infection with the scrapie agent, though these mice did not develop clinical symptoms and replication of infectivity was reduced as compared to wild-type mice [25]. In view of these data *in vivo*, the apparent inability of GPI-deficient PrP to support amplification of PrPres in PMCA reactions is highly intriguing. Our present data suggest that the conversion reaction of the anchorless PrP may require specific microenvironment and/or cofactors that are present *in vivo*, but are lost in homogenized samples used in PMCA reactions. One critical factor in this regard may be the interactions of the protein (*via* the anchor or directly) with biological membranes. In the case of GPI-containing PrP; these surface interactions (occurring largely through the anchor) are likely to be partially preserved even upon tissue homogenization in the presence of Triton X-100 micelles, allowing for highly efficient PrPres amplification by PMCA. In the absence of the GPI anchor, however, nonspecific surface interactions are likely to be compromised much more readily. It appears that in the latter case, these interactions and a conversion–conducive environment can be mimicked to a certain degree *in vitro* only in the presence of acidic detergents such as SDS.; consistent with a highly basic character of the prion protein.

The present findings are also relevant to understanding the origin of prion infectivity in body fluids such as urine and amniotic fluid [26]. Recent data indicate that PrP present in these fluids is both N-terminally truncated as well as deficient in the GPI anchor (Notari, S., Gambetti, P. and Chen, S., unpublished data). We found out that, at least under certain conditions, PrP lacking the native GPI anchor is a poor substrate for PrPres amplification suggests that PrP^{Sc} associated with prion infectivity detected in urine [27] may be derived from other sources than the delipidated PrP normally present in this body fluid.

4.4 RESULTS

The standard PMCA protocol [7] involves suspension of normal brain homogenate (containing PrP^{C}) in PBS containing 1 percent Triton X-100, 0.15 M NaCl and 4 mM EDTA. A minute quantity of PrP^{Sc} from brain of TSE-infected animals is then added, and the mixture is subjected to multiple rounds of cyclic amplification by incubation and sonication. Consistent with the previous report [7], we found that this procedure results in very efficient conversion of Syrian hamster PrP^{C} to PK-resistant form (PrPres), with the electrophoretic profile of newly generated PrPres faithfully replicating that of input PrP^{Sc}. Here we have made a systematic effort to adopt this PMCA protocol to bacterially expressed rShaPrP. To this end, we dissolved rShaPrP in the standard PMCA buffer, added small quantity of 263K SBHand subjected the samples to cycles of incubation and sonication. However, these experiments consistently failed to produce measurable quantities of rShaPrP-derived PrPres. Since additional cellular cofactors may be required for prion protein conversion [10, 14], we repeated these experiments using the PMCA buffer supplemented with 10 percent brain homogenate from PrP-knockout mice. However, also in this case no measurable conversion of rShaPrP could be detected, even upon increasing the number of PMCA cycles to 80 and using different concentrations of rShaPrP (1–25 μg/ml).

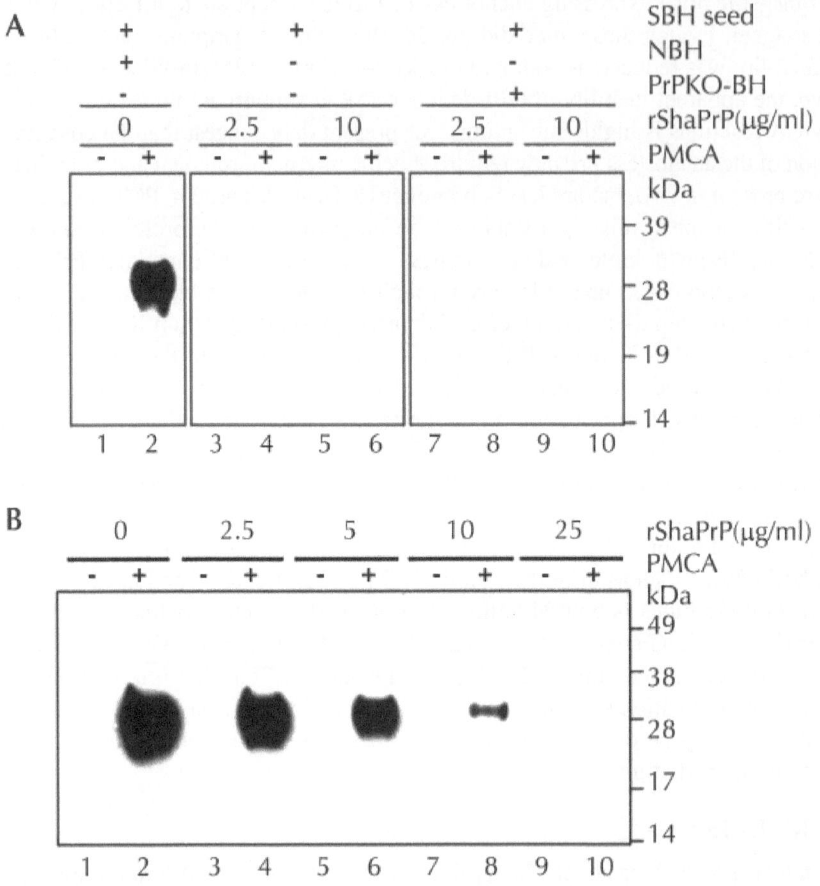

FIGURE 1 Application of PMCA procedure to recombinant Syrian hamster PrP (rShaPrP).

Lanes 1 and 2. Control experiment in which PMCA procedure was performed using 10 % normal hamster brain homogenate (NBH) in the presence of scrapie hamster brain homogenate (SBH) seed (1:100 ratio). Lanes 3–10. Experiments with rShaPrP. The recombinant protein was added to PMCA buffer containing SBH seed only (lanes 3–6) or SBH seed plus 10 % normal brain homogenate from PrP-knockout mice (PrP-KO-BH) (lanes 7–10), and the mixtures were immediately frozen (PMCA-, lanes 3, 5, 7, 9) or subjected to 24 cycles of PMCA (PMCA+, lanes 4, 6, 8, 10). An apparent lack of amplification of PrPres using rShaPrP as a substrate was also observed upon increasing the number of PMCA cycles to 48 or 80. (B) Effect of rShaPrP on the amplification of PrPres by PMCA using PrP^C present in NBH as a substrate. rShaPrP (0–25 μg/ml) was added to PMCA buffer containing SBH seed and NBH (1:100 ratio), and the mixtures were immediately frozen (PMCA-, lanes 1, 3, 5, 7, 9) or subjected to 24 cycles of PMCA (PMCA+, lanes 2, 4, 6, 8, 10). Both in panels A and B.; the samples were treated with PK.; and the blots were probed with anti-PrP 3F4 antibody.

PrP^{Sc} present in SBH seed was not detectable in Western blots shown in panels A and B.; but could be visualized using longer exposure times.

Next we tested whether rShaPrP can interfere with PMCA conversion of brain-derived PrP^C. Consistent with previous data [15], we found that the recombinant protein has a strong inhibitory effect, essentially completely blocking formation of PrPres at concentrations above ~10 μg/ml (i.e. corresponding to ~4-fold excess to PrP^C). Thus, rPrP appears to be recognized by PrP^{Sc}, competing for the same binding sites with brain-derived PrP^C. However, unlike PrP^C; it appears to lack the ability to propagate the structure of the PrP^{Sc} template.

The apparent inability of rShaPrP to convert to PrPres under standard PMCA conditions is both unexpected and intriguing, especially since tertiary and secondary structures of rPrP appear to be identical to those of brain-derived PrP^C [16]. An important difference between these two proteins, however, is the lack of N-glycosylation and the GPI anchor in bacterially expressed rPrP. The first of these factors appears to have little effect on prion protein conversion *in vitro*, as indicated by recent data showing that unglycosylated PrP^C (obtained by treatment of brain PrP^C with PNGase F) can be used as a substrate in PMCA reactions to produce infectious, PrPres-containing prions [17]. On the other hand, the role of the GPI anchor in these conversion reactions has not yet been explored.

To assess the importance of GPI anchor in PrPres formation *in vitro* by PMCA we have prepared NBH pretreated with PI-PLC. Treatment with this enzyme results in a somewhat slower migration of PrP^C during electrophoresis, indicating the removal of GPI anchor. This effect is best evident when electrophoretic profiles are compared using proteins that have been deglycosylated by PNGase F treatment. The abnormal electrophoretic behavior of delipidated PrP^C (i.e. slower rather than faster migration when compared with the GPI anchor-containing protein) is consistent with previous reports [13, 18]. No such shift in electrophoretic profile was observed in control samples treated with PI-PLC that had been inactivated by heating at 100°C.

(A) Western blot analysis of normal hamster brain homogenate (NBH) preincubated in PMCA buffer alone (lane 1) and after treatment with heat-inactivated PI-PLC (lane 2) or active PI-PLC (lane 3). Lanes 4–6 show Western blot of the same samples after deglycosylation with PNGase F. The blots were probed with anti-PrP 3F4 antibody. The concentration of PI-PLC was 1 unit/ml, and the samples were incubated with the enzyme for 2 hr at 37°C with constant shaking. (B) Slot blot analysis of NBH preincubated in PMCA buffer (lane 1) and after treatment with heat-inactivated PI-PLC (lane 2) or active PI-PLC (lane 3). The blot was probed with anti-PrP 3F10 antibody.

Intriguingly, the electrophoretic band corresponding to PI-PLC-treated sample was much less intense, corresponding to ~25–30% of that in non-treated samples. This effect was consistently observed regardless whether Western blots were probed with 3F4 or 3F10 antibody. Such an apparent "loss" of band intensity in Western blots upon PrP^C treatment with PI-PLC was noted previously and attributed to differences in the efficiency of transfer and/or binding to nitrocellulose membrane between GPI anchor-containing and delipidated PrP^C [13]. As this effect could interfere with our assessment of the efficiency of PrPres formation by PMCA, the samples used for Western blotting were reanalysed by slot blotting, a technique that does not include the transfer step. As shown in Figure 2(B). In this case there was no difference in the intensity of bands corresponding to PI-PLC-treated and untreated PrP^{C}. Thus, using our protocol slot blotting is better suited for quantitative comparison of samples containing native and delipidated PrP^{C}.

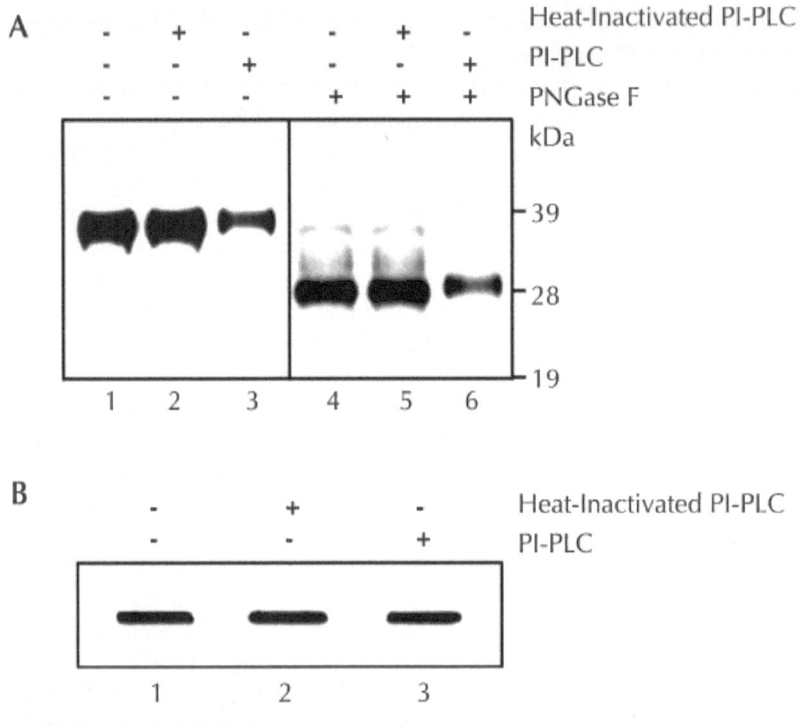

FIGURE 2 Removal of the GPI anchor from PrP^{C} by PI-PLC treatment.

In contrast to robust PrP^{Sc}-templated PMCA conversion of native PrP^{C} to PrPres, no amplification of PrPres could be detected by Western blotting when PrP^{C} without the GPI anchor (i.e. pretreated with active PI-PLC) was used as a substrate. To verify that this was not due to poor detection of delipidated PrP^{C} in Western blotting analysis as discussed earlier, the presence of PK-resistant form of PrP in PMCA-derived samples was further tested by slot blotting. As shown in Figure 3(B) also using this technique we were unable to detect any PrPres material in the product of PMCA reaction performed on PI-PLC treated samples, again in contrast to the presence of large quantities of such material in samples containing PrP^{C} with the native GPI anchor. Next we performed three more rounds of serial PMCA; mixing in each round fresh PI-PLC-treated NBH with a small aliquot (1:10 ratio) of the product of the previous round reaction. Also in this case, we failed to detect any amplification of PrPres in samples containing delipidated PrP^{C} in contrast to robust serial amplification observed using native (i.e. PI-PLC untreated) brain homogenate as a substrate. Altogether, these data clearly demonstrate that the GPI anchor plays an important role in prion protein conversion *in vitro*, and that the removal of this anchor by PI-PLC treatment greatly diminishes—if not completely eliminates—the ability of PrP^C to act as a substrate for amplification of PrP^{Sc} conformation by a standard PMCA procedure.

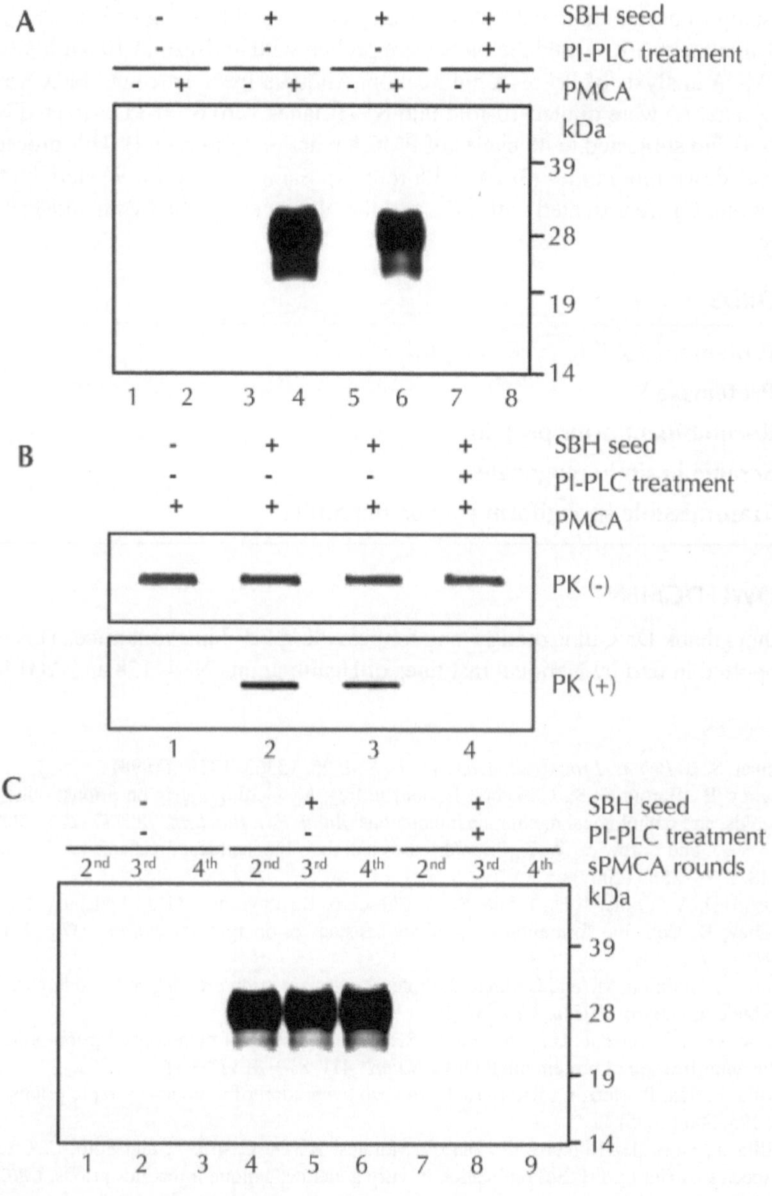

FIGURE 3 Amplification of PrPres by PMCA using as a substrate PrP^C present in normal hamster brain homogenate (NBH) before and after removal of the GPI moiety.

(A) Western blot analysis of PrPres amplification. NBH pretreated with PMCA buffer (lanes 3, 4), heat-inactivated PI-PLC (lanes 5, 6), or active PI-PLC (lanes 7, 8) was mixed with scrapie brain homogenate (SBH) seed. In a control experiment, NBH was used in the absence of SBH seed (lanes 1 and 2). The samples were immediately frozen (PMCA−, lanes 1, 3, 5 and 7) or subjected to 48 cycles of PMCA reaction (PMCA+, lanes 2, 4, 6 and 8). (B) Slot blot analysis of PrPres amplification. The PMCA samples described in panel A were subjected to slot blotting before (PK−) and after PK treatment (PK+), and the blots were probed with anti-PrP 3F10 antibody. (C) Serial PMCA analysis of PrPres amplification. Aliquots from 1st round PMCA reaction (see panel A) were diluted 10-fold into NBH (lanes 1–6) or PI-PLC-treated NBH (lanes 7–9) and subjected to 48 cycles of PMCA reaction (2nd round). This procedure was repeated two more times (3rd and 4th rounds). Samples used for Western blotting (panels A and C) were treated with PK.; and the blots were probed with anti-PrP 3F4 antibody.

KEYWORDS

- **Protein misfolding cyclic amplification**
- **Proteinase K**
- **Recombinant prior protein**
- **Scrapie brain homogenate**
- **Transmissible spongiform encephalopathies**

ACKNOWLEDGMENT

The authors thank Dr. Quingzhong Kong for providing PrP-knockout mice. This work was supported in part by National Institutes of Health grants NS44158 and AG14359.

REFERENCES

1. Prusiner, S. B. *Prions. Proc. Natl. Acad. Sci. U. S. A.* **95**, 13363–13383 (1998).
2. Caughey, B., Baron, G. S., Chesebro, B., and Jeffrey, M. Getting a grip on prions. oligomers, amyloids, and pathological membrane interactions. *Annu. Rev. Biochem.* **78**, 177–204 (2009).
3. Cobb, N. J. and Surewicz, W. K. Prion diseases and their biochemical mechanisms. *Biochemistry*, **48**, 2574–2585 (2009).
4. Kocisko, D. A., Come, J. H., Priola, S. A., Chesebro, B., Raymond, G. J., Lansbury, P.T., and Caughey, B. Cell-free formation of protease-resistant prion protein. *Nature.*, **370**, 471–474 (1994).
5. Hill, A. F., Antoniou, M., and Collinge, J. Protease-resistant prion protein produced in vitro lacks detectable infectivity. *J. Gen. Virol.* **80** (Pt 1), 11–14 (1999).
6. Saborio, G. P., Permanne, B., and Soto, C. Sensitive detection of pathological prion protein by cyclic amplification of protein misfolding. *Nature*, **411**, 810–813 (2001).
7. Castilla, J., Saa, P., Hetz, C., and Soto, C. In vitro generation of infectious scrapie prions. *Cell*, **121**, 195–206 (2005).
8. Castilla, J., Gonzalez-Romero, D., Saa, P., Morales, R., De Castro, J., and Soto, C. Crossing the species barrier by PrP(Sc) replication in vitro generates unique infectious prions. *Cell*, **134**, 757–768 (2008).

9. Castilla, J., Morales, R., Saa, P., Barria, M., Gambetti, and P., Soto, C. Cell-free propagation of prion strains. *EMBO. J.*, **27**, 2557–2566 (2008).
10. Deleault, N. R., Harris, B. T., Rees, J. R., and Supattapone, S. Formation of native prions from minimal components in vitro. *Proc. Natl. Acad. Sci. U. S. A*, **104**, 9741–9746 (2007).
11. Morillas, M., Swietnicki, W., Gambetti, P., and Surewicz, W. K. Membrane environment alters the conformational structure of the recombinant human prion protein. *J. Biol. Chem.*, **274**, 36859–36865 (1999).
12. Smirnovas, V., Kim, J. I., Lu, X., Atarashi, R., Caughey, B., and Surewicz, W. K. Distinct structures of scrapie prion protein (PrP^{Sc})-seeded versus spontaneous recombinant prion protein fibrils revealed by H/D exchange. *J. Biol. Chem.*, (2009), in press.
13. Nishina, K. A. and Supattapone, S. Immunodetection of glycophosphatidylinositol-anchored proteins following treatment with phospholipase C. *Anal. Biochem.*, **363**, 318–320 (2007).
14. Deleault, N. R., Lucassen, R. W., and Supattapone, S. RNA molecules stimulate prion protein conversion. *Nature*, **425**, 717–720 (2003).
15. Bieschke, J., Weber, P., Sarafoff, N., Beekes, M. Giese, A., and Kretzschmar, H. Autocatalytic self-propagation of misfolded prion protein. *Proc. Natl. Acad. Sci. U. S. A.* **101**, 12207–12211 (2004).
16. Hornemann, S., Schorn, C., and Wuthrich, K. NMR structure of the bovine prion protein isolated from healthy calf brains. *EMBO. Rep.*, **5**, 1159–1164 (2004).
17. Piro, J. R., Harris, B.T., Nishina, K., Soto, C., Morales, R., Rees, J. R., and Supattapone, S. Prion protein glycosylation is not required for strain-specific neurotropism. *J. Virol.*, **83**, 5321–5328 (2009).
18. Stahl, N., Borchelt, D .R., Hsiao, K., and Prusiner, S .B . Scrapie prion protein contains a phosphatidylinositol glycolipid. *Cell.*, **51**, 229–240 (1987).
19. Atarashi, R., Moore, R. A., Sim, V. L., Hughson, A. G., Dorward, D. W., Onwubiko, H. A., Priola, S. A., and Caughey, B. Ultrasensitive detection of scrapie prion protein using seeded conversion of recombinant prion protein. *Nat. Methods.*, **4**, 645–650 (2007).
20. Baron, G. S., Wehrly, K., Dorward, D. W., Chesebro, B., and Caughey, B. Conversion of raft associated prion protein to the protease-resistant state requires insertion of PrP-res (PrP(Sc)) into contiguous membranes. *EMBO. J.*, **21**, 1031–1040 (2002).
21. Kirby, L., Birkett, C. R., Rudyk, H., Gilbert, I. H., and Hope, J. *In vitro* cell-free conversion of bacterial recombinant PrP to PrPres as a model for conversion. *J. Gen. Virol.*, **84**, 1013–1020 (2003).
22. Caughey, B. Prion protein interconversions. *Philos. Trans. R. Soc. Lond., B. Biol. Sci.*, **356**, 197–200 (2001).
23. Caughey, B. and Raymond, G. J. The scrapie-associated form of PrP is made from a cell surface precursor that is both protease- and phospholipase-sensitive. *J. Biol. Chem.* **266**, 18217–18223 (1991).
24. McNally, K. L., Ward, A. E., and Priola, S. A. Cells expressing anchorless prion protein are resistant to scrapie infection. *J. Virol.*, **83**, 4469–4475 (2009).
25. Chesebro, B., et al. Anchorless prion protein results in infectious amyloid disease without clinical scrapie. *Science* **308**, 1435–1439 (2005).
26. Xiao, X., et al. Failure to detect the presence of prions in the uterine and gestational tissues from a Gravida with Creutzfeldt–Jakob disease. *Am. J. Pathol.*, **174**, 1602–1608 (2009).
27. Gregori, L., Kovacs, G. G., Alexeeva I., Budka, H., and Rohwer, R. G. Excretion of transmissible spongiform encephalopathy infectivity in urine. *Emerg. Infect., Dis.*, **14**, 1406–1412 (2008).

5 The Octarepeat Region of the Prion Protein

Alice Y. Yam, Carol Man Gao, Xuemei Wang, Ping Wu, and David Peretz

CONTENTS

5.1 INTRODUCTION

Prion diseases are fatal neurodegenerative disorders characterized by misfolding and aggregation of the normal prion protein PrP^C. Little is known about the details of the structural rearrangement of physiological PrP^C into a still-elusive disease-associated conformation termed PrP^{Sc}. Increasing evidence suggests that the amino-terminal octapeptide sequences of PrP (huPrP, residues 59–89), though not essential, play a role in modulating prion replication and disease presentation.

5.1.1 Methodology/Principal Findings

Here, we report that trypsin digestion of PrP^{Sc} from variant and sporadic human Creutzfeldt-Jakob Disease (CJD) results in a disease-specific trypsin-resistant PrP^{Sc} fragment including amino acids 49–231, thus preserving important epitopes such as the octapeptide domain for biochemical examination. The immunodetection analyses reveal that several epitopes buried in this region of PrP^{Sc} are exposed in PrP^C.

We conclude that the octapeptide region undergoes a previously unrecognized conformational transition in the formation of PrP^{Sc}. This phenomenon may be relevant to the mechanism by which the amino terminus of PrP^C participates in PrP^{Sc} conversion, and may also be exploited for diagnostic purposes.

Prion diseases are fatal neurodegenerative disorders characterized by dementia, motor dysfunction, and spongiform degeneration of the brain [1]. Propagation of the infectious prion particle is attributable to a conformational conversion of the widely expressed normal prion protein PrP^C into an abnormal, infectious conformation termed PrP^{Sc}. The PrP^{Sc}, unlike the physiological protein PrP^C; exists predominantly in an aggregated form and is partially resistant to protease digestion [2]. Digestion with Proteinase K (PK) leaves behind a core particle termed PrPres (for resistant PrP) or PrP^{27-30} (for 27–30 kDa PK-resistant fragments) that consists of the carboxy-terminal two-thirds of the protein. Consequently, the PK-labile amino-terminus has been suggested to be solvent-accessible and largely unstructured in the context of PrP^{Sc} as it is in PrP^C [3, 4]. Furthermore, because PrP^{27-30} remains infectious and because of the scarcity of tools to isolate full-length PrP^{Sc}, the amino-terminus has remained relatively unexamined in the context of aggregated PrP.

The amino-terminal tail of PrP is known to contain an array of five almost identical octapeptide sequences, also termed octarepeats, that has been reported to play a role in copper binding and homeostasis [5], as well as in protection from oxidative stress (reviewed in [6]). Importantly, mutations that result in expansion of the octarepeats have been linked to familial CJD. Documented cases of familial CJD report insertions of 2–9 octapeptide sequences [7-9], whose impact are recapitulated in transgenic mouse models [10]. Likewise, transgenic mice expressing PrP that lack all five octapeptide sequences appear to be impaired in propagating PrP^{Sc}, as these mice have longer incubation periods before they become symptomatic, lower prion titers, reduced amounts of PrPres, and no observable histopathology [11, 12]. *In vitro* studies also support a role of the octarepeats in PrP^{Sc} replication as octapeptide insertions or deletions affect the rate and propensity of oligomerization for recombinant PrP [13, 14]. As such, understanding the function of the amino-terminal portion of PrP is critical for understanding propagation of prion diseases.

Despite the importance of the amino-terminus, previous prion research has primarily focused on PK-treated PrP^{Sc}, partly because of the difficulties associated with separating PrP^{Sc} from PrP^{C} in infectious samples. Other biochemical means of protein enrichment such as antibody immunoprecipitation are largely ineffective at separating PrP^{C} from PrP^{Sc} due to their sequence identity. The PK digestion has been employed to circumvent this issue, but at the cost of removing amino-terminal sequences and decreasing yields of PrP^{Sc} and scrapie-associated infectivity [15]. To address this issue, we tried employing the more specific enzyme trypsin, rather than PK to distinguish PrP^{C} from PrP^{Sc}. We found that indeed trypsin cleavage significantly digested PrP^{C} but retained the majority of PrP^{Sc}, thus providing a means to separate the isoforms while maintaining the octarepeat sequence. Using this technique, we found the octarepeat sequence had multiple epitopes exposed in PrP^{C} but not PrP^{Sc}, suggesting that there is a conformational transition in this region during the conversion of PrP^{C} to PrP^{Sc}. Given the putative role of the octarepeat in infectivity, this novel structural change may provide clues to the mechanism of PrP^{Sc} replication.

5.2 MATERIALS AND METHODS

5.2.1 Tissue Samples

The CJD brain homogenates (BHs) were acquired from the NIBSC CJD Resource Centre and correspond to variant CJD (vCJD) MM (NHBY0/0014. in Figures 1B, 22–3, and S1), vCJD MM (NHBY0/0003. in Figure 1C only), sporadic CJD (sCJD) MM (NHBX0/0001), and sCJD MV (NHBX0/0004) strains with the indicated reference codes and genotypes at polymorphic residue 129 [16] Normal BH was retrieved from the tissue bank of the Swiss National Reference Centre of Prion Diseases (Zürich, Switzerland). All 10% BHs (w/v) were homogenized in 0.25 M sucrose. Protein concentrations were measured by BCA Assay (Pierce, Rockford, IL) with most 10% BHs being –10 mg/ml.

5.2.2 Antibodies

The POM2, POM17, POM1, and POM19 mouse monoclonal antibodies were obtained from the laboratory of Dr. Adriano Aguzzi at the Institute of Pathology at the University Hospital of Zürich. POM2 recognizes the sequence QPHGG (G/S)W whereas POM17 and POM1 recognize the helix one region of PrP (144–155) and compete with 6H4 antibodies for binding [17, 18]. POM19 recognizes residues 121–134 and 218–221 of the murine PrP sequence, but does not recognize the human PrP sequence. 3F4 (Covance, Princeton, NJ) recognizes residues 109–112 (MKHM) in the human PrP sequence. Anti-actin antibodies (clone C4, Millipore, Billerica, MA) recognize residues 50–70.

5.2.3 Western Blot Analysis

An amount of 25 μg of normal and infectious BH (2.5 μl of 10% BH) was digested with 50 μg/ml of trypsin or PK in 0.5 × TBS (25 mM Tris pH 7.5, 75 mM NaCl) with

0.5% Tween20, 0.5% TritonX-100, and 5 mM $CaCl_2$ for 1 hr at 37°C. Samples were separated by 12% sodium dodecyl sulfate polyacrylamide gel electrophoresis (SDS-PAGE) in parallel with 10 µg undigested sample, and immunoblotted with anti-PrP antibodies (3F4, POM2, POM17, or POM1) followed by an HRP-conjugated goat anti-mouse polyclonal antibody (Pierce). Chemiluminescent images were acquired *via* a Kodak Image Station 4000MM.

5.2.4 ELISA

An amount of 10% BH was diluted 10-fold into TBS with 2% Sarkosyl (TBSS) and digested with 0, 1, 10, or 100 µg/ml trypsin or PK for 1 hr at 37°C. Digestions were stopped by adding 2 mM PMSF and Complete Mini protease inhibitor cocktail (Roche, Indianapolis, IN) in four volumes of TBS. The samples were then detected by direct ELISA. Briefly, approximately 500 nl of digested 10% BH was centrifuged at 14,000 rpm for 30 min at 4°C. The PrP^{Sc} pellets were denatured in 6M GdnSCN; diluted with an equal volume of 0.1 M $NaHCO_3$ pH 8.9, and passively coated to ELISA plates overnight. The plates were washed and blocked in 0.1× BlockerCasein in TBS (Pierce) and coated PrP was detected *via* 0.1 µg/ml 3F4, POM2 or POM17 antibodies and an alkaline phosphatase (AP)-conjugated goat anti-mouse polyclonal antibody (Pierce). Samples were analyzed by ELISA in triplicate and washed six times with TBS 0.05% Tween20 between antibody incubations. Finally, LumiphosPlus substrate (Lumigen, Southfield, MI) with 0.05% SDS was added to the wells and incubated for 30 min at 37°C before the luminescence was measured *via* a Luminoskan luminometer (Thermo Electron Corporation, Waltham, MA).

5.2.5 Epitope Exposure of PrP^{Sc}

An amount of 10% BH diluted 10-fold into TBSS was digested with 50 µg/ml trypsin or PK for 1 hr at 37°C. Digestions were halted and samples were centrifuged as described earlier. The PrP^{Sc} pellets were resuspended in 0.1 M $NaHCO_3$ pH 8.9, and passively coated to ELISA plates overnight. The next day, coated proteins were denatured with increasing concentrations of GdnHCl (0–5 M) for 15 min at 37°C.; and detected by direct ELISA using 0.5 µg/ml 3F4 or 0.1 µg/ml POM2 antibodies as described earlier.

5.2.6 Immunoprecipitation of Native or Denatured PrP^{Sc}

An amount of 10% BH diluted 10-fold into TBSS was digested with 50 µg/ml trypsin, PK, or nothing for 1 hr at 37°C. Digestions were halted and samples were centrifuged as described earlier. PrP^{Sc} pellets were then resuspended in 0 or 5M GdnHCl, and adjusted to a final concentration of 0.1M GdnHCl in TBS with 1% Triton X-100 and 1% Tween 20. PrP was subsequently immunoprecipitated with POM2, 3F4, or POM17-conjugated to Protein G Dynal beads and analyzed by POM1 immunoblot (PrP^{Sc} from 75 µg BH was precipitated with 10 µg antibody). The POM19 immunoprecipitations served as a negative control as POM19 does not recognize human PrP. Approximately 40% of an equivalent sample was loaded to assess the input for each immunoprecipitation.

5.3 DISCUSSION

5.3.1 A Structural Change in the Octarepeat Region of PrP^{Sc} and Prion Propagation

Despite extensive debate, the importance of the PrP octapeptide sequences in PrP^{Sc} pathophysiology remains unclear. Initially, the region was thought to be unnecessary since PK-digested PrP^{Sc} from tissue samples still proved to be infectious. However, emerging evidence suggests that the octapeptide sequences may play a role in propagating prions. For instance, insertion of 1–9 octapeptides into the PrP sequence has been linked to several cases of familial CJD [9, 29, 30]. The effects of this octarepeat expansion have been recapitulated in transgenic mice carrying additional octapeptide sequences that demonstrate clinical (ataxia), histopathological (neuronal apoptosis and accumulation of aggregated PrP), and biochemical (protease-resistant PrP) hallmarks of prion disease [10]. While transgenic mice expressing PrP lacking the octapeptide sequences remain susceptible to scrapie infection, these mice appear to be impaired in PrP^{Sc} propagation as they exhibit longer incubation times, no histopathology, and significantly reduced prion titers and protease-resistant PrP [11]. In vitro aggregation studies also support involvement of the octarepeat region in PrP oligomerization. For instance, expansion of the octarepeats increased the rate of oligomerization for recombinant PrP [13], while deletion of the octarepeats resulted in reduced oligomerization [14]. These data together demonstrate that the octarepeat region is not required for disease progression, but that it may facilitate prion propagation and/or pathogenesis.

Given that the octarepeats are largely PK-sensitive, it has been presumed that the region contributes little structure to PrP^{Sc}. Here, by using trypsin digestion of PrP^{Sc} to preserve most of the PrP sequence (estimated 49–231) we have revealed that there is indeed a structural transition in the octarepeat region in the conversion of PrP^{C} to PrP^{Sc}. The structural change we measured upon denaturation (5–10 fold increase in detected PrP) mimics the structural changes observed within the PK-resistant core [22, 24, 31].

Several possibilities exist for how the octarepeat region may affect prion disease pathology. First, the octapeptide sequences may modulate the toxicity of PrP^{Sc}. Currently, the loss of infectivity resulting from PK digestion is largely attributed to the digestion of some proportion of PrP^{Sc} (90–231) [15, 17], but it is possible that infectivity is lost with the PK-labile amino-terminus as well. Second, the octapeptide sequence may be prone to self-association and cause multimerization of PrP^{C}, a process that, even if not strictly required for generation of prions, could accelerate conversion to PrP^{Sc}. In vitro binding and aggregation studies with octapeptide multimers of varying lengths support this theory [32, 33]. The finding that the octarepeats undergo a structural transition in the conversion of PrP^{C} to PrP^{Sc} open up another possibility that structural components of the octarepeat region, either alone or in conjunction with the primary sequence, may enhance the ability of PrP^{Sc} to propagate.

5.3.2 PrP^{Sc} Isolation by Trypsin Digestion

We have developed a novel method of purifying PrP^{Sc} from infectious samples that preserves the majority of the PrP sequence. Future studies comparing the infectious titers of PK- and trypsin-digested samples would allow the contribution of the octarepeats to infectivity to be directly measured. This method could also be used as an enhanced diagnostic tool. Sandwich ELISA was found to yield superior detection by preserving multiple POM2 epitopes. Because its cognate epitope is repeated on PrP.; POM2 can dock to PrP through multiple interactions, yielding extremely high avidities in the femtomolar range [18]. This peculiarity renders trypsinization, followed by ELISA, one of the most sensitive methods for detecting authentic PrP^{Sc}. The sensitivity gained by additional POM2 epitopes is also apparent in the amount of PrP captured by POM2 immunoprecipitation. Chemical denaturation of PrP exposed additional epitopes, enhancing the amount of PrP captured several fold over a single POM2, 3F4 or POM17 epitope. In addition, since PK-digested PrP^{Sc} has a high propensity to fibrillize in solution [34], preservation of full-length PrP^{Sc} may enhance the solubility of the aggregate, making it more amenable to structural studies—for example, by solid-state nuclear magnetic resonance spectroscopy which necessitates large amounts of highly purified analytes. Moving forward, we now have the ability to purify PrP^{Sc} that closely approaches unmodified, physiological aggregates. This new tool may help to dissect the role of the octarepeat region in prion disease.

5.4 RESULTS

5.4.1 Trypsin Digests PrP^C while Preserving PrP^{Sc}

To characterize the physicochemical properties of the PrP^{Sc} octarepeat region, we needed to preserve amino-terminal sequences while still removing PrP^C from CJD samples. Given the nonspecific nature of PK digestion, we chose to digest the samples with a more specific enzyme, trypsin, which cleaves only on the C-terminal side of lysine and arginine residues. The predicted tryptic map for the human PrP sequence shows that the majority of cleavage sites are within the PK-resistant core with a few additional cleavage sites between residues 23–49 of the mature PrP sequence and no sites within the octarepeat region (residues 51–91). To confirm this prediction, we digested BHs from several CJD strains in parallel with PK as a control. Immunoblot analysis of the digests indicated that trypsin-resistant PrP fragments (expected to be ~49–231) had higher mass than PK-resistant fragments (90–231) with a substantial fraction of molecules preserved. Mass spectrometry of human recombinant PrP_{23-231} validated that trypsin could cleave at all the sites predicted by the PrP sequence, thus confirming that the tryptic sites were conformationally protected in PrP^{Sc}. Further examination revealed that digestion with trypsin but not PK preserved the PrP octapeptide repeats recognized by the POM2 antibody. Importantly, preservation of this region significantly enhances PrP detection.

By contrast, PrP^C was digested by both proteases, and was no longer detected by POM2 and 3F4. However, when we immunoblotted the digests with POM17 or POM1 antibody (recognizing helix one, residues 144–155), we observed a trypsin-resistant fragment of 25 kDa. Given that the structure of PrP^C is composed of a globular domain (122–231) following an unstructured amino-terminus [19, 20], the 25 kDa trypsin-resistant fragment likely included the globular domain as it was the appropriate molecular weight and detected by POM1 and POM17 but not POM2 or 3F4 antibodies. Interestingly, this would suggest that the globular domain is tightly folded and that the tryptic sites within the POM17/POM1 epitope (R148 and R151) are not accessible for cleavage, contrary to how they appear in the solution structure [20]. To better understand the resistance of this domain to digestion, we digested both normal and vCJD BHs with increasing amounts of trypsin. The amino-terminal sequences of PrP^C were extremely labile as no PrP was detected by either POM2 or 3F4 at any of the trypsin concentrations tested, but a PrP^C 25 kDa fragment and PrP^{Sc} from vCJD BH were resistant to digestion. A survey of other non-CJD BHs confirmed the presence of a 25 kDa PrP^C tryptic fragment, indicating that the fragment was not unique to that sample. However, as most of studies of PrP^{Sc} utilized 3F4 and POM2 epitopes, PrP^C trypsin-resistant fragments did not impact analysis.

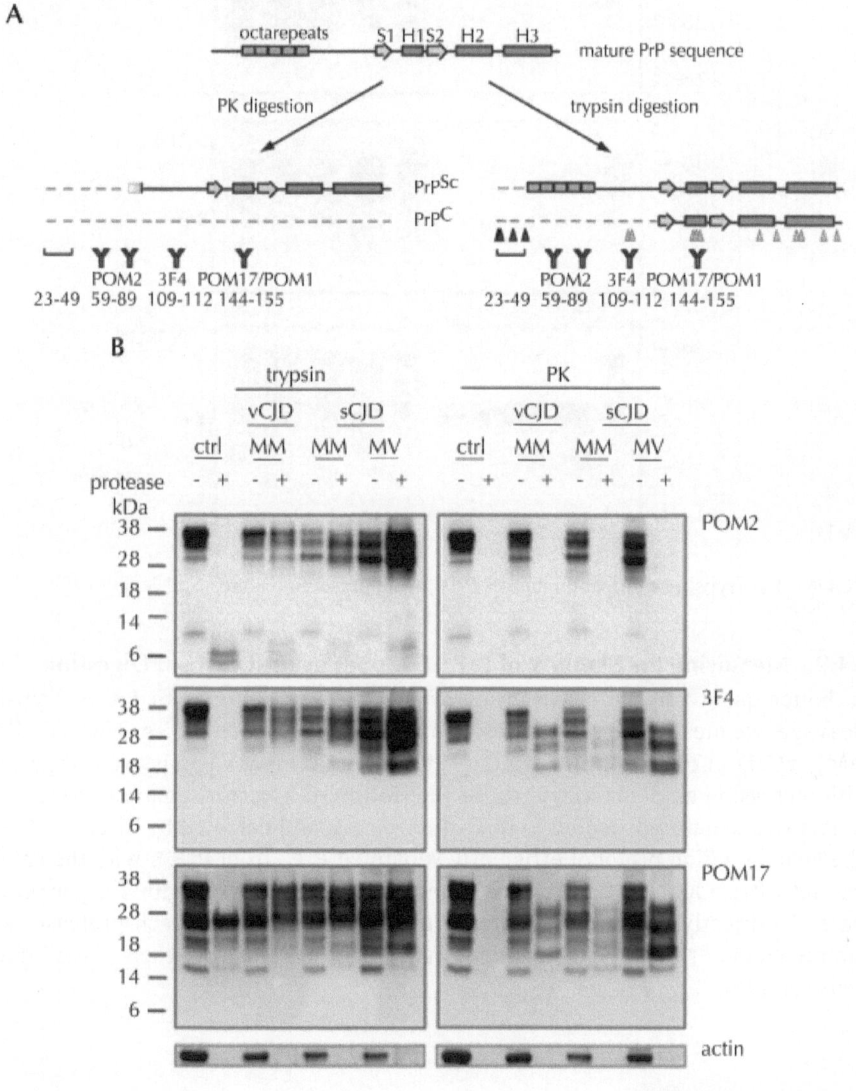

FIGURE 1 *(Continued)*

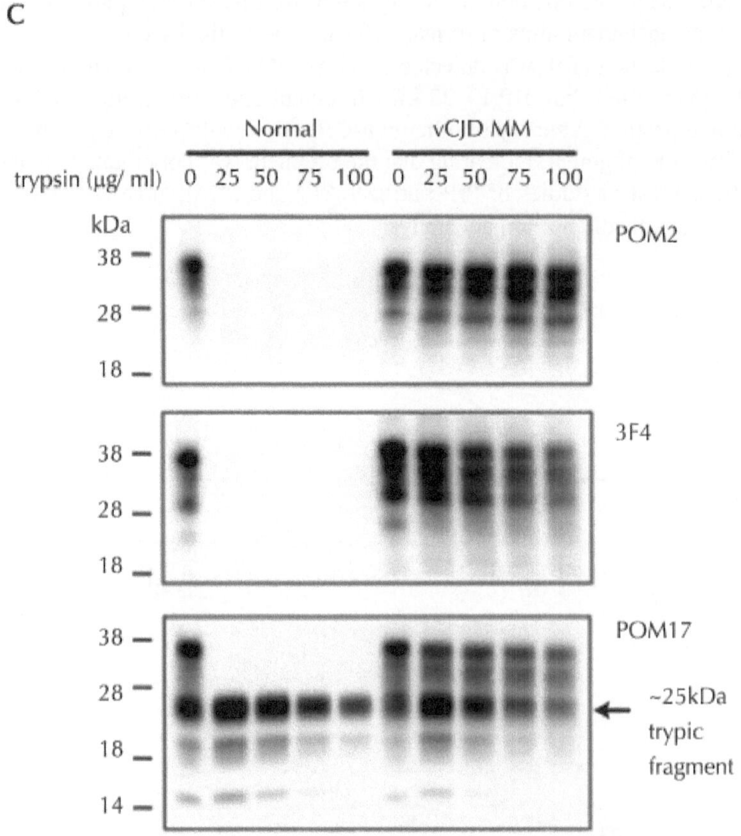

FIGURE 1 Trypsin significantly digests PrP^{C} while preserving almost full-length PrP^{Sc}.

5.4.2 Measuring the Stability of PrP^{Sc} Epitopes against Trypsin Digestion

To better quantitate the sensitivities of specific PrP epitopes to PK or trypsin cleavage, we measured protease-resistant epitopes by direct ELISA. Normal (129 MM), vCJD (129 MM), and sCJD (129 MM and 129MV) BHs were digested with increasing amounts of trypsin or PK; followed by centrifugation. PrP in the pellets was denatured, coated to microtiter plates, and detected by direct ELISA. The centrifugation protocol efficiently separated PrP^{C} from PrP^{Sc}, with the small amount of residual PrP^{C} efficiently removed *via* proteolysis with 1 µg/ml protease. To directly compare the amount of PrP^{Sc} preserved by each protease, we subtracted the PrP^{C}-derived signal detected in normal BH from each of the infectious samples.

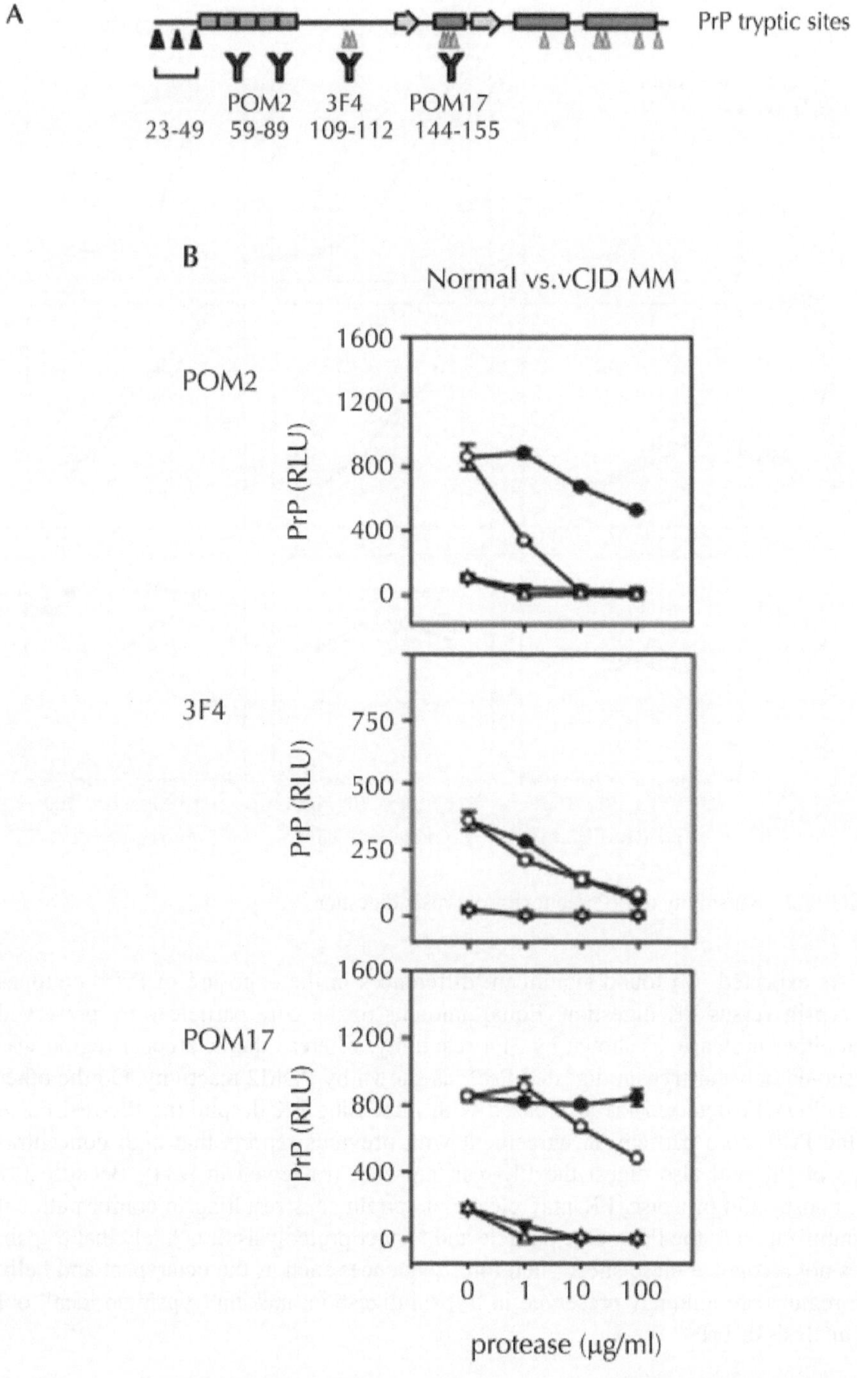

FIGURE 2 *(Continued)*

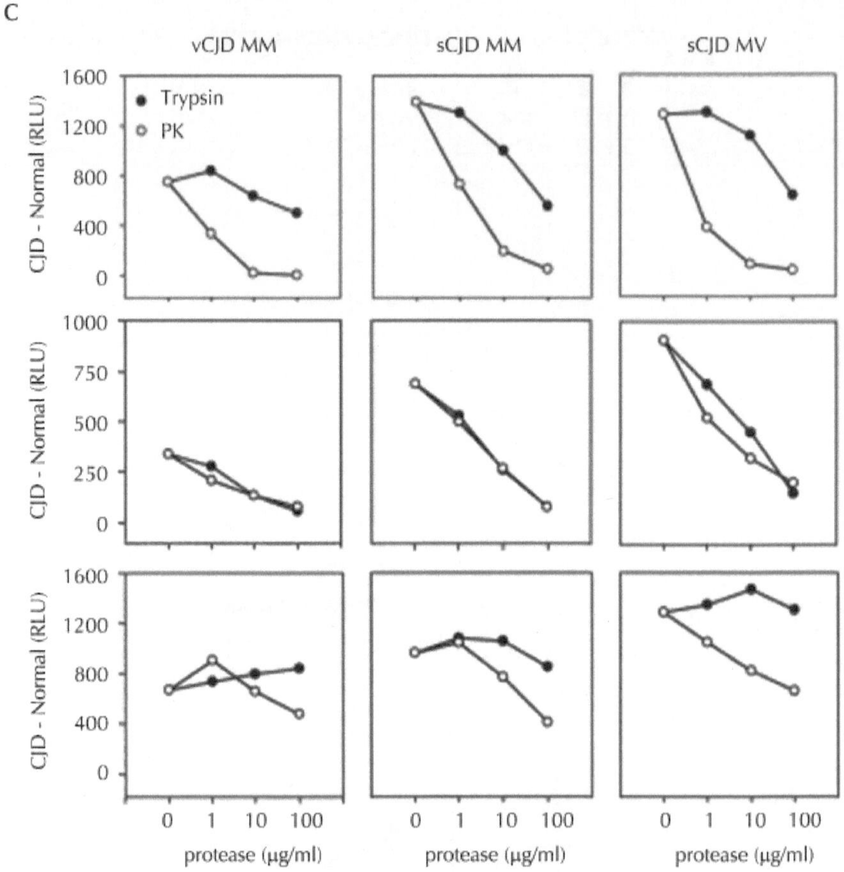

FIGURE 2 Sensitivity of PrP^{Sc} epitopes to trypsin digestion.

As expected, we found significant differences in the exposure of PrP^{Sc} epitopes to trypsin versus PK digestion. Equal amounts of the core particle were preserved with either protease, as shown by 3F4 reactivity, whereas the octarepeat region was preserved only for trypsin digested PrP^{Sc} as shown by POM2 reactivity. On the other hand, POM17-detected PrP decreased with increasing PK despite the PK-resistance of the PrP^{Sc} core particle, in agreement with previous reports that high concentrations of PK will also digest the PK-resistant core (reviewed in [21]). Because PK is a nonspecific protease, PK may cleave at certain sites resulting in conformational destabilization of the PrP^{Sc} core particle and further proteolysis. It is likely that trypsin does not recognize these sites. Therefore, sequences such as the octarepeat and helix one regions are uniquely preserved in trypsin digestion, making trypsin an ideal tool for analysis of PrP^{Sc}.

5.4.3 The Octarepeat Region Undergoes a PrP^{Sc}-specific Conformational Change

Despite extensive studies of PrP variants bearing mutations of the octarepeat region, understanding of the properties of this domain remains unclear. Preservation of the region by digestion with trypsin allowed us to study its structure by an epitope protection approach. Digested PrP^{Sc} from CJD BHs was coated directly to ELISA plates and probed with various antibodies to identify conformationally hidden epitopes. The stability of these conformations could be assessed by treatment with increasing amounts of denaturant Guanidine Hydrochloride, (GdnHCl) and looking for recovery of immunoreactivity. This ELISA-based approach allowed us to survey several different strains in both a high throughput and quantitative manner. Importantly, maximal antibody binding was observed for PrP^{C} at all concentrations of GdnHCl surveyed, demonstrating that the epitopes examined are accessible in the native conformation of PrP^{C} and that guanidine treatment of the passively-coated material did not strip PrP from the plate.

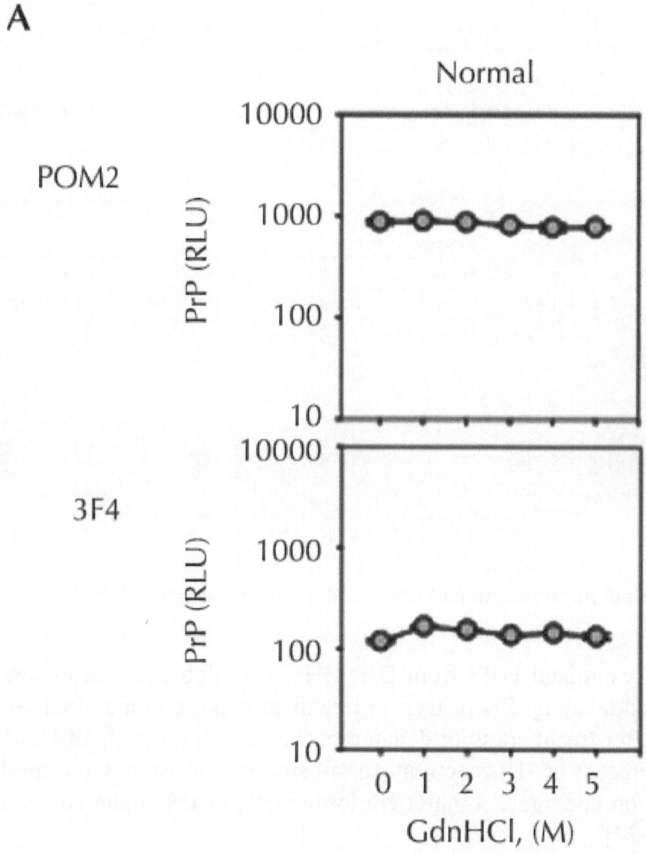

FIGURE 3 *(Continued)*

B

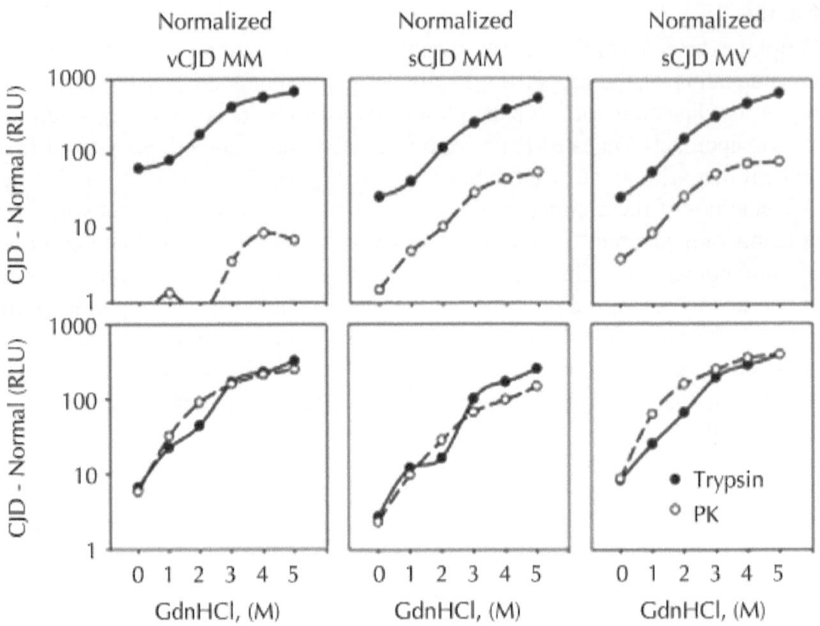

C

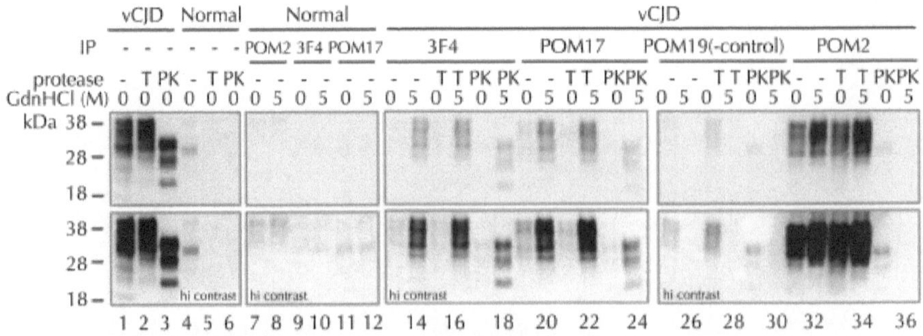

FIGURE 3 Epitope presentation of native and chemically denatured PrP^{Sc}.

We next examined PrP^{Sc} from CJD BHs. The high endogenous levels of PrP^{C} were digested away by PK or trypsin treatment and the immunoreactivity of PrP^{Sc} was tested after treatment with denaturant. Increasing amounts of GdnHCl yielded dramatic increases in 3F4 reactivity for all strains, consistent with previous reports that this region undergoes a major conformational change upon conversion of PrP^{C} to PrP^{Sc} [22-24].

Next, we examined the conformational stability of the octarepeat region, which has been previously presumed to be unstructured in PrP^{Sc}. Consistent with this view, significant POM2 immunoreactivity was observed even under native PrP^{Sc} conditions after trypsin digestion. However, when PrP^{Sc} was unfolded with GdnHCl there was a 5–10-fold increase in POM2 immunoreactivity, suggesting that this region indeed is buried in PrP^{Sc} and undergoes a conformational change upon conversion of PrP^{C} to PrP^{Sc}. As expected, significantly different levels of PrP^{Sc} were detected after PK digestion. No POM2 signal was observed for PK-digested PrP^{Sc} from vCJD BH; in agreement with the complete digestion of the octapeptides in this strain [25, 26]. Many, if not all, sCJD isolates are reported to contain both type 1 and 2 PrP^{Sc} [27, 28], thus retaining a mixture of PrP^{Sc} protease-resistant fragments with some containing a single octapeptide (type 1) after PK digestion [25], that could be detected by POM2. Interestingly, PK-resistant PrP containing a single octapeptide had a similar conformational stability as the trypsin-digested PrP,which retains all octarepeats.

We further confirmed these findings by immunoprecipitation studies for the vCJD sample, which appeared to contain the most homogeneous PrP^{Sc} by quantitative ELISA assays. Normal and vCJD BHs were digested, PrP^{Sc} was centrifuged, and PrP^{Sc} pellets were treated with or without denaturant, and immunoprecipitated with antibodies recognizing the various epitopes of interest. The immunoprecipitates were then examined by POM1 immunoblot. As expected, antibodies with epitopes buried within the protease-resistant core (3F4 and POM17) immunoprecipitated denatured but not native PrP^{Sc} relative to negative controls.

Next, we examined the ability of POM2 to bind the octapeptide repeats. Consistent with ELISA studies, no PrP was immunoprecipitated from PK-digested PrP^{Sc}. On the other hand, PrP^{Sc} from trypsin-digested BHs could be immunoprecipitated in agreement with the finding that one or more of the octapeptides are accessible.

5.5 CONCLUSION

The octarepeats were excised only at high concentrations of trypsin but rapidly disappeared with PK digestion. POM17 detection of the PrP^{Sc} core particle was consistent with results *via* 3F4 detection and confirmed that the vast majority of helix one was resistant to trypsin digestion.We observed by ELISA; a larger proportion of PrP could be immunoprecipitated *via* POM2 antibody when PrP^{Sc} was first denatured with guanidine treatment. Thus, the increased yield of POM2 immunoprecipitation upon denaturation of PrP^{Sc} supports the idea that the octapeptide domain undergoes a structural transition upon formation of PrP^{Sc}.

KEYWORDS

- **Creutzfeldt-Jakob disease (CJD)**
- **Immunoblot**
- **Prion protein**
- **Proteinase K**
- **Trypsin**

ACKNOWLEDGMENT

We thank Dazhi Tang for mass spectrometry analyses, Drs. Anthony Lau and Cleo Salisbury for helpful discussions and comments on the manuscript, Grace Ching for her assistance, and Drs. Philip Minor (National Institute for Biological Standards and Control, Potters Bar, United Kingdom) and Adriano Aguzzi (Institute of Pathology at the University Hospital of Zürich) for generously providing samples and for critical reading of the manuscript.

COMPETING INTERESTS

The authors of this work were or are employed by Novartis Vaccines and Diagnostics, Inc. This work is also part of a pending patent application, for which there is no direct financial benefit to the authors. This does not alter the authors adherence to all the PLoS ONE policies on sharing data and materials.

FUNDING

This work was funded by Novartis Vaccines and Diagnostics, Inc. The authors are or were employed by the funder and, as such, the funders played a role in the study design, data collection and analysis, decision to publish, and preparation of the manuscript.

REFERENCES

1. Aguzzi, A. and Calella, A. M. Prions: protein aggregation and infectious diseases. *Physiol. Rev.* **89**, 1105–1152 (2009).
2. Bolton, D. C., McKinley, M. P., and Prusiner S. B. Identification of a protein that purifies with the scrapie prion. *Science* **218**, 1309–1311 (1982).
3. Donne, D. G., Viles, J. H., Groth, D., Mehlhorn, I., James, T. L. et al. Structure of the recombinant full-length hamster prion protein PrP (29–231). The N terminus is highly flexible. *Proc. Natl. Acad. Sci. U. S. A.* **94**, 13452–13457 (1997).
4. Novitskaya, V., Makarava, N., Bellon, A., Bocharova, O. V., Bronstein, I. B. et al. Probing the conformation of the prion protein within a single amyloid fibril using a novel immunoconformational assay. *J. Biol. Chem.* **281**, 15536–15545 (2006).
5. Aguzzi, A., Baumann, F., and Bremer, J. The prion's elusive reason for being. *Annu. Rev. Neurosci.* **31**, 439–477 (2008).
6. Millhauser, G. L. Copper and the prion protein: methods, structures, function, and disease. *Annu. Rev. Phys. Chem.* **58**, 299–320 (2007).
7. Yanagihara, C., Yasuda, M., Maeda, K., Miyoshi, K., and Nishimura, Y. Rapidly progressive dementia syndrome associated with a novel four extra repeat mutation in the prion protein gene. *J. Neurol. Neurosurg. Psychiatry* **72**, 788–791 (2002).
8. Campbell, T. A., Palmer, M. S., Will, R. G., Gibb, W. R., Luthert P. J. et al. A prion disease with a novel 96-base pair insertional mutation in the prion protein gene. *Neurology*, **46**, 761–766 (1996).
9. Pietrini, V., Puoti, G., Limido, L., Rossi, G., Di Fede, G. et al. Creutzfeldt-Jakob disease with a novel extra-repeat insertional mutation in the PRNP gene. *Neurology*, **61**, 1288–1291 (2003).
10. Chiesa, R., Piccardo, P., Ghetti, B., and Harris D.A. Neurological illness in transgenic mice expressing a prion protein with an insertional mutation. *Neuron* **21**, 1339–1351 (1998).
11. Flechsig, E., Shmerling, D., Hegyi, I., Raeber, A.J., Fischer, M. et al. Prion protein devoid of the octapeptide repeat region restores susceptibility to scrapie in PrP knockout mice. *Neuron* **27**, 399–408 (2000).

12. Shmerling, D., Hegyi, I., Fischer, M., Blattler, T., Brandner, S. et al. Expression of amino-terminally truncated PrP in the mouse leading to ataxia and specific cerebellar lesions. *Cell* **93**, 203–214 (1998).
13. Moore, R. A., Herzog, C., Errett, J., Kocisko, D. A., Arnold, K. M. et al. Octapeptide repeat insertions increase the rate of protease-resistant prion protein formation. *Protein Sci.* **15**, 609–619 (2006).
14. Frankenfield, K. N., Powers, E. T., and Kelly, J. W. Influence of the N-terminal domain on the aggregation properties of the prion protein. *Protein Sci.* **14**, 2154–2166 (2005).
15. McKinley, M. P., Bolton, D. C., and Prusiner, S. B. A protease-resistant protein is a structural component of the scrapie prion. *Cell* **35**, 57–62 (1983).
16. Cooper, J. K., Ladhani, K., and Minor, P. D. Reference materials for the evaluation of pre-mortem variant Creutzfeldt-Jakob disease diagnostic assays. *Vox Sang.* **92**, 302–310 (2007).
17. Polymenidou, M., Stoeck, K., Glatzel, M., Vey, M., Bellon, A. et al. Coexistence of multiple PrP^{Sc} types in individuals with Creutzfeldt-Jakob disease. *Lancet Neurol.* **4**, 805–814 (2005).
18. Polymenidou, M., Moos, R., Scott, M., Sigurdson, C., Shi, Y. Z. et al. The POM monoclonals: a comprehensive set of antibodies to non-overlapping prion protein epitopes. *PLoS One* **3**, e3872 (2008).
19. Riek, R., Hornemann, S., Wider, G., Billeter, M., Glockshuber, R. et al. NMR structure of the mouse prion protein domain PrP (121–321). *Nature* **382**, 180–182 (1996).
20. Zahn, R., Liu, A., Luhrs, T., Riek, R., von Schroetter, C. et al. NMR solution structure of the human prion protein. *Proc. Natl. Acad. Sci. U. S. A* **97**, 145–150 (2000).
21. Aguzzi, A. and Polymenidou, M. Mammalian prion biology: one century of evolving concepts. *Cell* **116**, 313–327 (2004).
22. Peretz, D., Williamson, R. A., Matsunaga, Y., Serban, H., Pinilla, C. et al. A conformational transition at the N terminus of the prion protein features in formation of the scrapie isoform. J *Mol Biol.* **1997**, 273:614–622.
23. Leclerc, E., Peretz, D., Ball, H., Sakurai, H., Legname, G. et al. Immobilized prion protein undergoes spontaneous rearrangement to a conformation having features in common with the infectious form. *Embo J.* **20**, 1547–1554 (2001).
24. Safar, J., Wille, H., Itri, V., Groth, D., Serban, H. et al. Eight prion strains have PrP(Sc) molecules with different conformations. *Nat. Med.* **4**, 1157–1165 (1998).
25. Minor, P., Newham, J., Jones, N., Bergeron, C., Gregori, L. et al. Standards for the assay of Creutzfeldt-Jakob disease specimens. *J. Gen. Virol.* **85**, 1777–1784 (2004).
26. Collinge, J., Sidle, K. C., Meads, J., Ironside, J., Hill, A. F. Molecular analysis of prion strain variation and the aetiology of 'new variant' CJD. *Nature* **383**, 685–690 (1996).
27. Cali, I., Castellani, R., Alshekhlee, A., Cohen, Y., Blevins, J. et al. Co-existence of scrapie prion protein types 1 and 2 in sporadic Creutzfeldt-Jakob disease: its effect on the phenotype and prion-type characteristics. *Brain* **132**, 2643–2658 (2009).
28. Notari, S., Capellari, S., Langeveld, J., Giese, A., Strammiello, R. et al. A refined method for molecular typing reveals that co-occurrence of PrP(Sc) types in Creutzfeldt-Jakob disease is not the rule. *Lab. Invest.* **87**, 1103–1112 (2007).
29. Rossi, G., Giaccone, G., Giampaolo, L., Iussich, S., Puoti, G. et al. Creutzfeldt-Jakob disease with a novel four extra-repeat insertional mutation in the PrP gene. *Neurology* **55**, 405–410 (2000).
30. Duchen, L. W., Poulter, M., Harding, A. E. Dementia associated with a 216 base pair insertion in the prion protein gene. Clinical and neuropathological features. *Brain* **116**(Pt 3), 555–567 (1993).
31. Serban, D., Taraboulos, A., DeArmond, S. J., Prusiner, S. B. Rapid detection of Creutzfeldt–Jakob disease and scrapie prion proteins. *Neurology* **40**, 110–117 (1990).
32. Leliveld, S. R., Dame, R. T., Wuite, G. J., Stitz, L., and Korth, C. The expanded octarepeat domain selectively binds prions and disrupts homomeric prion protein interactions. *J. Biol. Chem.* **281**, 3268–3275 (2006).

33. Dong, J., Bloom, J. D., Goncharov, V., Chattopadhyay, M., Millhauser, G. L. et al. Probing the role of PrP repeats in conformational conversion and amyloid assembly of chimeric yeast prions. *J. Biol. Chem.* **282**, 34204–34212 (2007).
34. McKinley, M. P., Meyer, R. K., Kenaga, L., Rahbar, F., Cotter, R. et al. Scrapie prion rod formation in vitro requires both detergent extraction and limited proteolysis. *J. Virol.* **65**, 1340–1351 (1991).

6 Prion Protein Self-Peptides

Alan Rigter, Jan Priem, Drophatie Timmers-Parohi, Jan PM Langeveld, Fred G van Zijderveld, and Alex Bossers

CONTENTS

6.1 INTRODUCTION

Molecular mechanisms underlying prion agent replication, converting host-encoded cellular prion protein (PrP^C) into the scrapie associated isoform (PrP^{Sc}), are poorly understood. Selective self-interaction between PrP molecules forms a basis underlying the observed differences of the PrP^C into PrP^{Sc} conversion process (agent replication). The importance of previously peptide-scanning mapped ovine PrP self-interaction domains on this conversion was investigated by studying the ability of six of these ovine PrP based peptides to modulate two processes, PrP self-interaction and conversion.

Three peptides (octarepeat, binding domain 2 and C-terminal) were capable of inhibiting self-interaction of PrP in a solid-phase PrP peptide array. Three peptides (N-terminal, binding domain 2, and amyloidogenic motif) modulated prion conversion when added before or after initiation of the prion protein misfolding cyclic amplification (PMCA) reaction using brain homogenates. The C-terminal peptides (core region and C-terminal) only affected conversion (increased PrPres formation) when added before mixing PrP^{C} and PrP^{Sc}, whereas the octa repeat peptide only affected conversion when added after this mixing.

This study identified the putative PrP core-binding domain that facilitates the PrP^{C}-PrP^{Sc} interaction (not conversion), corroborating evidence that the region of PrP containing this domain is important in the species-barrier and/or scrapie susceptibility. The octarepeats can be involved in PrP^{C}-PrP^{Sc} stabilization, whereas the N-terminal glycosaminoglycan binding motif and the amyloidogenic motif indirectly affected conversion. Binding domain 2 and the C-terminal domain are directly implicated in PrP^{C} self-interaction during the conversion process and may prove to be prime targets in new therapeutic strategy development, potentially retaining PrP^{C} function. These results emphasize the importance of probable PrP^{C}-PrP^{C} and required PrP^{C}-PrP^{Sc} interactions during PrP conversion. All interactions are probably part of the complex process in which polymorphisms and species barriers affect TSE transmission and susceptibility.

Transmissible spongiform encephalopathies (TSEs) are fatal neurodegenerative disorders characterized by accumulation of the pathological isoform of prion protein mainly in tissues of the central nervous system. Formation of this pathological isoform is a posttranslational process and involves refolding (conversion) of the host-encoded prion protein (PrP^{C}) into a pathological isoform partially protease resistant PrP^{Sc} (derived from scrapie) or PrPres (PK-resistant PrP) [1]. The molecular mechanisms involved in PrP^{C} to PrP^{Sc} conversion are poorly understood, but polymorphisms in both PrP isoforms have been shown to be of importance in both interspecies and intraspecies transmissibilities [2]. The formation of PrP^{Sc} aggregates probably requires self-interactions of PrP^{C} molecules as well as with PrP^{Sc} [3, 4]. Thus binding and conformational changes are essential events in this conversion process.

The cell-free conversion of PrP^{C} provides a valuable *in vitro* model in which relative amounts of produced PrPres reflect important biological aspects of TSEs at the molecular level [5, 6]. A recent and very sensitive *in vitro* conversion system is the protein misfolding cyclic amplification (PMCA) assay [7-10], which has been shown to amplify minute amounts of PrP^{Sc} from a variety of sources including sheep scrapie [10]. The effects of single polymorphisms and species-barriers in PrP^{C} or PrP^{Sc} on PrP conversion can largely explain differences in susceptibility and transmissibility in sheep scrapie [5, 11-13]. Even though these polymorphisms are involved in modulation of disease development they do not seem to affect the initial binding of PrP^{C} to PrP^{Sc} [14] and do not seem to directly modulate PrP^{C}-PrP^{Sc} binding. Furthermore, in a recent peptide-array mapping study of ovine PrP^{C} we concluded that these polymorphisms are not part of the identified PrP binding domains likely to be involved in PrP self-interaction [15]. However, this does not exclude these polymorphisms from posing indirect effects on binding behavior of PrP^{C} to PrP^{Sc} and other possible chaperoning molecules. In that peptide-array binding study we unequivocally demonstrated that ovine PrP binds with PrP derived (self) amino acid sequences (sequence specific) separate from the polymorphic scrapie susceptibility determinants [15]. It remains to be elucidated whether the determined amino acid sequences play a role prior or during conversion in the self-interaction of PrP^C molecules and/or in the interactions of PrP^C with PrP^{Sc}. Simultaneously, whether these amino acid sequences play a role in the processes underlying PrP conversion needs to be elucidated. In the current study, we selected several ovine PrP sequence derived synthetic peptides to study not only their capacity to affect PrP binding to a solid-phase (PrP) peptide-array but also their potential modulating effect on PrP^C to PrP^{Sc} conversion.

6.2 METHODS

6.2.1 MBP-PrP Construction, Expression, and Purification

The mature part of sheep PrP (ARQ) open reading frame (ORF) was cloned into the pMAL Protein Fusion and Purification System (New England Biolabs) [15], resulting in the maltose binding protein (MBP) fusion to the N-terminus of PrP (MBP-PrP). The MBP-PrP was expressed and purified by affinity chromatography as described in the manual of the pMAL Protein Fusion and Purifications System (method I.; New England Biolabs) To improve binding of MBP-PrP and to prevent formation of interchain disulfide upon lysis (as suggested in the protocol), β-mercaptoethanol was added. Quantity and quality of the eluted MBP-PrP was determined before use in the peptide-array by SDS-PAGE (12% NuPAGE; Invitrogen). After separation the gel was either stained with Sypro Orange (total protein stain, Molecular Probes) or analysed by Western blotting and immunodetection of MBP-PrP with polyclonal antiserum R521-7 specific for PrP [71].

6.2.2 Peptides

Peptides were synthesized with an acetylated N-terminus and an amidated C-terminus as described before [71]. The synthesized peptides were purified by high performance liquid chromatography using mass spectrometric analysis for identification. The resulting purified peptides were at least 90 percent pure. All peptides dissolved well in water and solutions were stored frozen. Sequential properties like iso-electric point were calculated with the Peptide Property Calculator made available online in the tools section of Innovagen.

6.2.3 Peptide-Array Analysis

The synthesis of complete sets of overlapping 15-mer peptides were carried out on grafted plastic surfaces, covering the ovine PrP amino acid sequence of mature PrP (residues 25–234 [72]). The coupling of the peptides to the plastic surface consisting of 455-well credit-card size plastic (minicard) and subsequent ELISA analyses including subsequent background correction, relative density value calculation and binding pattern interpretation were performed as described before [15, 73]. This study also showed that linking of MBP to PrP did not have any disadvantageous effects and therefore its properties are indicative for PrP. Peptide blocking studies were performed by pre-incubating the MBP-PrP with molar excesses of prion peptides before incubating the PrP-peptide mixture as the antigen on the minicard. The binding to the peptide-array was considered relevant when at least at three consecutive peptides optical density values of at least 2.5 times the background were observed.

6.2.4 Protein Misfolding Cyclic Amplification Assay

The protein misfolding cyclic amplification (PMCA) assay first described by Saborio et al. [9] and has been shown applicable to amplify PrP^{Sc} from different sources [10, 43]. In short, a 10% brain homogenate from a (confirmed) scrapie-positive sheep (SPH) was diluted with 50100 times in 10% (confirmed) scrapie-negative sheep brain homogenate (SNH) after which the reaction was subjected to one round of sonication-incubation cycles (24 hr, 48 cycles). To test the influence of the peptides on conversion, the peptides were either added after combining the scrapie positive- and negative brain homogenates (supplemented PMCA) or peptide was pre-incubated with the scrapie-negative homogenate (pre-incubated PMCA) before addition of the scrapie-positive brain homogenate. The total amount of PrP^{C} in the PMCA reaction was calculated based on the quantification of the amount of PrP^{C} in brain tissue [74]. The amount of peptide needed for a reaction was calculated based on the total amount of PrP^{C} in the reaction, adjusted for the size difference between peptide and PrP^{C}; so that the amount of peptide added represents various molar excesses of peptide molecules relative to the total number of PrP^{C} molecules present in the reaction. The amount of formed PrPres was determined after proteinase K digestion (100 μg/ml) of the PMCA reaction and analysis by SDS-PAGE and subsequent Western blotting. The standard detection of PrPres was performed with monoclonal antibody 9A2 [75], except for PMCA reactions containing peptide SAU14, which contains the binding epitope of 9A2. In this case we used a proven combination of monoclonal antibodies L42 [76] and Sha31 [77]. To determine whether the determined amounts of PrPres were significantly different in comparison to its corresponding standard PMCA-assay reaction, an unpaired Student's *t*-test was performed and the *p*-value calculated.

6.3 DISCUSSION

In a previous study [15] we showed that ovine PrP binds to itself and mapped several domains using ovine (and bovine) prion protein derived peptide-arrays, yielding a PrP-specific binding pattern for soluble (monomeric) PrP that could be blocked by several PrP-specific monoclonal antibodies. The current study shows that the different PrP derived synthetic peptides exert different effects on binding in a peptide-array assay and on conversion of PrP^{C} to PrPres in the PMCA assay. In order to better interpret the PMCA data, one needs to consider the effects of sonication in the PMCA. It may be expected that after the first incubation cycle, sonication simply results in shearing the elongated PrP^{Sc}, which releases the peptide (Figure). This results in multiple seeds for the following incubation cycle so that after the first sonication cycle, conditions for both the pre-incubated and the supplemented PMCA can be considered identical. This is however not in agreement with the obtained results. Therefore, we have devised a more intricate schematic (Figure 2) to better account for the observed differences between the pre-incubated and supplemented PMCA for some of the peptides.

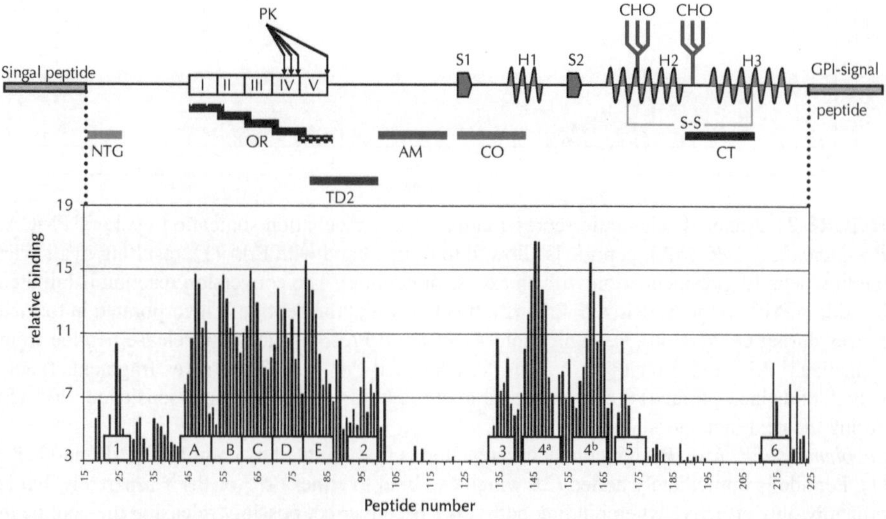

FIGURE 1 Schematic representation of one incubation/sonication cycle of PMCA. The peptide is allowed to form a bond with PrP^C (1), resulting in all PrP^C binding peptide(s) (pre-incubation with an excess of peptide). The conversion reaction is initiated by adding PrP^{Sc} to the 'sensitized' PrP^C (2). It is possible that the peptide remains associated with the formed PrPres during conversion (3). Sonication (4) shears PrPres, but does not necessarily release peptide from 'sensitized' PrP^C and probably also not from the elongated PrP^{Sc} (#). The PrPres fragments are in turn capable of recruiting and converting 'sensitized' PrP^C (5) during the next incubation cycle.

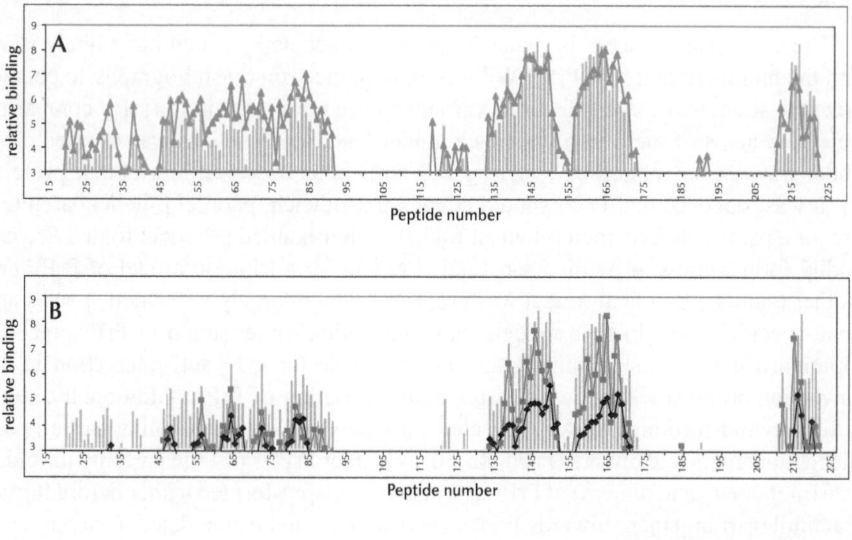

FIGURE 2 *(Continued)*

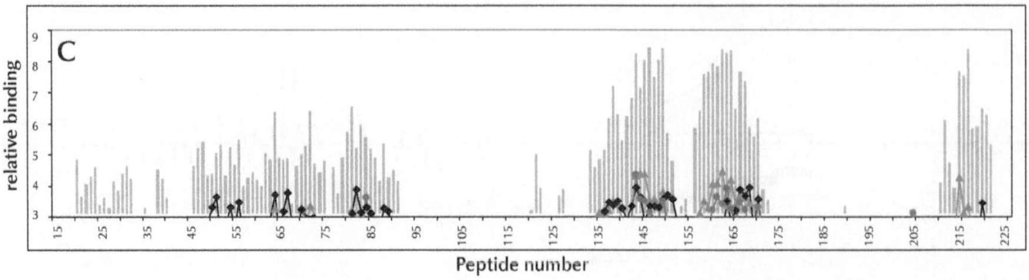

FIGURE 2 Adapted schematic representation of one incubation/sonication cycle of PMCA. *Pre-incubated PMCA* **(A)**. peptide is allowed to form a bond with PrP^C (1), resulting in all PrP^C binding peptide (pre-incubation with an excess of peptide). The conversion reaction is initiated by adding PrP^{Sc} to the 'sensitized' PrP^C (2). It is likely that the peptide is incorporated in formed PrPres during conversion (3). Sonication (4) shears PrPres, but does not release peptide from 'sensitized' PrP^C and probably not from the elongated PrP^{Sc} (#). The PrPres fragments (some with peptide incorporated) are in turn capable of recruiting and converting 'sensitized' PrP^C (5) during the next incubation cycle.
Supplemented PMCA **(B)**. the conversion reaction is first initiated by combining PrP^C and PrP^{Sc} (1). Peptide is immediately added (2), which can bind to either PrP^C or PrP^{Sc} separately, but is probably only effective when binding both. PrP^C is converted, possibly releasing the peptide in the process (3). Sonication (4) shears PrPres, probably releasing the peptide if it is still bound after step 3. The PrPres fragments (unlikely with peptide incorporated, #) are in turn capable of recruiting and converting PrP^C (5) during the next incubation cycle. Even though PrP^C.; PrPres and peptide are all present after sonication, it is likely that fist the PrP^C-PrP^{Sc} complex is formed (1) before interaction with peptide (2) occurs. However, if the peptide is capable of 'sensitizing' PrP^C.; the reaction proceeds as described for the pre-incubated PMCA reaction (6). After addition of all ingredients in the first cycle the pre-incubated and supplemented PMCA reactions will have both unbound and bound peptide available for the following sonication-incubation cycles.

The conversion process is a succession of distinct steps, which most likely starts with the multimerisation of PrP^C. Wille et al. used electron crystallography to characterize the structure of two infectious variants of the prion protein [44]. By comparing projection maps of these two variants a model featuring β-helices was devised. This model was further refined by studying 119 all-β-folds observed in globular proteins [3]. It was proposed that PrP^{Sc} should adopt a β-sandwich, parallel β-helical architecture, or a parallel left-handed β-helical fold. This left-handed β-helical folded PrP can readily form trimers, providing a natural template for a trimeric model of PrP^{Sc} and another (similar) β-helical model was proposed, which largely explained species and strain-specificity [4]. In both models, oligomerisation/trimerisation of PrP^C precedes initiation of conversion, implying an important role for PrP^C self-interaction in the conversion processes. Our data fits this multimerisation of PrP^C. Additionally, study of the amyloid-forming pathway revealed a pre-amyloid state containing partially unfolded monomers and dimers (PrP^i) [45-50]. Whether PrP^i is just the partially unfolded state (monomers and dimers) of PrP^C or whether it is a pre-formed trimer before further structural rearrangement towards PrP^{Sc} occurs remains to be elucidated. Conversion is initiated by recruitment of PrP to PrP^{Sc} after which PrP is (further) rearranged to adopt the tertiary structure of the PrP^{Sc} seed. The elongated PrP^{Sc} is in turn capable to recruit and convert further PrP. Taking these studies into account and their implications for conversion allows for a more detailed interpretation of the data presented in this study. Herein the peptide-array data is indicative for the effects on soluble PrP (PrP^C-PrP^C interaction), whereas the PMCA assay may be indicative for effects on PrP interactions (self- and PrP^C-PrP^{Sc} interaction) as well as interactions with chaperoning or inhibiting molecules. All six PrP-derived peptides tested affected conversion in either the supplemented and/or pre-incubated PMCA assay, whereas binding to the peptide-array was only completely abolished by three peptides [OR, TD2, and CT]. Taken together, this study shows that the previously determined self-interaction domains of PrP^C are of importance at several different phases in the conversion reaction.

Ovine peptide-array analysis previously revealed two high binding areas within PrP [15] of which the first high binding area encompasses the octarepeats, more specifically the consensus domain P(H)GG. This study showed that the octarepeat peptide (OR) was capable of blocking the binding pattern of PrP to the peptide-array, probably as a result of peptide-induced changes in the tertiary structure of the N-terminal tail and thus affecting PrP^C self-interaction. Only in the supplemented PMCA assay, a dose dependant and significant increase in PrPres is observed. The octa repeats can modulate [51-55] but are not a necessity for the molecular processes underlying conversion [21, 56]. The interaction between PrP^C and PrP^{Sc} seems almost instantaneous [14], which would leave the peptide free to interact with the PrP^C-PrP^{Sc} complex as a whole or with co-factors present in the homogenates. Because, the octarepeat stabilizes the interaction of PrP^C with the LRP-LR receptor [57], it seems that peptide OR indirectly affects the conversion process, either by affecting PrP^i stability/formation or by stabilizing PrPi interaction with PrP^{Sc}. Furthermore, di-peptide containing the octarepeat self-aggregates into nanometric fibrils [58] and these may also be formed in the supplemented PMCA assay.

The combined with our data, we propose that the flexible N-terminal tail containing the octarepeat region stabilizes PrP^C-PrP^{Sc} interaction during conversion and that free peptide OR forms nanometric fibrils mimicking and increasing PrP^C-PrP^{Sc} stabilization, thus aiding subsequent conversion. Also part of the peptide-array first high binding area is binding domain 2 ([102-WNK-104], Figure 3) and the data presented here shows that pre-incubation of the peptide TD2 (containing [102-WNK-104]) with PrP abolished binding of PrP to the peptide-array. This indicates the importance of this domain in PrP^C self-interaction. Increased PrPres production was observed in both the supplemented -and the pre-incubated PMCA-assay. The mechanism by which the peptide stimulates PrPres formation may simply be due to peptide enhanced interaction between separate PrP molecules. Alternatively, peptide TD2 could aid unfolding and/or refolding of PrP^C during conversion, binding domain 2 [102-WNK-104], together with the amyloidogenic motif, is part of the region of PrP^C that is partially unfolded and refolded during oligomerization of PrP^C into a β-sheet-rich soluble isoform of PrP [47].

The second high binding area in the peptide-array contains the domain [140-PLIHFGNDY-148] (domain 3, Figure 3). This study shows that the peptide CO containing the domain [140-PLIHFGNDY-148] does not affect binding of PrP to the peptide-array at all, thereby ruling out direct involvement in PrP^C self-interaction. Several studies have established that polymorphisms at sheep PrP amino acid position 136, 154 and 171 surrounding [140-PLIHFGNDY-148] are most relevant in differential TSE susceptibility [5, 59-63] and that stability of this PrP region is a crucial determinant in whether PrP^C is converted [40, 64] (affecting species-barrier and/or scrapie susceptibility). Therefore peptide CO induced PrPres formation in the pre-incubated PMCA assay is either due to this core region peptide facilitating binding of PrP^C to PrP^{Sc} or the peptide affects the stability of this region of PrP^C (interacting with the 'self-domain' or another domain of PrP) thereby facilitating refolding of PrP.

The N-terminal peptide NTG (containing the glycosaminoglycan binding motif) only moderately affects binding of PrP^C throughout the peptide-array, suggesting that interaction of the peptide with PrP^C results either in slight changes in the tertiary structuring affecting solubility of PrP^C or in diminished availability of the previously determined domains [15] for interaction with the peptide-array. Intriguingly, peptide NTG induces PrPres formation in both the supplemented (dose dependant) and pre-incubated (dose optimum) PMCA assay. The glycosaminoglycan heparan sulphate proteoglycan (HSPG) and pentosan polysulphate (PPS) stimulate PrPres formation *in vitro* and suggests that free glycosaminoglycans acted as a contact-mediator allowing interaction of PrP^C and PrP^{Sc} [65]. The N-terminal peptide NTG likely indirectly affects *in vitro* conversion either by mimicking glycosaminoglycan binding to domain [27-RPKPGGG-33] or by recruiting glycosaminoglycans onto PrP^C, facilitating conversion of PrP^C into new PrP^{Sc} after seeding. These studies and our PMCA assay data strongly implicate glycosaminoglycans as an important cofactor in the conversion process.

The ability of peptide AM; which encompasses the amyloidogenic motif [116-AGAAAAGA-123], to moderately block binding of PrP to the peptide-array was somewhat surprising, since we previously showed that the amyloidogenic motif was not involved in PrP self-interaction [15]. Peptide AM mainly inhibits binding of PrP to the peptides covering the N-terminal part of the mature PrP protein, suggesting that peptide AM interacts with one (or more) of the other previously determined binding domains. In contrast to earlier reports [41, 66, 67], we observed that peptide AM (containing the amyloidogenic motif) slightly but significantly increased PrPres formation in both the pre-incubated and supplemented PMCA. However, all these inhibiting peptides contained two or more additional amino acids of the putative aggregation sites (flanking the amyloidogenic motif) implicated in aggregation/oligomerisation [68], suggesting inhibition by these peptides is due to interference with aggregation/oligomerization. Additionally, differences between the used conversion systems (i.e. availability of cofactors) are likely to play a role as well.

The peptide AM used in this study specifically focuses only on the amyloidogenic motif. Our data suggests that peptide AM interacts with the N-terminal tail of PrP^C (octarepeat motif or [102-WNK-104]), probably altering its tertiary structure and facilitating the proposed stabilizing effect of the N-terminal tail. Alternatively, peptides containing only the amyloidogenic motif are also capable of forming a β-sheet rich layer at the water-air interface when sonicated [69] and peptide AM may form a β-sheeted backbone that interacts with the PrP^C-PrP^{Sc}

complex, mimicking and/or complementing the proposed stabilizing effect of the N-terminal tail.

Peptide CT overlaps most of the third alpha helix of PrP^C as well as the second glycosylation site and the second cysteine involved in the di-sulphide bridge formed in PrP^C. The capacity to completely block PrP binding to the peptide-array suggests that the domain [225-SQAY-228] is of importance in PrP^C self-interaction. This study shows a slight significant increase in PrPres formation when peptide CT is pre-incubated with scrapie negative brain homogenate. This contradicts results using a similar peptide capable of inhibiting cell free conversion [41]. However, this inhibiting peptide is four amino acids larger than peptide CT; which may account for the difference in effects and/or it may just be due to the differences in experimental technique between the cell free conversion and the PMCA assay. This seems to be corroborated by the observation that in the supplemented PMCA (setup closest resembling conditions in cell free conversion [41]) peptide CT seems to slightly inhibit PrPres formation albeit not significantly. Fibrillization of a human PrP peptide fragment is hindered by disulfide bridge formation between two peptides [70] or when an additional disulfide bridge is introduced [34], which indicates that peptide CT (when pre-incubated with PrP^C) likely compromises the disulfide bridge, destabilizing PrP^C; which consequently promotes trimerisation or formation of a conversion intermediate and thus facilitating conversion.

In the PMCA-assay all peptides revealed an inducing effect on PrPres formation in the supplemented and/or pre-incubated PMCA-assay. The aforementioned possible explanations for these effects have been discussed for each peptide. However it cannot be ruled out that the peptides may have had an opposite effect, instead of interacting with PrP; peptide may have interacted with possible conversion inhibitory factors present in the homogenate, thus indirectly allowing conversion to take place more efficiently. Identifying these possible 'natural' inhibitory factors may prove an alternative line of investigation towards the underlying mechanisms involved in prion replication and may provide additional targets for future prion therapy.

6.4 RESULTS

As previously determined the recombinant ovine PrP yielded a reproducible sequence specific binding pattern with amino acid sequences using a solid-phase array of overlapping 15-mer peptides encompassing the complete ovine or bovine amino acid sequence (peptide-array). Roughly this pattern breaks down into two high binding areas containing two and three consensus domains respectively, combined with some lower binding domains (Figure 3). Based on the interaction domains extrapolated from this binding pattern as well as properties reported in literature, the following six ovine PrP regions were selected for peptide blocking studies. The sequences of these peptides represented structural properties of PrP as explained hereafter (summarized in Table 1 and mapped in Figure 3). Peptide NTG, spanning the amino acids (AAs) at the N-terminal part of the mature PrP^C, including the glycosaminoglycan (GAG) binding motif KKRPK [16] and binding domain 1

[27-RPKPGGG-33] (ovine numbering used throughout, [15]), **Peptide OR,** spanning the octarepeat AA motif [QPHGGGWG.; AA 54-94] of the N-terminal region (PrP self-interaction was mapped to AA motif P(H)GG [15]), which is probably involved in a range of interactions [17-31] of which metal-binding is the best characterized, **Peptide TD2**, which overlaps the limiting region containing strain and species dependant variable sites for proteinase K trimming of PrP^{Sc} [32, 33] and spanning binding domain 2 ([102-WNK-104], ovine numbering used throughout) of the first high binding area [15], **Peptide AM,** which includes the amyloidogenic motif (AGAAAAGA) of PrP that did not exhibit any binding in the peptide-array [15], **Peptide CO,** encompassing amino acids of the core region of PrP spanning from the first β-sheet onto the first α-helix. The peptide includes binding domain 3 [140-PLIHFGNDYE-149] and is immediately adjacent to binding domain 4[a] of the second high binding area [15], These domains are also important in PrP^{Sc} conformation-specific immuno-precipitation [34-40], and **Peptide CT,** spanning the C-terminal AA's covering part of the third helix, partially covering low binding domain 6 [192-TTTTKGENFT-202] and almost identical to a peptide capable of inhibiting cell-free conversion [41].

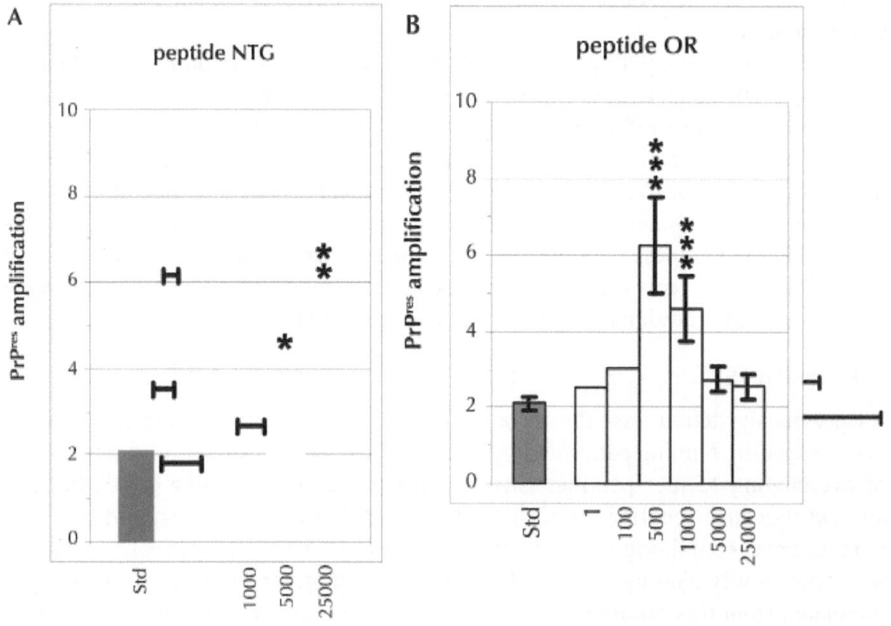

FIGURE 3 *(Continued)*

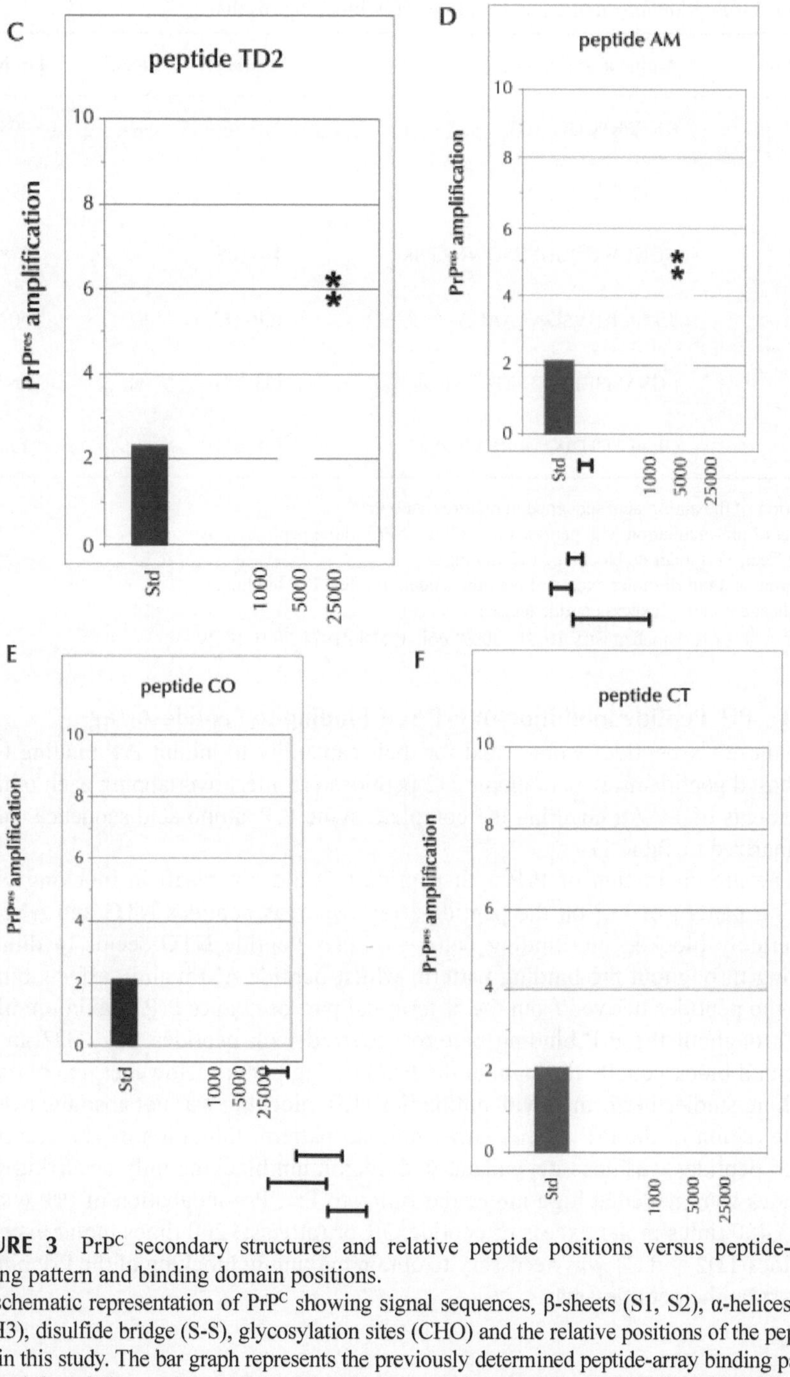

FIGURE 3 PrP^C secondary structures and relative peptide positions versus peptide-array binding pattern and binding domain positions.

The schematic representation of PrP^C showing signal sequences, β-sheets (S1, S2), α-helices (H1, H2, H3), disulfide bridge (S-S), glycosylation sites (CHO) and the relative positions of the peptides used in this study. The bar graph represents the previously determined peptide-array binding pattern [15] and the relative positions of the determined interaction domains. **1** [27-RPKPGGG-33], **A-E** octarepeat motif [PxGG], **2** [102-WNK-104], **3** [140-PLIHFGNDY-148], **4**^{a} [152-YYR-154], **4**^{b} [165-YYR-167], **5** [177-NFV-179] and **6** [225-SQAY-228]. All numbering used is for sheep PrP.

TABLE 1 Peptide information and peptide-array blocking results.

Peptide	Amino acid sequence	Position[1]	Block[2]	Equiv.[3]
NTG	KKRPKPGGGWNT	25–36	+/–	403
OR	GQPHGGGWGQ	61–95	++	396
TD2	GGGGWGQGGSHSQWNKPSK	89–107	++	198
AM	KTNMKHVAGAAAAGA	109–123	+/–	502
CO	LGSAMSRLPLIHFGNDYEDR	133–151	no	499*
CT	GENFTETDIKIMERVVEQMC	198–217	++	198

[1] Position of the amino acid sequence in mature ovine PrP^{C}.
[2] Effect of pre-incubation with peptide on binding of PrP to the peptide-array.
(−no effect, +/−moderate blocking, ++blocking).
[3] Minimal amount of molar excess of peptide needed to affect PrP binding.
(*Highest amount of excess peptide tested).
Rigter et al. BMC Biochemistry 10, 29 (2009) doi:10.1186/1471-2091-10-29.

6.4.1 PrP Peptide Inhibition of PrP Self-Binding to Peptide-Array

First these six peptides were tested for their capability to inhibit PrP binding to the PrP based peptide-array containing 242 peptides (15-mer) overlapping each other by increments of 1 AA.; covering the complete ovine PrP amino acid sequence (results summarized in Table 1).

The pre-incubation of PrP with peptide CO did not result in blocking of the binding pattern of PrP on the peptide-array, whereas peptides NTG and AM only moderately blocked the binding pattern of PrP. Peptide NTG seems to diminish binding throughout the binding pattern, whilst peptide AM mainly affects binding with the peptides derived from the N-terminal part of mature PrP. Maximum blocking throughout the PrP binding pattern occurred with peptides OR, TD2 and CT which all block equally throughout the PrP binding pattern. However, in contrast to blocking studies performed with antibodies [15], blocking was not absolute over the whole region of the PrP peptide-array binding pattern. Inhibition by the aforementioned peptides was dose-dependant, with maximum blocking only occurring when peptides were added at high molecular ratios to PrP. Pre-incubation of PrP with (at least) 400 times molar excess of peptide OR or (at least) 200 times molar excess of peptides TD2 and CT was necessary to obtain maximum blocking of the PrP binding pattern on the peptide-array.

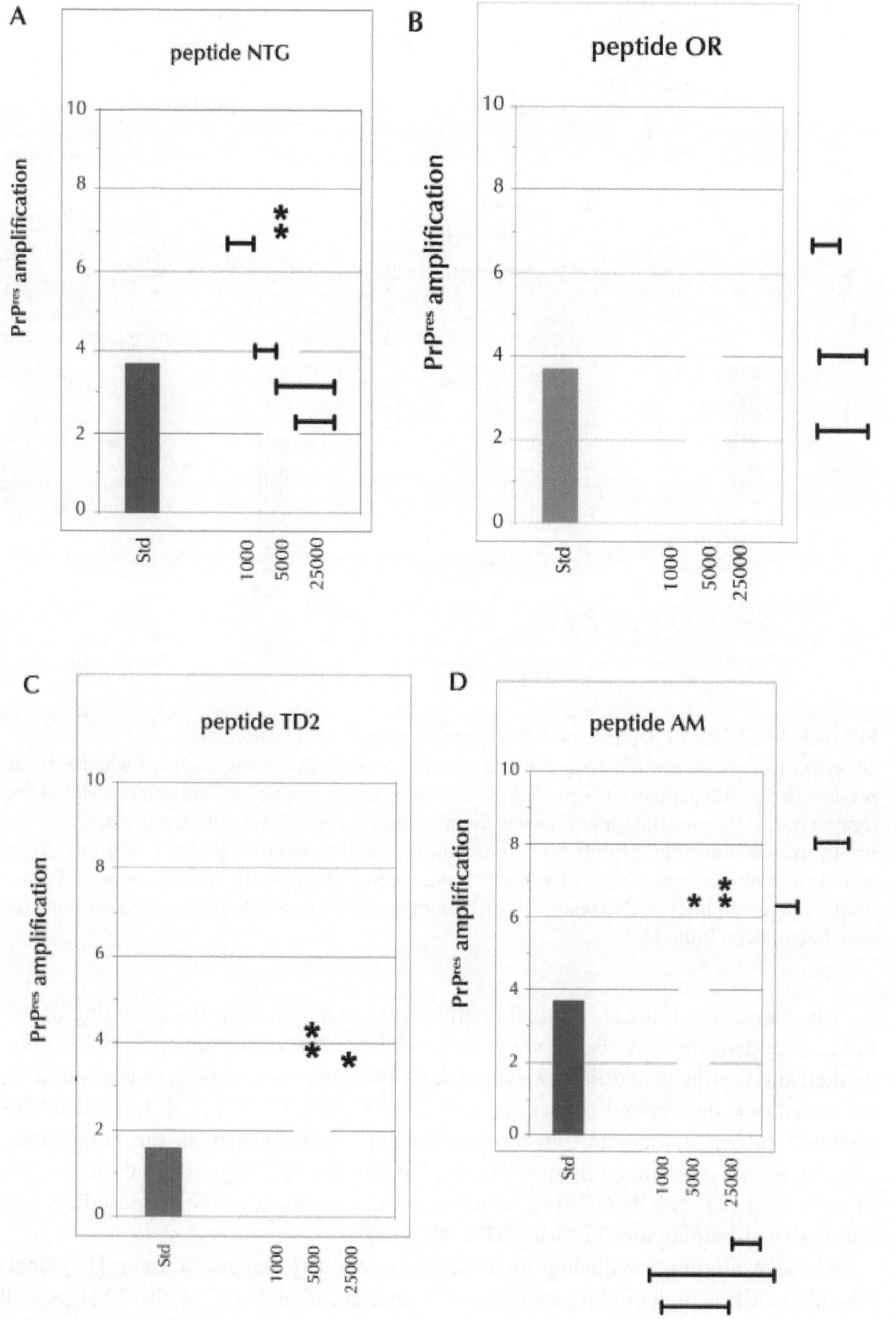

FIGURE 4 *(Continued)*

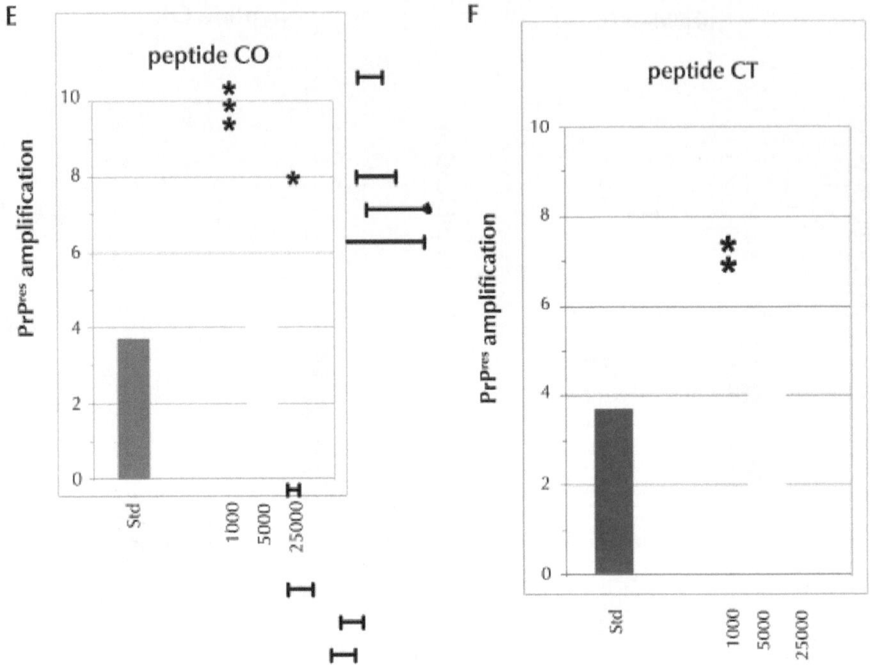

FIGURE 4 Effects of peptides on PrP to peptide-array binding pattern.
No effect (A) on the peptide-array binding pattern and intensities was observed when PrP was pre-incubated with peptide CO (green line) compared to non-peptide pre-incubated PrP binding (grey bars). Only moderate inhibition of the binding pattern (B) was observed when PrP was pre-incubated with either peptide NTG (blue line) or peptide AM (orange line), whereas almost complete inhibition was observed when PrP was pre-incubated with either peptide OR (blue line), TD2 (green line) or CT (orange line). Minimal peptide concentration needed for blocking was determined (Table 1).

The capability of the N-terminal peptide NTG to moderately block binding of PrP to the peptide-array and also moderately modulated PrPres formation in the supplemented and pre-incubated PMCA assay (described later) necessitates re-evaluation of the previously determined consensus domain [33-GWNTG-37] [15]. Instead of this common consensus domain two binding domains seem present in the N-terminus. Binding motifs determined by motif-grafted antibodies [42] suggests that the domain of interest is [27-RPKPGGG-33], which encompasses most of the proposed glycosaminoglycan-binding motif [25-KKRPK-29] [16].

These results confirm the importance of the previously mapped domains [15] located within the first high binding area [PHGG] (octarepeat) and [102-WNK-104], as well as the importance of the C-terminal low binding domain [192-TTTTKGENFT-202] in

PrP self-interaction. To a lesser extent the involvement of the N-terminal glycosaminoglycan binding motif contained within domain [27-RPKPGGG-33] and amyloidogenic motif [116-AGAAAAGA-123] in interaction is confirmed. Interestingly peptide CO, encompassing the previously mapped domain [140-PLIHFGNDYE-149] did not influence binding of PrP to the peptide-array.

6.4.2 PrP Peptides Modulation of PrPres Formation in the PMCA-Assay

The peptides analysed in the prion protein peptide-array were also studied for their modulating capacity in the sheep PrP protein misfolding cyclic amplification (PMCA) assay [9, 10, 43] using sheep brain homogenates from confirmed scrapie-positive and scrapie-negative sheep in one round of sonication cycles. To test the influence of the peptides on conversion, peptides were either added after combining the scrapie positive- and negative brain homogenates (peptide supplemented PMCA) or alternatively peptide was added first to the scrapie-negative brain homogenate before addition of the scrapie-positive material (peptide pre-incubated PMCA). This allowed us to assess if the effect of the peptides on conversion was dependent on the first rapid interaction between PrP^{C} and PrP^{Sc} [14] or not. Peptide was added in several molar ratios, relative to the calculated total amount of PrP^{C} present in the reaction. PrP^{Sc} specific proteinase-K (PK) resistant fragments were quantified by Western blotting.

6.4.3 Peptide Supplemented PMCA

The addition of peptide after mixing scrapie positive- and negative brain homogenates resulted in a dose dependant increase of PK resistant PrP (PrPres) after sonication for four [NTG, OR, TD2, and AM] of the six peptides tested (results summerized in Table 2). In general, the amount of newly formed PrPres roughly ranged between 2 fold (standard reaction) up to 8 fold as compared to the input amount of PrP^{Sc}. Addition of a large molar excess of peptide NTG resulted in a significant increase in PrPres formation in a dose dependant manner at molar excesses of 5.000 ($p = 0.0161$) and 25.000 ($p = 0.0007$). Addition of peptide OR resulted in the largest increase of PrPres. However, in contrast to the other peptides the effect was inversely dose dependant and significant at molar excesses of 500 ($p < 0.0001$) and 1000 $p = 0.0001$). Molar excesses below 500 of peptide or did not result in significant increase of PrPres as with higher molar excess of this peptide. Addition of peptide TD2 resulted in a slight to moderate increase in PrPres formation. PrPres increase was significant only at a molar excess of 25.000 ($p = 0.0027$). Addition of peptide AM resulted in a moderate and significant increase of PrPres, but only when peptide was added at a molar excess of 5.000 ($p = 0.0012$). Addition of either peptide CO or CT did not significantly influence PrPres formation at the tested molar excesses. However it does seem that these peptides slightly inhibit PrPres formation at molar excesses of 1000 and 5000, respectively.

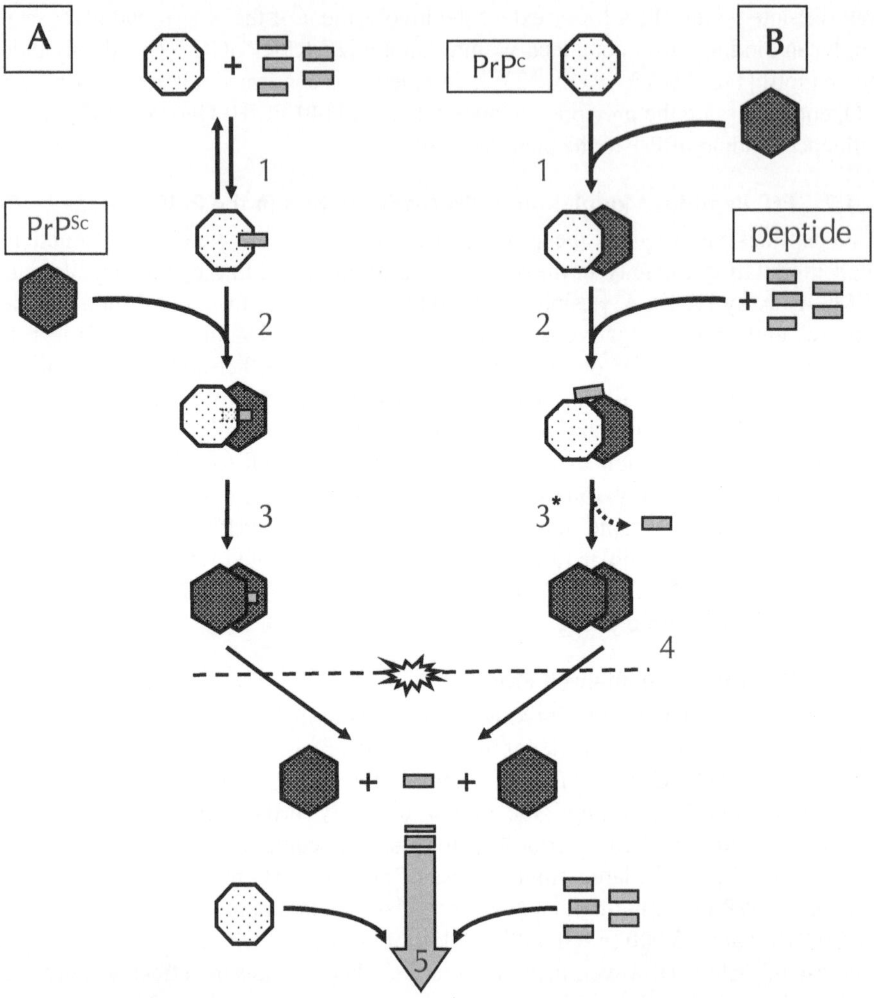

FIGURE 5 Effects of peptide supplementation on PrPres formation in ovine PMCA.
Bar graphs depicting PrPres formation of a standard PMCA reaction compared to a PMCA reaction supplemented with different molar excess amounts of peptide NTG (A), peptide OR (B), peptide TD2 (C), peptide AM (D), peptide CO (E) and peptide CT (F). For each peptide its specific corresponding standard is depicted (grey bar). In order to determine the optimal molar excess of octarepeat peptide or in the supplemented PMCA; a single test with peptide OR added at molar excess 1, 10, 100 and 250 was performed. The amounts of PrPres in these reactions were comparable to the standard. As an example results at molar excess 1 and 100 are depicted in graph B. The number of independent measurements (n), the median and s.e.m. for each peptide and molar excess are summarized in Table 2. An unpaired Student's t-test was performed to determine whether PrPres formation with peptide was significantly different to the corresponding standard PrPres formation. The p-values are listed in Table 2 and significant differences are marked, p-values between 0.05 and 0.001 with *, p-values between 0.001 and 0.0001 with ** and p-values = 0.0001 with ***.

6.4.4 Peptide Pre-incubated PMCA

Pre-incubating scrapie-negative brain homogenate with peptide before initiating conversion with scrapie-positive brain homogenate, surprisingly resulted in increased formation of proteinase-K resistant PrP (PrPres) after sonication for five [NTG, TD2, AM, CO, and CT] of the six peptides tested (results summerized in Table 2) generally in a dose dependant manner. However, compared to the peptide supplemented PMCA assay results differed for several of the peptides. Pre-incubation with peptide NTG resulted in a slight but significant increase of PrPres at the optimal molar excess of 5.000 ($p = 0.0076$). In contrast to peptide supplemented PMCA reactions, pre-incubation with peptide OR did not significantly affect PrPres formation at any of the molar excesses tested. Pre-incubation with peptide TD2 induced a moderate but significant increase of PrPres at molar excesses 5.000 ($p = 0.0048$) and 25.000 ($p = 0.0113$), with an apparent optimum at 5.000. Whereas peptide AM induced amplification of PrPres optimally at molar excess 5.000 in the supplemented PMCA assay, pre-incubation with this peptide resulted in a slight but significant dose dependant increase in PrPres at molar excesses 5.000 ($p = 0.0362$) and 25.000 ($p = 0.0044$). Pre-incubation peptide CO resulted in a significant moderate or slight increase in PrPres at molar excesses 1.000 ($p < 0.0001$) and 25.000 ($p = 0.0496$). However, increase of PrPres at molar excess of 1.000 was larger and unmistakably more significant than at the higher molar excess, suggesting an inverse dose dependant increase of peptide induced PrPres formation. Finally, pre-incubation of SNH with peptide CT also resulted in a significant inverse dose dependant increase of PrPres at the molar excess of 1.000 ($p = 0.0006$).

TABLE 2 Peptide modulation of supplemented- and pre-incubated PMCA assay.

Complemented PMCA-assay						Pre-incubated PMCA-assay						
Peptide	**m.e.**[1]	**median**	**s.e.m.**	**n**[2]	**sign.**[3]	**median**	**s.e.m.**	**n**[2]	**sign.**[3]	**s.e.m.**	**n**[2]	**sign.**[3]
NTG	1000	2.07	0.24	3		4.15	0.60	5	0.60	5		
	5000	3.56	0.73	3	*	5.91	0.76	5	**	0.76	5	**
	25000	4.68	1.26	3	**	4.13	0.63	5	0.63	5		
OR	500	6.26	1.27	4	***	§*n.t.*	§*n.t.*					
	1000	4.61	0.87	5	***	3.85	0.54	4	0.54	4		
	5000	2.74	0.68	4		4.44	0.85	4	0.85	4		
	25000	2.57	0.33	2		3.99	1.19	4	1.19	4		

TABLE 2 *(Continued)*

Complemented PMCA-assay						**Pre-incubated PMCA-assay**						
Peptide	**m.e.**[1]	**median**	**s.e.m.**	**n**[2]	**sign.**[3]	**median**	**s.e.m.**	**n**[2]	**sign.**[3]	**s.e.m.**	**n**[2]	**sign.**[3]
TD2	1000	3.71	0.73	8	n.q.	1.73	0.35	4	0.35	4		
	5000	3.00	0.93	8		3.15	0.33	4	**	0.33	4	**
	25000	4.63	0.62	8	**	2.91	0.34	4	*	0.34	4	*
AM	1000	1.92	0.27	3		4.36	0.28	7	0.28	7		
	5000	4.03	0.34	3	**	5.38	0.76	7	*	0.76	7	*
	25000	2.36	0.46	3		5.63	0.50	7	**	0.50	7	**
CO	1000	1.56	0.02	2		8.59	0.46	5	***	0.46	5	***
	5000	2.02	0.58	2		4.95	1.43	5	1.43	5		
	25000	1.97	0.22	2		6.09	1.53	5	*	1.53	5	*
CT	1000	1.80	0.21	2		6.21	0.39	5	**	0.39	5	**
	5000	1.41	0.29	2		5.23	1.76	5	1.76	5		
	25000	2.85	1.12	2		4.77	0.92	5	0.92	5		

[1]Amount of molar excess of peptide compared to PrP^C tested in the PMCA assay.

[2]Numbers of independent measurements performed.

[3]Comparison of conversion ratios of the peptide supplemented -or pre-incubated PMCA and their corresponding standards using the unpaired Student's *t*-test (*p*-values <0.05 are considered statistically different). Significantly different values are marked, *p*-values between 0.05 and 0.001 with *, *p*-values between 0.001 and 0.0001 with ** and *p*-values ≤0.0001 with ***. The *p*-value of a comparison that was not quite significantly different is marked with n.q.

[§]n.t. = not tested.

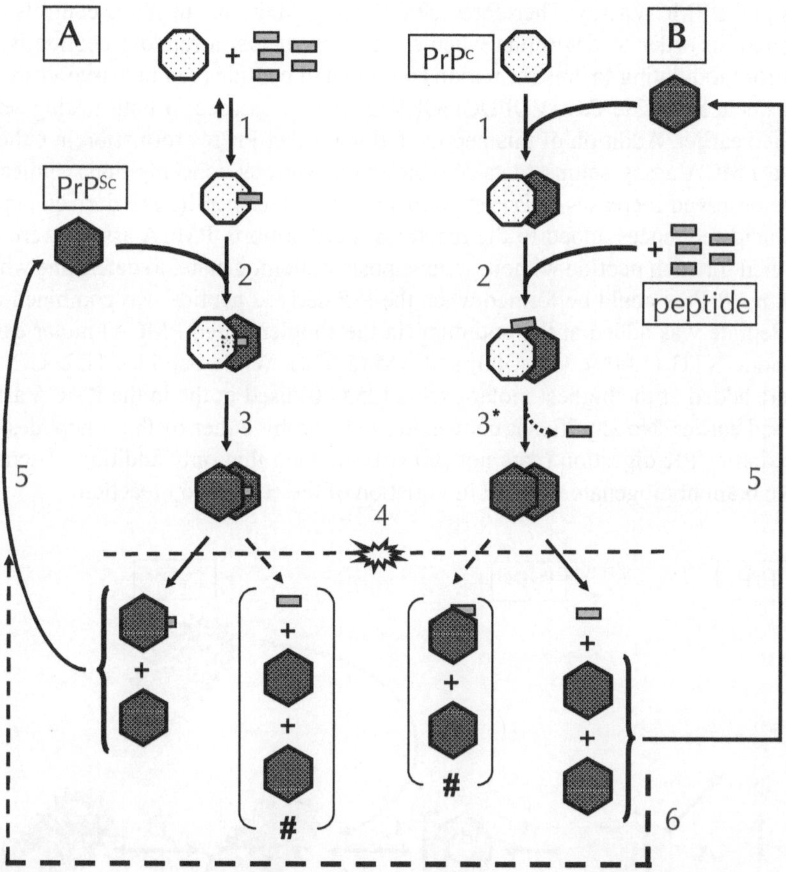

FIGURE 6 Effects of peptide pre-incubation on PrPres formation in ovine PMCA. Bar graphs depicting PrPres formation of a standard PMCA reaction compared to a PMCA reaction in which the scrapie negative homogenate was pre-incubated with peptide with different molar excess amounts of peptide NTG (A), peptide or (B), peptide TD2 (C), peptide AM (D), peptide CO (E) and peptide CT (F) before adding scrapie positive homogenate. For each peptide its specific corresponding standard is depicted (grey bar). The number of independent measurements (n), the median and s.e.m. for each peptide and molar excess are summarized in Table 2. An unpaired Student's *t*-test was performed to determine whether the PrPres formation with peptide was significantly different to the corresponding standard PrPres formation. The *p*-values are listed in Table 2 and significant differences are marked, *p*-values between 0.05 and 0.001 with *, *p*-values between 0.001 and 0.0001 with ** and *p*-values ≤ 0.0001 with ***.

6.4.5 PMCA Assay Negative Controls

Even though each PMCA assay setup revealed at least one PrP specific peptide incapable of modulating PrPres formation, additional negative controls were performed. To rule out the possibility that factors other than PrP sequence specificity could be responsible for the observed results, the isoelectric point, net charge and average hydrophilicity were determined for each peptide (data not shown). Comparison revealed no clear correlation between these non-sequential features and the observed effects in the PMCA assay. Therefore, the following additional negative controls were performed, in order to determine whether addition of just a random peptide is sufficient for modulating PrPres formation an unrelated peptide (canine parvo virus specific sequence peptide DGAVQPDGGQPAVRNER) was used in both testing setups described earlier. Addition of this peptide did not affect PrPres formation in either of the two PMCA assay setups (data not shown) at various concentrations, indicating that the observed increases in PrPres were a result of the specific PrP derived peptide amino acid sequences added to the reactions. Furthermore, PMCA assays were also performed for each peptide without scrapie positive homogenate, to determine whether *de novo* PrPres could be formed when the PrP derived peptide was combined with PrP^C. Peptide was added at the optimum (in the supplemented PMCA) molar excess for peptide NTG (1.000), OR (500) and AM (5.000). While peptides TD2, CO, and CT were added at the highest molar excess (25.000) used in the in the PMCA assays described earlier. No significant conversion induced by either of these peptides was detected after PK digestion (data not shown), showing that only addition of scrapie-positive brain homogenate resulted in initiation of the conversion reaction.

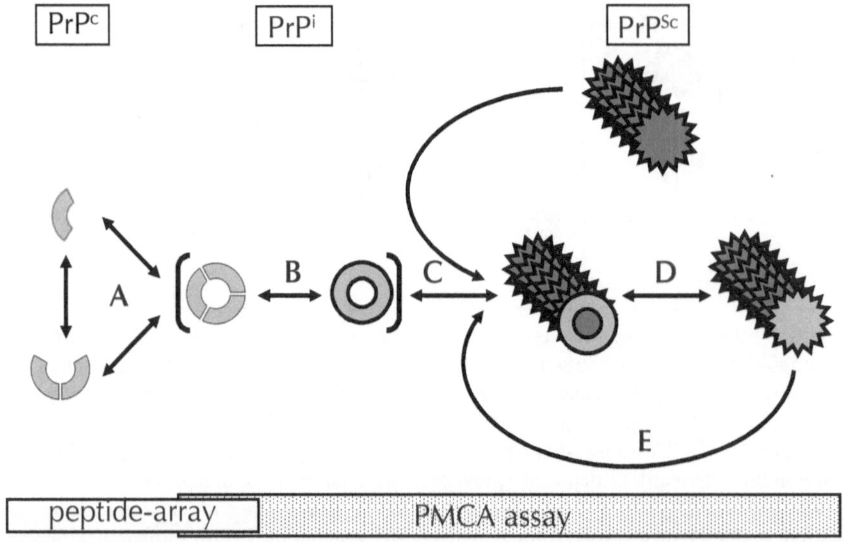

FIGURE 7 Schematic representation of likely steps in prion protein conversion.
Preceding conversion a multimer of PrP^C molecules is formed [A]. Possibly a dimer is formed and both monomeric and dimeric PrP is partially refolded before forming the structural subunit of a fibril, trimeric PrP. PrP^C may however first form a trimer before partially unfolding and/or refolding. Partial unfolding/refolding is at the basis of forming an intermediate isoform of PrP (PrP^i) [B]. PrP^i can subsequently interact with the conversion seed (PrP^{Sc}) [C], which allows conversion of PrP into PrP^{Sc}. Several PrP^i may have to be 'stacked' onto the conversion seed before conversion of the first PrP^i bound can occur. During conversion PrP^i is further refolded into PrP^{Sc}, its tertiary structure adapted to that of the conversion seed (D). The newly elongated PrP^{Sc} in turn can act as a seed for further conversion of PrP (E). The bars beneath the schematic denote for which conversion processes the peptide-array results (white bar) and PMCA assay results (shaded bar) are indicative.

6.5 CONCLUSION

The binding domains found for ovine PrP^C using a prion protein peptide-array are primarily indicative of prion protein self-interaction. Apparently several specific self-interactions between individual PrP molecules occur, which include both PrP^C-PrP^C as well as PrP^C-PrP^{Sc} interactions. The data presented here imply an influence of binding domain [140-PLIHFGNDY-148] on the stability of the region of PrP previously determined to be involved in the species-barrier and/or susceptibility to scrapie. Furthermore our data indicates a stabilizing function for the octarepeats region (N-terminal tail) in PrP^C-PrP^{Sc} interaction and thus improving subsequent conversion. The data further suggests that the N-terminal glycosaminoglycan-binding motif [27-RPKP-GGG-33] affects the conversion process indirectly, and implicates glycosaminoglycans as an important cofactor in prion disease pathogenicity. Peptide AM containing the amyloidogenic motif indirectly affects conversion either by aiding and/or complementing the proposed stabilizing function of the N-terminal tail of PrP^C. Finally, the data implicates direct involvement of the two binding domains [102-WNK-104] and [225-SQAY-228] in self-interaction between PrP^C molecules preceding binding to PrP^{Sc} and subsequent conversion. Therefore, these two domains may prove prime targets for development of new therapeutic strategies. Our results emphasize the importance of the stability of the PrP^C-PrP^C and PrP^C-PrP^{Sc} interactions in PrP conversion, which is an essential determinant in the effects of disease-associated mutations, as well as the species-barrier. The focusing on the (stabilizing) self-interaction domains of PrP and the subsequent conversion processes may lead to further therapeutic strategies with the possibility to leave the physiological function of the prion protein unaffected.

KEYWORDS

- **Glycosaminoglycan**
- **Peptides**
- **Western blotting**
- **Scrapie**
- **Sonication**

AUTHORS' CONTRIBUTIONS

Alan Rigter and Alex Bossers conceived the study and together with JPML were responsible for study design and coordination. All PMCA assays were performed by JP and all peptide-arrays were performed at Pepscan presto B.V. by DPT. Supporting experiments and data-analysis were performed by Alan Rigter, AR, JPML, and AB were responsible for data interpretation. The experiments for the study were facilitated by grants under the supervision of AB, JPML, and FGvZ. AR drafted the manuscript and AB and JPML critically read the manuscript before submission.

ACKNOWLEDGMENT

We thank Drs. Jaques Grassi (SPI.; CEA) and Martin Groschup (FLI, Reims, GE) for generously supplying the monoclonal antibodies Sha31 and L42. This work was supported by grant 903-51-177 from the Dutch Organization for Scientific Research (NWO), by a grant from the Dutch Ministry of Agriculture, Nature Management and Fisheries (LNV) and by EU NeuroPrion project STOPPRIONs FOOD-CT-2004-506579.

REFERENCES

1. DeArmond, S. J. and Prusiner, S. B. Perspectives on prion biology, prion disease pathogenesis, and pharmacologic approaches to treatment. *Clin. Lab. Med.* **23**(1), 1–41 (2003).
2. Bossers, A., Rigter, A., de Vries, R., and Smits, M. A. *In vitro* conversion of normal prion protein into pathologic isoforms. *Clin. Lab. Med.* **23**(1), 227–247 (2003).
3. Govaerts, C., Wille, H., Prusiner, S. B., and Cohen, F. E. Evidence for assembly of prions with left-handed beta-helices into trimers. *Proc. Natl. Acad. Sci. U. S. A.* **101**(22), 8342–8347 (2004).
4. Langedijk, J. P., Fuentes, G., Boshuizen, R., and Bonvin, A. M. Two-rung model of a left-handed beta-helix for prions explains species barrier and strain variation in transmissible spongiform encephalopathies. *J. Mol. Biol.* **360**(4), 907–920 (2006).
5. Bossers, A., de Vries, R., and Smits, M. A. Susceptibility of sheep for scrapie as assessed by *in vitro* conversion of nine naturally occurring variants of PrP. *J. Virol.* **74**(3), 1407–1414 (2000).
6. Kocisko, D. A., Priola, S. A., Raymond, G. J., Chesebro, B., Lansbury, P. T. Jr., and Caughey, B. Species specificity in the cell-free conversion of prion protein to protease-resistant forms. A model for the scrapie species barrier. *Proc. Natl. Acad. Sci. U. S. A.* **92**(9), 3923–3927 (1995).
7. Castilla, J., Saa, P., Morales, R., Abid, K., Maundrell, K., and Soto, C. Protein misfolding cyclic amplification for diagnosis and prion propagation studies. *Methods Enzymol.* **412**, 3–21 (2006).
8. Castilla, J., Saa, P., and Soto, C. Detection of prions in blood. *Nat. Med.* **11**(9), 982–985 (2005).
9. Saborio, G. P., Permanne, B., and Soto, C. Sensitive detection of pathological prion protein by cyclic amplification of protein misfolding. *Nature* **411**(6839), 810–813 (2001).
10. Soto, C., Anderes, L., Suardi, S., Cardone, F., Castilla, J., Frossard, M. J., Peano, S., Saa, P., Limido, L., Carbonatto, M., et al. Pre-symptomatic detection of prions by cyclic amplification of protein misfolding. *FEBS Lett.* **579**(3), 638–642 (2005).
11. Dubois, M. A., Sabatier, P., Durand, B., Calavas, D., Ducrot, C., and Chalvet-Monfray, K. Multiplicative genetic effects in scrapie disease susceptibility. *C. R. Biol.* **325**(5), 565–570 (2002). PubMed Abstract.
12. Tranulis, M. A. Influence of the prion protein gene, *Prnp*, on scrapie susceptibility in sheep. *Apmis* **110**(1), 33–43 (2002).
13. Sabuncu, E., Petit, S., Le Dur, A., Lan Lai, T., Vilotte, J. L., Laude, H., and Vilette, D. PrP polymorphisms tightly control sheep prion replication in cultured cells. *J. Virol.* **77**(4), 2696–2700 (2003).
14. Rigter, A. and Bossers, A. Sheep scrapie susceptibility-linked polymorphisms do not modulate the initial binding of cellular to disease-associated prion protein prior to conversion. *J. Gen. Virol.* **86**(Pt 9), 2627–2634 (2005).
15. Rigter, A., Langeveld, J. P., Timmers-Parohi, D., Jacobs J. G., Moonen, P. L., and Bossers, A. Mapping of possible prion protein self interaction domains using peptide arrays. *BMC Biochem.* **8**(1), 6 (2007).
16. Pan, T., Wong, B. S., Liu, T., Li, R., Petersen, R. B., and Sy, M. S. Cell-surface prion protein interacts with glycosaminoglycans. *Biochem. J.* **368**(Pt 1), 81–90 (2002).
17. Brown, D. R. Prion protein expression aids cellular uptake and veratridine-induced release of copper. *J. Neurosci. Res.* **58**(5) 717–725 (1999).
18. Brown, D. R. Role of the prion protein in copper turnover in astrocytes. *Neurobiol. Dis.* **15**(3), 534–543 (2004).

19. Brown, D. R., Clive C., and Haswell, S. J. Antioxidant activity related to copper binding of native prion protein. *J. Neurochem.* **76**(1), 69–76 (2001).
20. Dupuis, L., Mbebi, C., Gonzalez de Aguilar, J. L., Rene, F., Muller, A., de Tapia, M., and Loeffler, J. P. Loss of prion protein in a transgenic model of amyotrophic lateral sclerosis. *Mol. Cell. Neurosci.* **19**(2), 216–224 (2002).
21. Flechsig, E., Shmerling, D., Hegyi, I., Raeber, A. J., Fischer M., Cozzio, A., von Mering C., Aguzzi, A., and Weissmann, C. Prion protein devoid of the octapeptide repeat region restores susceptibility to scrapie in PrP knockout mice. *Neuron* **27**(2), 399–408 (2000).
22. Frankenfield, K. N., Powers, E. T., and Kelly, J. W. Influence of the N-terminal domain on the aggregation properties of the prion protein. *Protein Sci.* **14**(8), 2154–2166 (2005).
23. Klamt, F., Dal-Pizzol, F., Conte da Frota, M. J., Walz, R., Andrades M. E., da Silva, E. G., Brentani, R. R., Izquierdo, I., and Fonseca Moreira, J. C. Imbalance of antioxidant defense in mice lacking cellular prion protein. *Free Radic. Biol. Med.* **30**(10), 1137–1144 (2001).
24. Miele, G., Jeffrey, M., Turnbull, D., Manson, J., and Clinton, M. Ablation of cellular prion protein expression affects mitochondrial numbers and morphology. *Biochem. Biophys. Res. Commun.* **291**(2), 372–377 (2002).
25. Pauly, P. C., Harris, D. A. Copper stimulates endocytosis of the prion protein. *J. Biol. Chem.* **273**(50), 33107–33110 (1998).
26. Rachidi, W., Vilette, D., Guiraud, P., Arlotto, M., Riondel, J., Laude, H., Lehmann, S., and Favier, A. Expression of prion protein increases cellular copper binding and antioxidant enzyme activities but not copper delivery. *J. Biol. Chem.* **278**(11), 9064–9072 (2003).
27. Sakudo, A., Lee, D. C., Saeki, K., Nakamura, Y., Inoue, K., Matsumoto, Y., Itohara, S., and Onodera, T. Impairment of superoxide dismutase activation by N-terminally truncated prion protein (PrP) in PrP-deficient neuronal cell line. *Biochem. Biophys. Res. Commun.* **308**(3), 660–667 (2003).
28. Walz, R., Amaral, O. B., Rockenbach, I. C., Roesler, R., Izquierdo, I., Cavalheiro, E. A., Martins, V. R., and Brentani, R. R. Increased sensitivity to seizures in mice lacking cellular prion protein. *Epilepsia* **40**(12), 1679–1682 (1999).
29. Wong, B. S., Pan, T., Liu, T., Li, R., Gambetti, P., and Sy, M. S. Differential contribution of superoxide dismutase activity by prion protein *in vivo*. *Biochem. Biophys. Res. Commun.* **273**(1), 136–139 (2000).
30. Wong, E., Thackray, A. M., and Bujdoso, R. Copper induces increased beta-sheet content in the scrapie-susceptible ovine prion protein PrPVRQ compared with the resistant allelic variant PrPARR. *Biochem. J.* **380**(Pt 1), 273–282 (2004).
31. Zeng, F., Watt, N. T., Walmsley, A. R., and Hooper, N. M. Tethering the N-terminus of the prion protein compromises the cellular response to oxidative stress. *J. Neurochem.* **84**(3), 480–490 (2003).
32. Jacobs, J. G., Langeveld, J. P., Biacabe, A. G., Acutis, P. L., Polak, M. P., Gavier-Widen, D., Buschmann, A., Caramelli, M., Casalone, C., Mazza, M., et al. Molecular discrimination of atypical bovine spongiform encephalopathy strains from a geographical region spanning a wide area in Europe. *J. Clin. Microbiol.* **45**(6), 1821–1829 (2007).
33. Thuring, C. M., Erkens, J. H., Jacobs, J. G., Bossers, A., Van Keulen, L. J., Garssen, G. J., Van Zijderveld, F. G., Ryder, S. J., Groschup, M. H., Sweeney, T., et al. Discrimination between scrapie and bovine spongiform encephalopathy in sheep by molecular size, immunoreactivity, and glycoprofile of prion protein. *J. Clin. Microbiol.* **42**(3), 972–980 (2004).
34. Knowles, T. P. and Zahn, R. Enhanced stability of human prion proteins with two disulfide bridges. *Biophys. J.* **91**(4), 1494–1500 (2006).
35. Korth, C., Stierli, B., Streit, P., Moser, M., Schaller, O., Fischer, R., Schulz-Schaeffer, W., Kretzschmar, H., Raeber, A., Braun, U., et al. Prion (PrP^{Sc})-specific epitope defined by a monoclonal antibody. *Nature* **390**(6655), 74–77 (1997).
36. Kuczius, T., Grassi, J., Karch, H., and Groschup, M. H. Binding of N- and C-terminal anti-prion protein antibodies generates distinct phenotypes of cellular prion proteins (PrP^{C}) obtained from human, sheep, cattle and mouse. *FEBS J.* **274**(6), 1492–1502 (2007).

37. Martucci, F., Acutis, P., Mazza, M., Nodari, S. Colussi, S. Corona, C., Barocci, S., Gabrielli, A., Caramelli, M., Casalone, C., et al. Detection of typical and atypical BSE and scrapie prion strains by prion protein motif-grafted antibodies. *J. Gen. Virol.* (2009).
38. Moroncini, G., Kanu, N., Solforosi, L., Abalos, G., Telling, G. C., Head, M., Ironside, J., Brockes, J. P., Burton, D. R., and Williamson, R. A. Motif-grafted antibodies containing the replicative interface of cellular PrP are specific for PrP^{Sc}. *Proc. Natl. Acad. Sci. U. S. A.* **101**(28), 10404–10409 (2004).
39. Moroncini, G., Mangieri, M., Morbin, M., Mazzoleni, G., Ghetti, B., Gabrielli, A., Williamson, R. A., Giaccone, G., and Tagliavini, F. Pathologic prion protein is specifically recognized in situ by a novel PrP conformational antibody. *Neurobiol. Dis.* **23**(3), 717–724 (2006).
40. Paramithiotis, E., Pinard, M., Lawton, T., LaBoissiere, S., Leathers, V. L., Zou, W. Q., Estey, L. A., Lamontagne, J., Lehto, M. T., Kondejewski, L. H., et al. A prion protein epitope selective for the pathologically misfolded conformation. *Nat. Med.* **9**(7), 893–899 (2003).
41. Horiuchi, M., Baron, G. S., Xiong, L. W., and Caughey, B. Inhibition of interactions and interconversions of prion protein isoforms by peptide fragments from the C-terminal folded domain. *J. Biol. Chem.* **276**(18), 15489–15497 (2001).
42. Solforosi, L., Bellon, A., Schaller, M., Cruite, J. T., Abalos, G. C., and Williamson, R. A. Toward molecular dissection of PrP^{C}-PrP^{Sc} interactions. *J. Biol. Chem.* **282**(10), 7465–7471 (2007).
43. Soto, C., Saborio, G. P., and Anderes, L. Cyclic amplification of protein misfolding. application to prion-related disorders and beyond. *Trends Neurosci.* **25**(8), 390–394 (2002).
44. Wille, H., Michelitsch, M. D., Guenebaut, V., Supattapone, S., Serban, A., Cohen, F. E., Agard, D. A., and Prusiner, S. B. Structural studies of the scrapie prion protein by electron crystallography. *Proc. Natl. Acad. Sci. U. S. A.* **99**(6), 3563–3568 (2002).
45. Apetri, A. C., Maki, K., Roder, H., and Surewicz, W. K. Early intermediate in human prion protein folding as evidenced by ultrarapid mixing experiments. *J. Am. Chem. Soc.* **128**(35), 11673–11678 (2006).
46. Apetri, A. C., Surewicz, K., and Surewicz, W. K. The effect of disease-associated mutations on the folding pathway of human prion protein. *J. Biol. Chem.* **279**(17), 18008–18014 (2004).
47. Gerber, R., Tahiri-Alaoui, A., Hore, P. J., and James, W. Oligomerization of the human prion protein proceeds via a molten globule intermediate. *J. Biol. Chem.* **282**(9), 6300–6307 (2007).
48. Hosszu, L. L., Trevitt, C. R., Jones, S., Batchelor, M., Scott, D. J., Jackson, G. S., Collinge, J., Waltho, J. P., and Clarke, A. R. Conformational properties of beta -PrP. *J. Biol. Chem.* (2009).
49. Jenkins, D. C., Sylvester, I. D., and Pinheiro, T. J. The elusive intermediate on the folding pathway of the prion protein. *FEBS J.* **275**(6), 1323–1335 (2008).
50. Stohr, J., Weinmann, N., Wille, H., Kaimann, T., Nagel-Steger, L., Birkmann, E., Panza G., Prusiner, S. B., Eigen, M., and Riesner, D. Mechanisms of prion protein assembly into amyloid. *Proc. Natl. Acad. Sci. U. S. A.* **105**(7), 2409–2414 (2008).
51. Campbell, T. A., Palmer, M. S., Will, R. G., Gibb, W. R., Luthert, P. J., Collinge, J. A prion disease with a novel 96-base pair insertional mutation in the prion protein gene. *Neurology* **46**(3), 761–766 (1996).
52. Cochran, E. J., Bennett, D. A., Cervenakova, L., Kenney, K., Bernard, B., Foster, N. L., Benson, D. F., Goldfarb, L. G., and Brown, P. Familial Creutzfeldt-Jakob disease with a five-repeat octapeptide insert mutation. *Neurology* **47**(3), 727–733 (1996).
53. Goldfarb, L. G., Brown, P., Little, B. W., Cervenakova, L., Kenney, K., Gibbs, C. J. Jr., and Gajdusek, D. C. A new (two-repeat) octapeptide coding insert mutation in Creutzfeldt-Jakob disease. *Neurology* **43**(11), 2392–2394 (1993).
54. Goldfarb, L. G., Brown, P., McCombie, W. R., Goldgaber, D., Swergold, G. D., Wills, P. R., Cervenakova, L., Baron, H., Gibbs, C. J Jr., and Gajdusek, D. C. Transmissible familial Creutzfeldt-Jakob disease associated with five, seven, and eight extra octapeptide coding repeats in the *PRNP* gene. *Proc. Natl. Acad. Sci. U. S. A.* **88**(23), 10926–10930 (1991).
55. Laplanche, J. L., Delasnerie-Laupretre, N., Brandel, J. P., Dussaucy, M., Chatelain, J., and Launay, J. M. Two novel insertions in the prion protein gene in patients with late-onset dementia. *Hum. Mol. Genet.* **4**(6), 1109–1111 (1995).

56. Lawson, V. A. Priola, S. A. Meade-White, K. Lawson, M., and Chesebro, B. Flexible N-terminal region of prion protein influences conformation of protease-resistant prion protein isoforms associated with cross-species scrapie infection in vivo and in vitro. *J Biol Chem*, **279**(14):13689-13695. (2004).
57. Hundt, C., Peyrin, J. M., Haik, S., Gauczynski, S., Leucht, C., Rieger, R., Riley, M. L., Deslys, J. P., Dormont, D., Lasmezas, C. I., et al. Identification of interaction domains of the prion protein with its 37-kDa/67-kDa laminin receptor. *EMBO J.* **20**(21), 5876–5886 (2001).
58. Madhavaiah, C. and Verma, S. Self-aggregation of reverse bis peptide conjugate derived from the unstructured region of the prion protein. *Chem. Commun.* (Camb) (6), 638–639 (2004).
59. Bossers, A., Belt, P., Raymond, G. J., Caughey, B., de Vries, R., and Smits, M. A. Scrapie susceptibility-linked polymorphisms modulate the in vitro conversion of sheep prion protein to protease-resistant forms. *Proc. Natl. Acad. Sci. U. S. A.* **94**(10), 4931–4936 (1997).
60. Bossers, A., Schreuder, B. E., Muileman, I. H., Belt, P. B., and Smits, M. A. PrP genotype contributes to determining survival times of sheep with natural scrapie. *J. Gen. Virol.* **77**(Pt 10), 2669–2673 (1996).
61. Goldmann, W., Hunter, N., Smith, G., Foster, J., and Hope, J. PrP genotypes and the Sip gene in Cheviot sheep form the basis for scrapie strain typing in sheep. *Ann. N. Y. Acad. Sci.* **724**, 296–299 (1994).
62. Hunter, N., Foster, J. D., Goldmann, W., Stear, M. J., Hope, J., and Bostock, C. Natural scrapie in a closed flock of Cheviot sheep occurs only in specific PrP genotypes. *Arch. Virol.* **141**(5), 809–824 (1996).
63. Surewicz, W. K., Jones, E. M., and Apetri, A. C. The emerging principles of Mammalian prion propagation and transmissibility barriers. Insight from studies in vitro. *ACC Chem. Res.* **39**(9), 654–662 (2006).
64. Wong, C., Xiong, L. W., Horiuchi, M., Raymond, L., Wehrly, K., Chesebro, B., and Caughey, B. Sulfated glycans and elevated temperature stimulate PrP(Sc)-dependent cell-free formation of protease-resistant prion protein. *EMBO J.* **20**(3), 377–386 (2001).
65. Chabry, J., Caughey, B., and Chesebro, B. Specific inhibition of in vitro formation of protease-resistant prion protein by synthetic peptides. *J. Biol. Chem.* **273**(21), 13203–13207 (1998).
66. Chabry, J., Priola, S. A., Wehrly, K., Nishio, J., Hope, J., and Chesebro, B. Species-independent inhibition of abnormal prion protein (PrP) formation by a peptide containing a conserved PrP sequence. *J. Virol.* **73**(8), 6245–6250 (1999).
67. Ziegler, J., Viehrig, C., Geimer, S., Rosch, P., and Schwarzinger, S. Putative aggregation initiation sites in prion protein. *FEBS Lett.* **580**(8), 2033–2040 (2006).
68. Satheeshkumar, K. S. and Jayakumar, R. Sonication induced sheet formation at the air-water interface. *Chem. Commun.* (Camb) (19), 2244–2245 (2002).<AQ5: Please supply volume number for Ref. 69, if necessary.>
69. Bosques, C. J. and Imperiali, B. The interplay of glycosylation and disulfide formation influences fibrillization in a prion protein fragment. *Proc. Natl. Acad. Sci. U. S. A.* **100**(13), 7593–7598 (2003).
70. van Keulen, L. J., Schreuder, B. E., Meloen, R. H., Poelen-van den Berg, M., Mooij-Harkes, G., Vromans M. E., and Langeveld, J. P. Immunohistochemical detection and localization of prion protein in brain tissue of sheep with natural scrapie. *Vet. Pathol.* **32**(3), 299–308 (1995).
71. Goldmann, W., Hunter, N., Martin, T., Dawson, M., and Hope, J. Different forms of the bovine PrP gene have five or six copies of a short, G-C-rich element within the protein-coding exon. *J. Gen. Virol.* **72**(Pt 1), 201–204 (1991).
72. Rigter, A., Priem, J., Timmers-Parohi, D., Langeveld, J. P., and Bossers, A. Mapping functional prion-prion protein interaction sites using prion protein based peptide-arrays. *Methods Mol. Biol.* **570**, 257–271 (2009).
73. Moudjou, M., Frobert, Y., Grassi, J., and La Bonnardiere, C. Cellular prion protein status in sheep. tissue-specific biochemical signatures. *J. Gen. Virol.* **82**(Pt 8) 2017–2024 (2001).

74. Langeveld, J. P., Jacobs, J. G., Erkens, J. H., Bossers, A., van Zijderveld, F. G., and van Keulen, L. J. Rapid and discriminatory diagnosis of scrapie and BSE in retro-pharyngeal lymph nodes of sheep. *BMC Vet. Res.* **2**, 19 (2006).
75. Harmeyer, S., Pfaff, E., and Groschup, M. H. Synthetic peptide vaccines yield monoclonal antibodies to cellular and pathological prion proteins of ruminants. *J. Gen. Virol.* **79**(Pt 4), 937–945 (1998).
76. Feraudet, C., Morel, N., Simon, S., Volland, H., Frobert, Y., Creminon, C., Vilette, D., Lehmann, S., and Grassi, J. Screening of 145 anti-PrP monoclonal antibodies for their capacity to inhibit PrP^{Sc} replication in infected cells. *J. Biol. Chem.* **280**(12), 11247–11258 (2005).

7 PrP^{Sc} Spreading Patterns and Prion Types

Wiebke M Wemheuer, Sylvie L Benestad, Arne Wrede, Wilhelm E Wemheuer, Bertram Brenig, Bjшrn Bratberg, and Walter J Schulz-Schaeffer

CONTENTS

7.1 INTRODUCTION

Scrapie in sheep and goats has been known for more than 250 years and belongs nowadays to the so-called prion diseases that also include for example bovine spongiform encephalopathy in cattle (BSE) and Creutzfeldt-Jakob disease in humans. According to the prion hypothesis, the pathological isoform (PrP^{Sc}) of the cellular prion protein (PrP^{C}) comprises the essential, if not exclusive, component of the transmissible agent. Currently, two types of scrapie disease are known—classical and atypical/Nor98 scrapie. In the present study we examine 24 cases of classical and 25 cases of atypical/Nor98 scrapie with the sensitive PET blot method and validate the results with conventional immunohistochemistry. The sequential detection of PrP^{Sc} aggregates in the CNS of classical scrapie sheep implies that after neuroinvasion a spread from spinal cord and obex to the cerebellum, diencephalon and frontal cortex via the rostral brainstem takes place. We categorize the spread of PrP^{Sc} into four stages. the CNS entry stage, the brainstem stage, the cruciate sulcus stage and finally the basal ganglia stage. Such a sequential development of PrP^{Sc} was not detectable upon analysis of the present atypical/Nor98 scrapie cases. PrP^{Sc} distribution in one case of atypical/Nor98 scrapie in a presumably early disease phase suggests that the spread of PrP^{Sc} aggregates starts in the di- or telencephalon. In addition to the spontaneous generation of PrP^{Sc}, an uptake of the infectious agent into the brain, that bypasses the brainstem and starts its accumulation in the thalamus, needs to be taken into consideration for atypical/Nor98 scrapie.

Scrapie in sheep and goats, which has been reported for more than 250 years [1], belongs to the transmissible spongiform encephalopathies (TSEs) - also known as prion diseases. This group of fatal diseases includes bovine spongiform encephalopathy (BSE) in cattle, chronic wasting disease (CWD) in deer and Creutzfeldt-Jakob disease (CJD) in humans. TSEs are characterized by the accumulation of protein aggregates, which are relatively stable against proteolysis. According to the prion hypothesis, a misfolded protein is the relevant part of the infectious agent [2]. It is widely accepted that this "proteinaceous infectious particle" is the pathological isoform of the physiological prion protein (PrP^{C}) which is encoded by a cellular gene [3]. Recently, it has been shown that infectivity can be generated from a synthetic misfolded form of the prion protein [4]. The depending on the kind of prion disease, the pathological prion protein (PrP^{Sc}) is detectable solely in the central nervous system (CNS) or may also be found in other tissues, especially in those of the lymphoreticular system (LRS) [5].

In the worldwide population of small ruminants, BSE and scrapie are considered to be the relevant TSEs affecting sheep and goats. Scrapie, however, is not a homogenous disease form, as demonstrated by the existence of several strains upon transmission to rodents [6] and the peculiar molecular properties of the sheep-passaged scrapie isolate CH1641 [7, 8]. The discovery of a novel type of scrapie in Norway in 1998 (Nor98) that was clearly distinguishable from all previously reported forms of scrapie [9], and that was soon after detected in several other countries, added to the diversity of this TSE [10]. In our present work we concentrate on scrapie field cases that include cases of "classical" scrapie as well as "atypical"/Nor98 scrapie. Obvious differences exist between the two scrapie forms with regard to the epidemiology of the disease and the properties of the proteinaceous particle. The latter include Western blot profiles and the stability against denaturation and proteases [11-13]. The two forms of sheep scrapie also differ with regard to the genotypes affected. Amino acids at codon 136 (A/V), 154 (H/R) and 171 (H/Q/R) are considered to markedly influence susceptibility to classical scrapie, the most susceptible alleles are $V_{136}R_{154}Q_{171}$ (VRQ) and $A_{136}R_{154}Q_{171}$ (ARQ), while the $A_{136}R_{154}R_{171}$ allele (ARR) seems to confer a certain resistance against the disease [14, 15]. Atypical/Nor98 scrapie affects a number of genotypes, including the ARR allele, and animals with the AHQ allele or a Phenylalanin (F) instead of Leucin (L) at codon 141 in the ARQ allele are proportionally overrepresented [16-18].

The results of a number of case reports and studies have shown that the deposition form and distribution of PrP^{Sc} aggregates in atypical/Nor89 scrapie sheep are clearly distinct from classical scrapie, immunohistochemical methods and recently the sensitive PET blot method have been used for the detection of PrP^{Sc} in the ovine brain [9, 19-23]. Formerly, the PET blot had only been used for the sensitive detection of PrP^{Sc} in extra-cerebral organs of classical scrapie sheep [24-27]. Surprisingly, the anatomical distribution of PrP^{Sc} in the ovine brain found in the literature is more thoroughly documented for atypical/Nor98 scrapie than for classical scrapie. Although, the pathogenesis of classical scrapie is well-studied [28, 29], detailed descriptions on how the infectious agent spreads once it has reached the brain seem to be lacking for both scrapie types. For classical scrapie, numerous reports exist on the different forms of PrP^{Sc} that can be found in the brain tissue and the presence of PrP^{Sc} aggregates in peripheral neural and non-neural tissues - at least in sheep carrying susceptible PrP genotypes. Also, the entry of the infectious agent into the CNS has been described thoroughly for field classical scrapie infections and has been shown to agree with the oral infection of sheep with BSE and scrapie as well as the oral infection of rodent models infected with scrapie [29-32]. The infectious agent apparently enters the CNS via the intermediolateral column of the thoracic spinal cord (Th_8–Th_{10} in natural scrapie infection) and the dorsal motor nucleus of the vagus nerve (DMNV) in the brainstem. Unfortunately, reports on the spread of ovine PrP^{Sc} from the brainstem into the brain are usually not very detailed. In atypical/Nor98 scrapie, most of the PrP^{Sc} load in affected sheep is found in the cerebellum and cerebrum. It still needs to be determined whether this novel disease is a sporadic prion disease or not. If sheep could acquire the disease from their environment, where would the infectious agent enter the CNS? The pattern of PrP^{Sc} deposition is apparently reproduced when atypical/Nor98 scrapie is transmitted from one sheep to another via intracerebral inoculation [33].

In this study, the PrP^{Sc} deposition pattern in the CNS of 24 classical and 25 atypical/Nor98 field scrapie sheep was determined using the sensitive and specific PET blot method. Different amounts of PrP^{Sc} in the CNS of classical scrapie have been assigned to different stages of PrP^{Sc} spread into the brain, depending on the affected neuroanatomical structures.

7.2 MATERIALS AND METHODS

7.2.1 Material

The brains and, if available, the spinal cords as well as lymphatic tissue (tonsils and/or retropharyngeal lymph nodes) were collected from 49 scrapie field cases and 6 further sheep from scrapie-free flocks as controls. Scrapie positivity was diagnosed either ante mortem by tonsil biopsy or post mortem using the respective methods stipulated by the EU VO999/2001 at that time (samples were collected during a time span of 12 years). The scrapie-positive group included 19 German and 5 Norwegian sheep diagnosed with classical scrapie and 24 Norwegian atypical/Nor98 scrapie cases, plus one German atypical/Nor98 case. The control group was made up of six German sheep derived from scrapie-free flocks. The PrP genotypes were determined either by PCR and melting curve analysis [34] or by automated sequencing as described previously [9]. Further information on the individual animals including age, breed, genotype, presence of clinical signs and availability of LRS and spinal cord is listed in Table 1.

7.2.2 Histopathology

1–3 μm-thick CNS/lymphatic tissue sections were cut, collected on silane-coated glass slides and stained with haematoxylin and eosin (H&E). Brain sections were also stained with Luxol Fast Blue then counterstained by periodic acid Schiff' reagent (LFB/PAS) for the orientation and discrimination of neuronal nuclei and neural tracts.

7.2.3 PET Blot

The PET blot procedure followed the protocol as described previously [23, 35] using the monoclonal antibody (mAb) P4 (R-Biopharm, Darmstadt, Germany), which had proved to give the best results regarding sensitivity and specificity for the detection of PrP^{Sc} in classical and atypical/Nor98 sheep scrapie [23]. In brief, immunolabeling of PrP^{Sc} was performed after a 1–3 μm tissue section had been placed on a nitrocellulose membrane (0.45 μm, Bio-Rad, Hercules, CA.; USA), which was then deparaffinized and rehydrated. This was followed by treatment with proteinase K (250 μg/mL, Sigma-Aldrich, MO.; USA) overnight at 56°C and the decontamination of the membranes in 4 M guanidine thiocyanate (GdnSCN) for 30 min. Membranes were blocked with 0.2% casein in PBS containing 1% Tween before the primary antibody (mAbP4) was applied 1:5000 in TBST. An alkaline phosphatase-coupled goat-anti-mouse antibody (Dako, Glostrup, Denmark) and the formazan-reaction with NBT/BCIP were used to visualize the result. Thorough rinsing of the membranes with TBST was required between the different steps.

7.2.4 Immunohistochemistry

Tissue sections on silane-coated glass slides were stained with one of the primary mAbs P4, L42 (R-Biopharm, Darmstadt, Germany), F89/160.1.5 (Veterinary Medical Research and Developement, Pullman, WA.; USA), and 12F10 (kindly provided by W Bodemer and D Motzkus, German Primate Center), which were used 1:500 in combination with an alkaline phosphatase-coupled goat anti-mouse antibody (Dako) and neufuchsine as chromogene as described previously [23]. Alternatively, a commercially available kit from Dako (Envision AEC.; Glostrup, Denmark) was applied by using mAb F89/160.1.5 at a dilution of 1:2000 in combination with the mAb 2G11 (1:200, kindly provided by J Grosclaude (INRA.; Jouy-en-Josas, France)).

TABLE 1 Genotype, age, breed, the presence of clinical signs and the availability of lymphatic tissue and spinal cord of the individual sheep.

Genotype	Breed	Age in months	Lymphatic tissue avail-able	Spinal cord vail-able	Clinical signs present	Spinal cord available	Clinical signs present	
Classical scrapie cases								
ARQ/ARQ	G.M./B.M. crossbreed	~42	Yes	Yes	No	Yes	Yes	No
ARQ/ARQ	G.M./B.M. crossbreed	~24	Yes	Yes	No	Yes	Yes	No
ARQ/ARQ	G.M./B.M. crossbreed	>48	Yes	Yes	No	Yes	Yes	No
ARQ/ARQ	Black headed Mutton	~72	Yes	Yes	No	Yes	Yes	No
ARQ/ARQ	German Merino	~60	Yes	Yes	Yes	Yes	Yes	Yes
ARQ/ARQ	German Merino	>48	Yes	Yes	Yes	Yes	Yes	Yes
ARQ/ARQ	G.M./B.M. crossbreed	>48	Yes	Yes	Yes	Yes	Yes	Yes
ARQ/ARQ	G.M./B.M. crossbreed	Unknown	Yes	No	Unknown	Yes	No	Unknown
ARQ/ARQ	G.M./B.M. crossbreed	~48	Yes	No	Unknown	Yes	No	Unknown
ARQ/ARQ	Black headed Mutton	Unknown	Yes	No	Unknown	Yes	No	Unknown
ARQ/ARQ	B.M./Mountain sheep crossbreed	27	Yes	Yes	Yes	Yes	Yes	Yes
ARQ/ARQ	G.M./B.M./Mountain sheep crossbreed	25	Yes	Yes	Yes	Yes	Yes	Yes
ARQ/ARQ	B.M./Mountain sheep crossbreed	Unknown	Yes	Yes	Yes	Yes	Yes	Yes

TABLE 1 *(Continued)*

Genotype	Breed	Age in months	Lymphatic tissue available	Spinal cord vailable	Clinical signs present	Spinal cord available	Clinical signs present	
ARQ/ARQ	B.M./Mountain sheep crossbreed	29	Yes	Yes	No	Yes	Yes	No
ARQ/ARQ	B.M./Mountain sheep crossbreed	38	Yes	Yes	Yes	Yes	Yes	Yes
VRQ/ARQ	Texel	Unknown	Yes	No	Unknown	Yes	No	Unknown
VRQ/ARQ	Norwegian pelt sheep	~24	Yes	No	Unknown	Yes	No	Unknown
VRQ/ARQ	Steigar sheep	~42	No	No	Unknown	No	No	Unknown
VRQ/ARQ	Texel	Unknown	Yes	No	Unknown	Yes	No	Unknown
VRQ/ARQ	Steigar sheep	Unknown	No	No	Unknown	No	No	Unknown
VRQ/ARQ	Texel/Mountain sheep crossbreed	32	Yes	Yes	Yes	Yes	Yes	Yes
VRQ/ARH	Texel	30	Yes	Yes	No	Yes	Yes	No
VRQ/ARH	Steigar sheep	Unknown	Yes	No	Unknown	Yes	No	Unknown
VRQ/ARH	Texel	Unknown	Yes	No	Unknown	Yes	No	Unknown
Atypical/Nor98 scrapie cases								
AFRQ/ARQ	Spæl sheep	~78	No	No	Uncertain	No	No	Uncertain
AFRQ/AHQ	Suffolk/Rygja/Steigar crossbreed	~72	No	Yes	Yes	No	Yes	Yes

TABLE 1 *(Continued)*

Genotype	Breed	Age in months	Lymphatic tissue avail-able	Spinal cord vail-able	Clinical signs present	Spinal cord available	Clinical signs present	
ARQ/AHQ	German Merino	Unknown	Yes	No	Unknown	Yes	No	Unknown
AHQ/AHQ	Steigar sheep	~48	Yes	Yes	Yes	Yes	Yes	Yes
AHQ/AHQ	Spæl sheep	~72	Yes	No	Yes	Yes	No	Yes
AHQ/AHQ	Norwegian white sheep	~84	Yes	No	Yes	Yes	No	Yes
AHQ/AHQ	Spæl sheep	~42	Yes	Yes	Uncertain	Yes	Yes	Uncertain
AHQ/AHQ	Spæl sheep	~48	Yes	No	Yes	Yes	No	Yes
AHQ/AHQ	Spæl sheep	~72	No	No	Yes	No	No	Yes
AHQ/ARH	Norwegian white sheep	~120	Yes	No	Yes	Yes	No	Yes
AHQ/AFRQ	Dala sheep	~84	No	Yes	Yes	No	Yes	Yes
AHQ/AFRQ	Steigar sheep	~60	No	No	Yes	No	No	Yes
AFRQ/AFRQ	Norwegian white sheep	~60	Yes	No	Uncertain	Yes	No	Uncertain
AFRQ/AFRQ	Dala sheep	~36	Yes	Yes	Yes	Yes	Yes	Yes
AFRQ/AFRQ	Norwegian white sheep	~78	Yes	No	Yes	Yes	No	Yes
AFRQ/AFRQ	Rygja/Dala crossbreed	~84	Yes	No	Uncertain	Yes	No	Uncertain
AFRQ/AFRQ	Dala sheep	~108	Yes	No	No	Yes	No	No
AHQ/ARR	Spæl sheep	~72	No	No	Yes	No	No	Yes

TABLE 1 *(Continued)*

Genotype	Breed	Age in months	Lymphatic tissue avail-able	Spinal cord vail-able	Clinical signs present	Spinal cord available	Clinical signs present	
AHQ/ARR	Norwegian white sheep	~96	Yes	No	No	Yes	No	No
AHQ/ARR	Dala sheep	~72	Yes	No	Uncertain	Yes	No	Uncertain
AHQ/ARR	Dala sheep	~84	No	Yes	No	No	Yes	No
ARR/AFRQ	Norwegian white sheep	~96	No	No	Unknown	No	No	Unknown
ARR/AFRQ	Norwegian white sheep	~66	Yes	No	No	Yes	No	No
ARR/ARR	Steigar	~84	No	Yes	No	No	Yes	No
Unknown	Rygja/Dala crossbreed	Unknown	No	No	No	No	No	No
Control cases								
ARQ/ARQ	Skudde	Unknown	Yes	No	No	Yes	No	No
ARR/ARH	Leine	30	Yes	No	No	Yes	No	No
ARR/ARR	German Merino	~132	Yes	No	No	Yes	No	No
ARR/ARR	Leine	~60	Yes	No	No	Yes	No	No
ARR/ARR	Leine	~96	Yes	No	No	Yes	No	No
ARR/ARR	Leine	>48	Yes	No	No	Yes	No	No

B.M. = blackheaded mutton, G.M. = German Merino.

Depending on the circumstances under which the samples were collected, the post mortem times of tissues varied between 2 hr and 4 days. Usually one half of the brain/tonsil/lymph node was fixed in 4% buffered formaldehyde, cut into slices and embedded in paraffin within 5 to 7 days, while the other half was frozen and stored at -80 °C.

7.2.5 Examination and Evaluation of Immunolabelled Sections

From each sheep all available sections of the CNS and the LRS were examined with the PET blot, and the intensity of the PrP^{Sc} staining as well as the forms and distribution of the PrP^{Sc} deposition were evaluated. The presence of PrP^{Sc} deposits and the deposition forms in CNS and LRS sections were verified by immunohistochemistry. This was usually done using either mAb P4 (German cases) or mAb F89/160.1.5 in combination with mAb 2G11 (Norwegian cases), but if considered necessary immunohistochemistry was repeated with further antibodies as stated above. The intensity of PrP^{Sc} deposits in the PET blots was evaluated on a scale of 0.5 to 4 (0 = no PrP^{Sc} deposits visible, 0.5 = very little indefinable deposits, 1 = very little distinct PrP^{Sc} deposits, 1.5 = little distinct PrP^{Sc} deposits, 2 = moderate PrP^{Sc} deposits, all deposition forms well distinguishable, 2.5 = moderate to pronounced PrP^{Sc} deposits, all deposition forms well distinguishable, 3 = pronounced PrP^{Sc} deposits, deposition forms partly interfere with each other, 3.5 = pronounced PrP^{Sc} deposits, deposition forms interfere with each other, 4 maximal PrP^{Sc} deposits, deposition forms interfere with each other). The value system of the scale itself was established and agreed on by two independent persons that routinely evaluate PET blots.

7.2.6 Western Blot Analysis

Ten percent tissue homogenates (wt/vol) were either prepared in PBS containing 0.5 percent desoxycholic acid sodium salt (DOC) using glass grinding tubes and pestles or 20 percent homogenates were obtained by the standard sampling procedure of the TeSeE Western Blot Kit (Bio-Rad, Hercules, CA.; USA).

Twenty percent homogenates were processed using the TeSeE sheep/goat Western Blot Kit according to the manufacturer's instructions. The antibody P4 was added at a dilution of 1:1000 to the primary antibody of the kit. Ten percent homogenates were subjected to a different protocol using homemade 15% acrylamid gels, a 0.45 μm nitrocellulose (NC) membrane (Bio-Rad) for semi-dry blotting and mAbP4 (1:2000). The membrane was treated with 4 M GdnSCN and blocked with 0.2% casein in PBS including 1% Tween for 30 min respectively before the primary antibody was applied overnight at 4 °C. An HRP-conjugated goat anti-mouse antibody (Dako, Carpintera, CA.; USA) and Super Signal Femto West Maximum Sensitivity Substrate (Perbio, Erembodegem, Belgium) were used to visualize the result on x-ray film. The molecular size of PrP^{Sc} was compared only within one system.

7.3 DISCUSSION

In this study, 24 cases of classical and 25 cases of atypical/Nor98 scrapie cases were examined with the PET blot method, focusing on the similarities and differences in the distribution of PrP^{Sc} deposits that were detectable with this method. Recently the PET blot has been shown to provide a sensitive and specific detection of PrP^{Sc} in both types of sheep scrapie in the same manner as had been previously shown for human, bovine and rodent neuronal and non-neuronal tissues [30, 35, 38-41]. The high sensitivity of this method allows PrP^{Sc} deposits to be detected even in FFI patients where conventional immunohistochemistry fails to detect them, and contrasts with Western blotting, which requires up to 1 g of tissue equivalent [42]. The PET blot provides, apart from its sensitivity and specificity, a good overview of where to find PrP^{Sc} in a brain section (Figure 2), as no counterstaining is necessary. The fine resolution of the immunolabeling gives a good impression of the structures that accumulate PrP^{Sc}, but the general delineation of the single cell is better with immunohistochemistry, which is why these two methods complement each other in a sensible way.

7.3.1 Neuroinvasion and Spread of PrP^{Sc} in the Ovine Brain

In this study, we also give a more detailed account of how the disease-associated PrP aggregates seem to spread in the CNS tissue of sheep infected with classical scrapie. The sequential detection of PrP^{Sc} aggregates in the CNS of classical scrapie sheep implies that a cell-to-cell spread takes place from the entry sites in the spinal cord and obex to the cerebellum, diencephalon and frontal cortex via the rostral brainstem. From these entry sites we conclude that the vagus nerve for the DMNV and sympathetic fibres for the spinal cord are the structures that transport the infectious agent to the CNS. This is very similar to the results obtained in hamsters after oral inoculation with the 263K scrapie strain [43]. The cerebellum may also receive PrP^{Sc} via the cerebellar tracts of the spinal cord. Noticeable perivascular PrP^{Sc} deposition in the brains of scrapie-affected sheep also raises the possibility that the infectious agent reaches the brain via the haematogenous route [44]. The distribution of brain metastases in humans reflects a haematogenous entry into the brain as it is proportional to the cerebral blood flow per area. From this one can conclude that a general PrP^{Sc} uptake from the blood would cause quite a different cerebral distribution pattern of PrP^{Sc} deposits than we observed [45]. There are three other possible explanations for the perivascular accumulation of PrP^{Sc} aggregates in classical scrapie. Cells of glial origin, for example microglia, might use the blood vessels as a structural lead for their movement and carry PrP^{Sc} molecules with them, possibly also distributing them among the astrocytes forming the blood brain barrier (BBB). This would be a vascular spread in the broader sense. As a second possibility, microglia cells that have incorporated PrP^{Sc} move to the blood vessels in order to dispose of the aggregates and this leads to a perivascular deposition of the aggregates. Another way for PrP^{Sc} to reach blood vessels could be that they spread via sympathetic nerve fibres of the Plexus nervorum perivascularis. Haematogenous neuroinvasion has also been discussed with regard to the circumventricular organs (CVOs) due to the fact that these are usually affected in scrapie-infected sheep and that they are not protected by the BBB [46]. The possibility that the CVOs might be in contact with PrP^{Sc} from the blood during the pathogenesis of the disease cannot be excluded, but our results argue against a major involvement of the CVOs in neuroinvasion. In a very early case the DMNV was affected, but the area postrema and further CVOs were devoid of PrP^{Sc}. This agrees again with the results obtained for the oral infection of hamsters with scrapie [30].

In contrast to the classical scrapie cases, a sequential development of PrP^{Sc} distribution cannot be seen upon analysis of the present atypical/Nor98 scrapie cases. PrP^{Sc} distribution in one sheep of presumably early disease phase suggests that the aggregation of PrP^{Sc} has its origin in the di- or telencephalon. A spontaneous genesis of misfolded PrP could arise in the cerebral cortex. On the other hand, an ascending spread of the infectious agent that bypasses the brainstem and enters the CNS via sensible nerve fibres should be taken into consideration, for example proprioceptive fibres [47] or the spinothalamic tract. This would lead to further spreading of PrP^{Sc} from the thalamic nuclei to the cerebellar and cerebral cortex and from these to the brainstem and spinal cord, for example via the corticospinal tract.

7.3.2 Where Does the Spread of Nor98- PrP^{Sc} Start in the Brain?

It has been speculated by Nentwig et al. [48] that the PrP^{Sc} deposits and histopathologic lesions in atypical/Nor98 scrapie possibly evolve from the cerebrum to the cerebellum and the brainstem, but according to their examination of six sheep brains, this concept would not explain the PrP^{Sc} distribution in one sheep where they found PrP^{Sc} mainly in the cerebellum. However, immunohistochemistry—as used by these authors—is sometimes not able to detect the fine reticular deposits, for example seen in the cortex of Creutzfeldt-Jakob disease type 1, especially in the rare VV1 subtype [49]. The sensitive PET blot method, in contrast, is able to visualize these reticular deposits [38]. PrP^{Sc} deposits in one case described by Nentwig et al. could have therefore simply been missed in the cortex by immunohistochemistry. If this proved to be correct, according to the argumentation of Nentwig et al., PrP^{Sc} deposition and histopathologic lesions could indeed evolve from the cerebrum into the cerebellum and the brainstem. The 15 whole brains of atypical/Nor98 scrapie sheep examined by Moore et al. [21] should accordingly represent more or less the final stage of disease, as PrP^{Sc} can be generally found in all parts of the brain, including the brainstem. In other reports on the occurrence of atypical/Nor98 scrapie, cases have been described in which no PrP^{Sc} was detectable by immunohistochemistry in the obex region at all, but in the cerebellum and cerebrum [50-54]. If the misfolding of PrP^{C} in atypical/Nor98 scrapie does really start in the cerebrum it is obvious why early stages are not present in the worldwide pool of preserved atypical/Nor98 brains, as only the sampling of brainstem and cerebellum is compulsory in small ruminants according to EU regulations. Thus the question of whether PrP^{Sc} accumulation might start sporadically in the cerebrum—and if so, at one or more sites at the same time?—cannot be resolved by this or any other current study using field cases of atypical/Nor98 scrapie. This situation is comparable to the one with CJD type 1, where a spontaneous misfolding of PrP^{C} in the cerebral cortex and a caudal spread from there is assumed, but not proven [55]. The incidence of atypical/Nor98 in sheep is higher than that of CJD in humans [17]. A case control study of atypical/Nor98 scrapie has shown that animal movement does not seem to be a factor for the transmission of atypical/Nor98 scrapie between flocks, thus if sheep were to acquire this prion disease from their environment, its contagiousness would indeed be very low [56]. It has been speculated that this might be due to the relatively low protease stability, which could also explain the lack of intracellular PrP^{Sc} deposits [33].

There are certainly small differences between the PrP^{Sc} distribution detected by Moore et al. [21] in their described atypical/Nor98 scrapie cases and the ones revealed here by the PET blot method. For instance, in the present atypical/Nor98 scrapie material, PrP^{Sc} was never detectable in the cerebellar nuclei. Also, the affected parts of the hippocampus appear to be different. This might be due to differences in the treatment of tissue, the methods and/or differences in the antibodies used (mAb2G11 versus mAbP4). As previously reported, perineuronal staining has also been detected for the substantia nigra in some atypical/Nor98 scrapie sheep using immunohistochemistry [11], whereas in our study only plaque-like PrP^{Sc} deposits could be seen in this neuroanatomical structure. Similarly, neuronal deposits could be found in many affected sites of classical scrapie, but in contrast to previous publications [22], neither PET blot nor immunohistochemistry revealed PrP^{Sc} in the Purkinje cells of the cerebellum. It is known that especially intraneuronal immunoreactivity needs to be interpreted with caution [10]. However, the congruence between previous reports on PrP^{Sc} deposition patterns and the present results is obvious.

7.4 RESULTS

7.4.1 Western Blot

In all sheep that had been classified as atypical/Nor98 scrapie cases, the characteristic small fragment of 11–12 kDa [9] was present in CNS tissue samples after proteinase K digestion. Hereof the typical triplet pattern of 18–30 kDa in all classical scrapie sheep was clearly distinguishable. We usually used CNS tissue for Western blotting to determine the molecular profile. Only in one sheep with classical scrapie PrP^{Sc} were amounts in the brainstem so minimal that lymphatic tissue was needed to perform a valid Western blot. To ensure that Western blot PrP^{Sc} patterns of different tissues were comparable in one sheep, lymphatic tissues of further sheep with classical scrapie (from this study) were examined as well.

7.4.2 PET Blot and Immunohistochemistry

Disease-associated prion protein could be identified in the CNS of all scrapie sheep with the PET blot and there was no PrP^{Sc} detectable in the tissues of the negative control group. As previously described, immunohistochemical methods were able to confirm the presence of PrP^{Sc} deposits in all sheep except for one atypical/Nor98 case, despite using of a panel of antibodies [23].

Immunolabeling with the PET blot method allowed the identification of a number of deposition forms of PrP^{Sc} in the CNS and all were confirmed by immunohistochemical methods.

As described before [23] PrP^{Sc} was detectable in the LRS tissue of all classical scrapie sheep where it was present, but in none of the atypical/Nor98 scrapie animals with available LRS tissue could PrP^{Sc} be found (for availability of lymphatic tissue see Table 1). Figures 1(a) and 1(b) show PET blot and immunohistochemical staining of the PrP^{Sc} aggregates in the follicle of a tonsil derived from a classical scrapie case.

The same lymph follicle in the tonsil of a sheep with classical scrapie is shown stained either with the PET blot method (a) or conventional immunohistochemistry (b) (both mAb P4, bars = 250 μm). In the cervical spinal cord segment of a sheep with atypical/Nor98 scrapie (c) stained with the PET blot method (mAb P4, bar = 800 μm) synaptic PrP^{Sc} aggregates are present in the substantia gelatinosa (vertical arrow) and granular PrP^{Sc} in the corticospinal tract (horizontal arrows). In the cerebellar cortex of a classical scrapie sheep stained with the PET blot (d) complex PrP^{Sc} deposits are visible. Glia-associated PrP^{Sc} deposits take a stellate form in the molecular layer (mAb P4, bar = 150 μm, M = molecular layer, P = layer of Purkinje cells, G = granular cell layer). In the majority of the atypical/Nor98 scrapie sheep the cerebellar cortices show a more intense staining of the molecular layer than the granular layer (e) (PET blot, mAb P4, bar 600 μm). Intraneuronal, perineuronal and glia-associated PrP^{Sc} aggregates (f) in the reticular formation of a classical scrapie sheep (PET blot, mAb P4, bar = 50 μm). Tissue sections derived from sheep with the genotypes VRQ/ARH (a, b), AHQ/AHQ (c, e) and ARQ/ARQ (d, f).

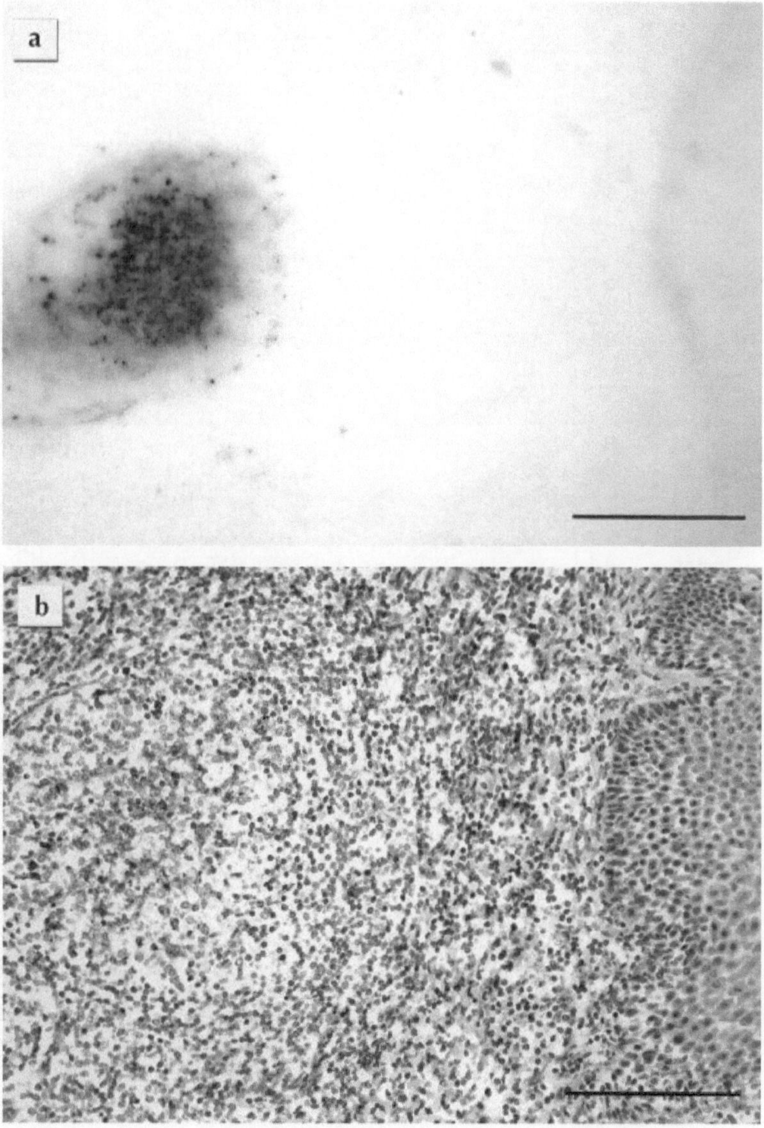

FIGURE 1 *(Continued)*

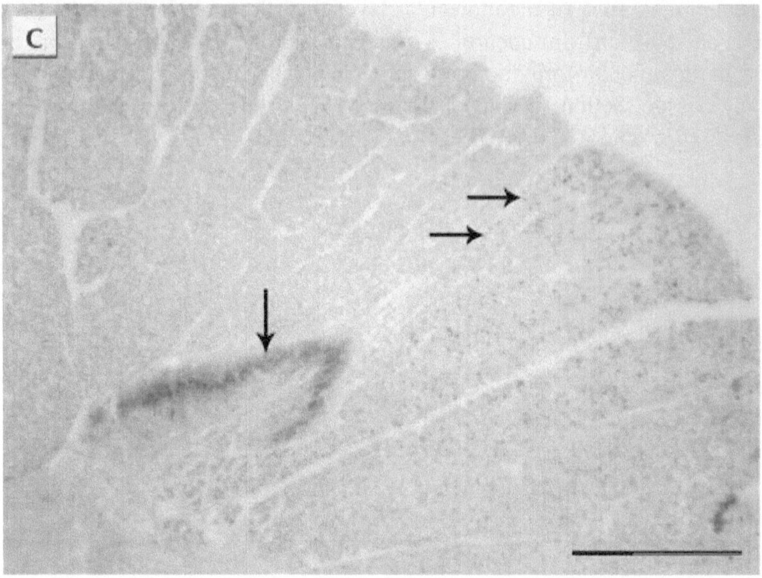

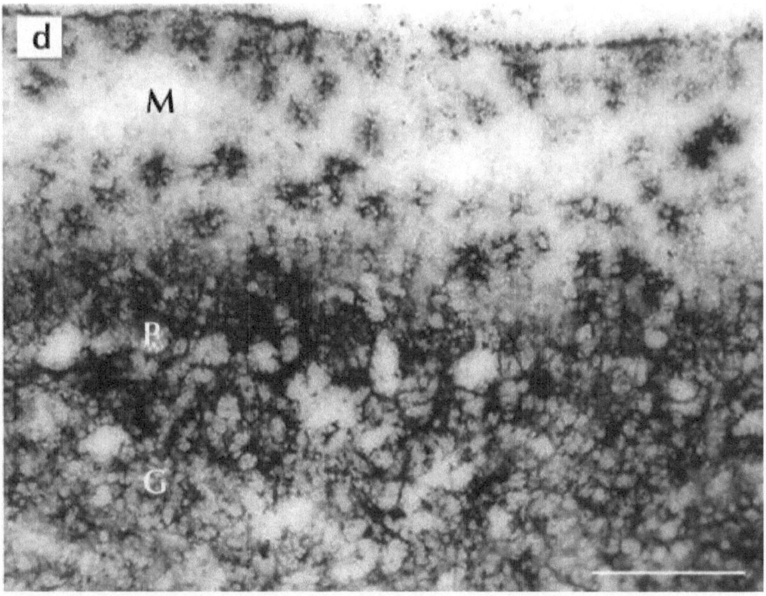

FIGURE 1 *(Continued)*

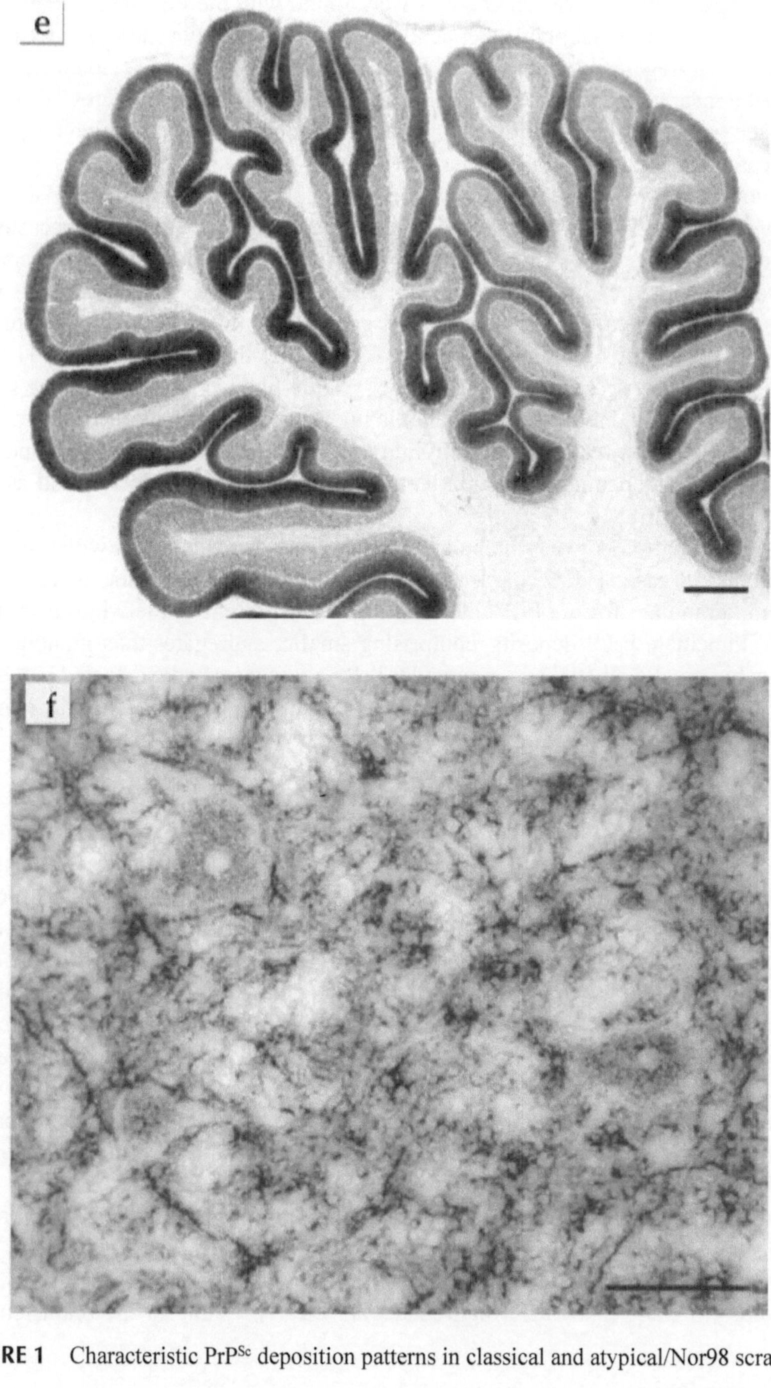

FIGURE 1 Characteristic PrP^{Sc} deposition patterns in classical and atypical/Nor98 scrapie.

7.4.3 Deposition Forms

Intra- and perineuronal PrP^{Sc} aggregates were found with the PET blot solely in classical scrapie, as were subpial, subenpendymal, and perivascular deposits. Extra neuronal PrP^{Sc} aggregates in the brains of sheep affected by classical scrapie often had a ramified appearance and were found in gray and white matter structures. They were addressed as glia-associated PrP^{Sc} aggregates and found to be relatively conspicuous in the cerebellar molecular layer where they took a stellate form [36].

In contrast, PrP^{Sc} aggregates found in the white matter of atypical/Nor98 scrapie sheep were always well-defined granules that varied a bit in size and were occasionally arranged like pearls on a string. The latter deposition form could also be observed in classical cases, but here also linear PrP^{Sc} was sometimes present. PrP^{Sc} deposits in the gray matter of atypical/Nor98 scrapie cases generally showed a fine granular pattern, also termed "synaptic/reticular" in human TSEs rather than "fine granular" [12, 37]. In some atypical/Nor98 scrapie cases, larger plaque-like aggregates could be seen in the substantia nigra, basal ganglia, thalamic nuclei and white matter. However, a differentiation between real plaques (amyloid) and plaque-like deposits is not possible with immunohistochemical detection methods or with the PET blot method as demonstrated before [38].

A discrimination between globular and punctuate deposits in the white matter of atypical/Nor98 cases [11, 21] was irreproducible with the PET blot, which is why the term "granular" for the PrP^{Sc} deposits that were present in the white matter was chosen. Punctuate PrP^{Sc} deposits, comprising smaller aggregates than granular PrP^{Sc} deposits but more defined than the reticular PrP^{Sc} aggregates, were detected in the gray matter of classical scrapie sheep. Small deviations in the composition of the complex deposition pattern could not be related to genotypes in the sheep examined.

7.4.4 Distribution of PrP^{Sc} in the CNS Sequential Appearance of PrP^{Sc} Distribution in the CNS of Classical Scrapie Sheep

To determine the sequential appearance of PrP^{Sc} in the CNS.; all field cases of classical scrapie were subjected to a thorough examination regarding the anatomical structures affected by PrP^{Sc} deposition. In the following, all cases were arranged according to the amount of PrP^{Sc} they had accumulated in total, and the occurrence of PrP^{Sc} in a panel of 127 neuroanatomical loci was compared between the cases. From this evaluation arose a classification of the classical scrapie cases into four stages of PrP^{Sc} spread in the CNS (see Figures 2–5). Criteria for these turned out to be certain neuroanatomical structures whose involvement marked a stage, meaning that the respective structure accumulated PrP^{Sc} aggregates (with a minimal score of 1) in all animals belonging to this stage and the following stage/stages. They are described in detail below and visualized in Figures 3 and 4.

The examined classical scrapie cases classified into four stages of PrP spread according to certain affected neuroanatomical sites (PET blots, mAb P4). In the CNS entry stage (a–d) only discrete PrP^{Sc} deposits are visible in the obex region, while in the brainstem stage (e–h) PrP^{Sc} aggregates are clearly visible in the brainstem and start to appear in more rostral structures. Once PrP^{Sc} deposits can be found in the deep cortical layers of the frontal cortex (i), the cruciate sulcus stage (i–l) is reached. In the basal ganglia stage, intense deposits in basal ganglia and thalamic nuclei can be found (m–p). Brain sections shown for the first, third and fourth stage derived from sheep with the genotype ARQ/ARQ while the sheep whose brain sections are depicted in the brainstem stage carried the genotype ARH/VRQ (bar = 5 mm).

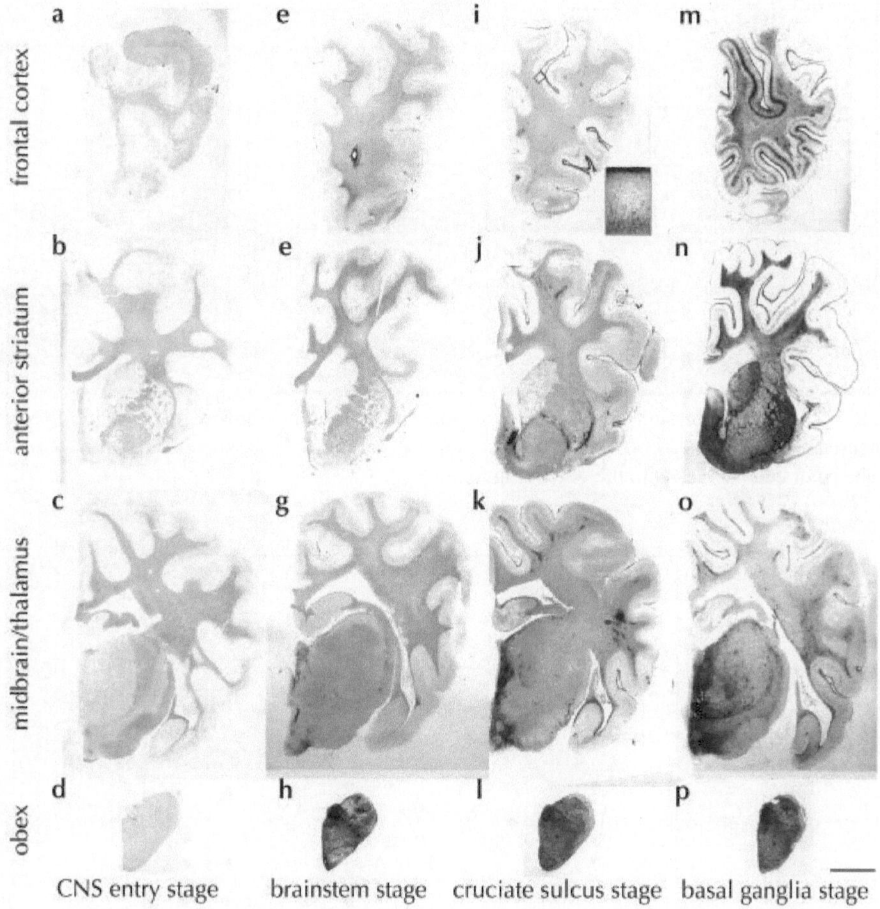

FIGURE 2 Classification of the PrP^{Sc} spread during disease development in classical scrapie.

Stage \ CNS structures	spinal cord	DMNV	brainstem	cerebellum	hypothalamus	amygdaloid nuclei	septal nuclei	cruciate sulcus	medial thalamic nuclei	caudate nucleus	putamen
CNS entry stage	X	X									
brainstem stage	X	X	X	X	X						
cruciate sulcus stage	X	X	X	X	X	X	X	X			
basal ganglia stage	X	X	X	X	X	X	X	X	X	X	X

FIGURE 3 Progression of classical scrapie in the brain shown for certain affected neuroanatomical sites.
The color code agrees with the one in Figure 5.

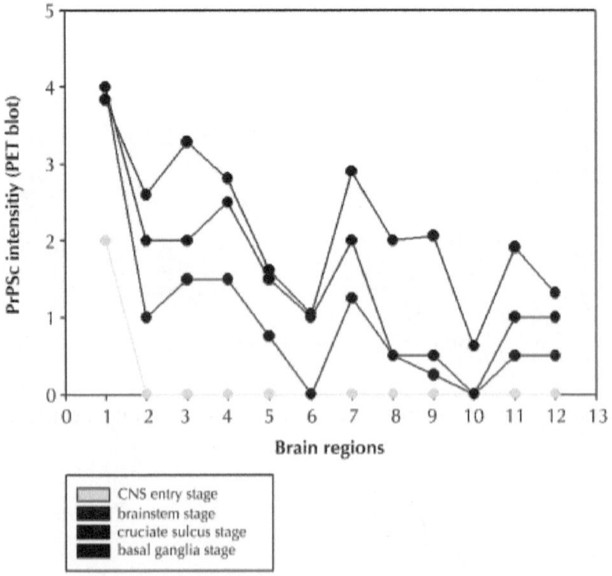

FIGURE 4 Accumulation of PrP^{Sc} in different brain regions during disease progression.

The four stages of the examined classical scrapie cases are depicted in four overlying graphs that illustrate how PrP^{Sc} aggregates (PET blot method, mAb P4) are increasingly accumulated in the brains from caudal (left) to rostral (right). Evaluation of PrP^{Sc} intensity was performed on a scale from 0.5 to 4 (see material and methods) and shown for the following brain areas. 1 dorsal motor nucleus of the vagus nerve (DMNV), 2 inferior olive, 3 dorsal tegmental nucleus, 4 cerebellar molecular layer, 5 cerebellar granular layer, 6 cerebral peduncle, 7 central gray (mesencephalon), 8 caudate nucleus, 9 ventral pallidum, 10 rostral commissure, 11 cruciate sulcus, 12 frontal white matter.

Neuroanatomical structures	Findings in classical scrapie	Findings in atypical/Nor98 scrapie
spinal cord		
substantia gelatinosa	very dense, complex deposits	often reticular deposits
cornu dorsale	complex deposits	–
cornu ventrale	complex deposits	–
cornu laterale (in the thoracic spinal cord)	complex deposits	–
dorsolateral tract	glia-associated deposits	often granular deposits
corticospinal (pyramidal) tract	glia-associated deposits	often granular deposits
cerebellar tract	glia-associated deposits	often granular deposits
brainstem		
dorsal nucleus of the vagus	very dense, complex deposits	–
spinal trigeminal nucleus	dense, complex deposits	often reticular deposits
hypoglossal nucleus	dense, complex deposits	–
olivary nucleus	weak to dense, complex deposits	seldom weak reticular deposits
dorsal tegmental nucleus	dense, complex deposits	–
reticular formation	principally glia-associated deposits	partly, granular deposits
pyramidal tract	principally glia-associated deposits	often, granular deposits
midbrain		
central grey	dense, complex deposits	–
oculomotor nucleus	dense, complex deposits	–
red nucleus	dense, complex deposits	–
substantia nigra	dense, complex deposits	partly reticular and plaque-like deposits
cerebral peduncle	glia-associated deposits	granular deposits
cerebellum		
cerebellar nuclei	complex deposits	–
cerebellar cortex – molecular layer	glia-associated, stellate deposits	dense, reticular deposits
cerebellar cortex – ganglionar layer	dense, perineuronal deposits	–
cerebellar cortex – granular layer	dense, complex deposits	dense deposits in the cerebellar glomeruli
cerebellar peduncles	glia-associated deposits	granular deposits
thalamus		
geniculate nuclei	often complex deposits	often reticular deposits
medial thalamic nuclei	often complex deposits	often reticular deposits, sometimes plaque-like
amygdaloid nuclei	mostly dense, complex deposits	partly reticular deposits
hypothalamic nuclei	dense, complex deposits	reticular deposits
internal capsule	glia-associated deposits	very dense, granular deposits
corpus callosum	glia-associated deposits	often dense, granular deposits
anterior striatum		
putamen	complex deposits	dense, reticular deposits
caudate nucleus	complex deposits	dense, reticular deposits
pallidum	complex deposits	partly weak, reticular deposits
comissura rostralis	weak glia-associated deposits	very dense, reticular deposits
septum	complex deposits, often very dense in the islands of Calleja	reticular and granular deposits
lateral olfactory tract	glia-associated deposits	granular deposits
frontal cortex		
frontal cortex	complex deposits in layers V and VI, cruciate sulcus exceedingly affected	evenly distributed reticular deposits (layer I and IV may be accentuated)
subcortical fibres	weak glia-associated deposits	very dense granular and reticular deposits
hippocampus formation		
hippocampus formation	complex, evenly distributed deposits, usually stronger in the hippocampus than in the dentate gyrus	reticular deposits, exceedingly in the Fissura hippocampi and the granular laye of the dentate gyrus

– : no PrP^{Sc} deposits detectable in any of the examined sheep

- CNS entry stage
- Brainstem stage
- Cruciate sulcus stage or
- Basal ganglia stage

FIGURE 5 Form and appearance of PrP^{Sc} deposits in classical and atypical/Nor98 scrapie sheep presented for representative CNS regions.

As a sequential development of PrP^{Sc} distribution could not be observed upon analysis of the present atypical/Nor98 scrapie cases, no coding colors were used for the results of this scrapie type in contrast to classical scrapie sheep. PrP^{Sc} deposits are detectable in the respective location in classical scrapie sheep belonging to the stages of spread indicated in the bottom of the figure using the same color code as in Figure 3.

7.4.5 CNS Entry Stage

One sheep showed only few discrete PrP^{Sc} deposits in the brain that were restricted to the dorsal motor nucleus of the vagus nerve (DMNV), the solitary tract nucleus and the spinal trigeminal tract in the brainstem. Further PrP^{Sc} aggregates could be detected in the substantia intermedialis lateralis and centralis of the thoracic spinal cord. This first stage, where PrP^{Sc} is detectable only in these CNS areas, can be considered the "CNS entry stage" in accordance with studies of other authors who have monitored the ascension of PrP^{Sc} from the intestines to the CNS [29, 30].

7.4.6 Brainstem Stage

In the second stage, all segments of the spinal cord and all nuclei of the obex region accumulate PrP^{Sc}, which also disseminates to the more rostral parts of the medulla, this may therefore be called "brainstem stage". In the caudal medulla the cellulae marginales and substantia gelatinosa of the spinal trigeminal tract nucleus show a very intense staining. The mesencephalon and thalamus display discrete PrP^{Sc} deposits, which are generally found to be subpial and/or perivascular while the mamillary body, habenular nuclei and the hypothalamic nuclei accumulate substantial amounts of PrP^{Sc}. The cerebellar nuclei accumulate PrP^{Sc} if the rostral medulla is largely involved and focal deposits of PrP^{Sc} are visible in the cerebellar cortex.

7.4.7 Cruciate Sulcus Stage

During the next stage, the mesencephalon, amygdaloid nuclei, septal nuclei, optic tract, cerebral peduncle, hippocampus formation, frontal cortex and subcortical white matter are increasingly affected. Regarding the frontal cortex, it is notably the sulcus cruciatus—and in a number of cases only this part of the cortex—that accumulates PrP^{Sc} in its deeper cortical layers. This stage is therefore designated "cruciate sulcus stage". PrP^{Sc} deposits in the cerebellar cortex are not yet evenly distributed.

7.4.8 Basal Ganglia Stage

In the final stage, PrP^{Sc} deposits can be seen also in the medial thalamic nuclei (mediodorsal, ventrolateral, ventral posterior and anterior group), the corpora geniculata and the basal ganglia. A positive staining for PrP^{Sc} in the latter determines a classical case in our definition for the "basal ganglia stage". The white matter also displays remarkable amounts of PrP^{Sc}, which are strongly linked to perivascular distribution.

All stages are depicted in Figures 2–4 and the stage at which PrP^{Sc} reaches a respective neuroanatomical site is indicated in Figure 5 using a color code. In the sheep examined in this study we could not find any influence of the different genotypes on the neuroanatomical distribution of PrP^{Sc} aggregates.

Comparison of PrP^{Sc} deposition patterns in classical and atypical/Nor98 scrapie

The PrP^{Sc} deposits in atypical/Nor98 scrapie cases were examined and evaluated in the same way as with the classical field cases. In contrast to the classical scrapie cases, differentiating distribution/spread stages of PrP^{Sc} in the CNS was not feasible with the atypical/Nor98 scrapie cases. In Figure 6, the same brain sections that illustrate the different stages of classical scrapie in Figure 2 are depicted for a case of atypical/Nor98 scrapie. In all atypical/Nor98 scrapie sheep, where brainstem material was available (n = 15) apart from one (see below), PrP^{Sc} aggregates were detectable in the rhombencephalon and mesencephalon. Regularly affected neuroanatomical structures were the spinal trigeminal nucleus, reticular formation, pyramid, pontine fibres, substantia nigra and cerebral peduncle. In the spinal cord the corticospinal tract and substantia gelatinosa accumulated PrP^{Sc} in most cases. Certain gray matter structures such as the DMNV, hypoglossal nucleus, dorsal tegmental nucleus, oculomotor nucleus, red nucleus and central gray of the mesencephalon, never displayed any PrP^{Sc} in the examined atypical/Nor98 scrapie cases. These listed neuroanatomical sites, however, accumulated large amounts of PrP^{Sc} in the respective stage of PrP^{Sc} distribution in the CNS of classical scrapie sheep as explained above (Figures 2–5). There were no PrP^{Sc} aggregates detectable in the cerebellar nuclei of the examined atypical/Nor98 scrapie cases, in contrast to the classical scrapie cases as described above. The synaptic or reticular PrP^{Sc} staining pattern in the cerebellar cortex of atypical/Nor98 scrapie sheep was in most cases more intense in the molecular than in the granular layer. Intra- and extracellular complex PrP^{Sc} aggregates in the cerebellar cortex of classical scrapie sheep were predominantly present in the granular layer and surrounding the Purkinje cells, the molecular layer displayed mainly glia-associated PrP^{Sc} deposits that took a stellate form. The cerebellar peduncles and white matter of the cerebellum itself showed PrP^{Sc} aggregates for both scrapie types. In the diencephalon of most atypical/Nor98 scrapie sheep, the corpora geniculata, medial thalamic nuclei and reticular nucleus accumulated PrP^{Sc} aggregates. In all atypical/Nor98 cases where the anterior striatum could be examined (n = 14), PrP^{Sc} deposits were also present in the caudate nucleus and putamen. The white matter of diencephalon and telencephalon showed PrP^{Sc} deposits in both types of sheep scrapie. In atypical/Nor98 scrapie, these were mainly confined to the subcortical fibres and certain white matter tracts, for example the corpus callosum or the commissura rostralis, while the distribution in classical scrapie was more disseminated.

Brain sections of an atypical/Nor98 scrapie case stained with the PET blot (mAb P4) have a different PrP^{Sc} distribution than the ones of classical scrapie cases as shown in Figure 2 (bar = 5 mm). Brain section derived from a sheep with the genotype ARQ/AHQ.

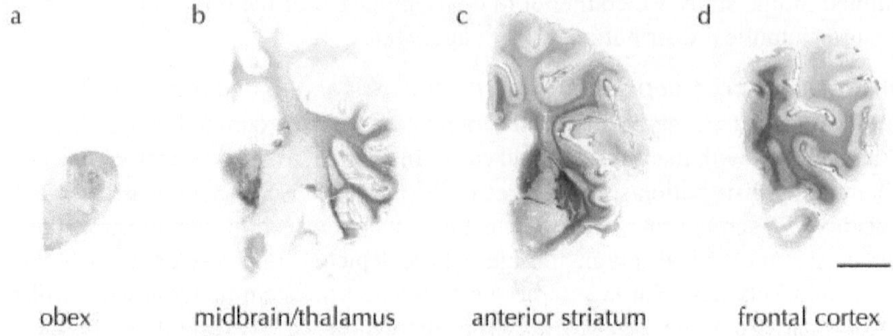

FIGURE 6 The PrP^{Sc} distribution in the brain of atypical/Nor98 scrapie.

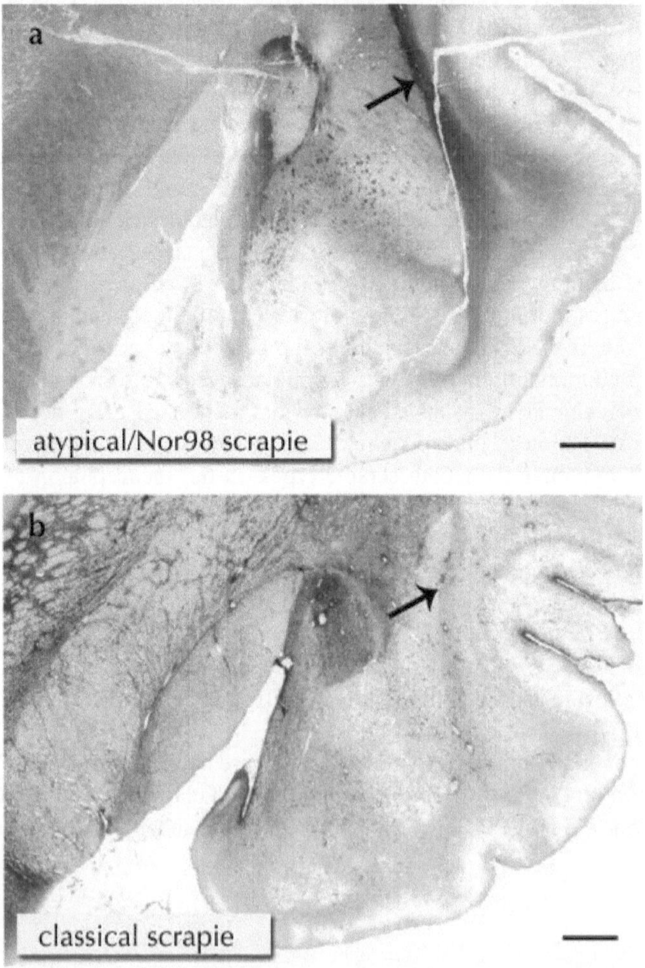

FIGURE 7 *(Continued)*

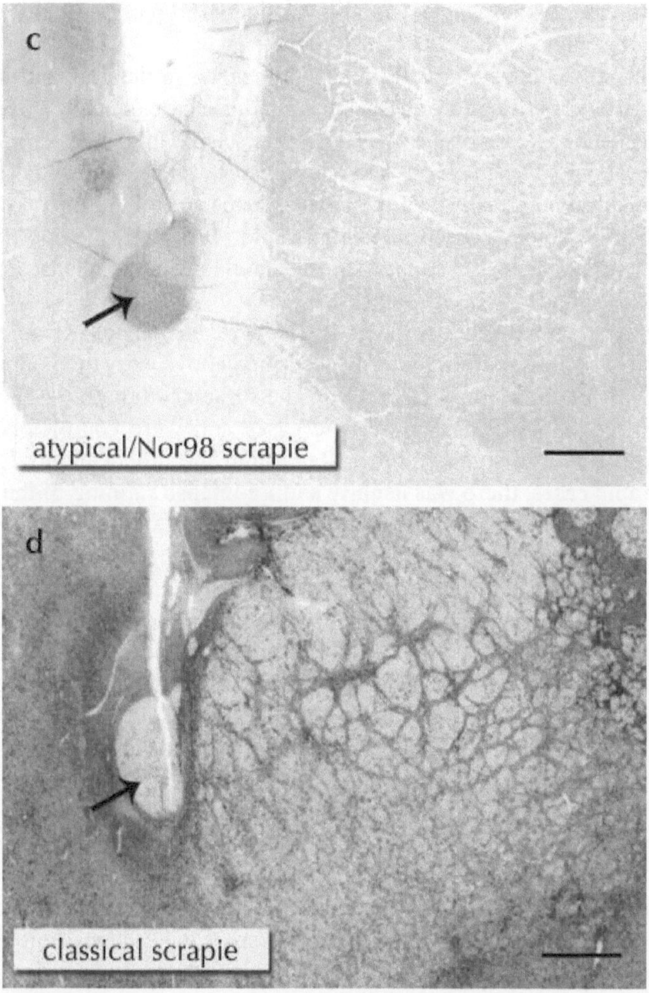

FIGURE 7 Differences in the neuroanatomical distribution of PrP^{Sc} deposit in atypical/Nor98 and classical sheep scrapie.

In atypical/Nor98 scrapie, white matter structures like the external capsule or rostral commissure contain substantially more PrP^{Sc} than the subcortical nuclei or basal ganglia respectively (a and c), whereas this is the reverse in classical scrapie (b and d). The external capsule (a and b) and the rostral commissure (c and d) are marked with arrows (mAb P4, bars = 1 mm). Tissue derived from sheep with the genotypes ARQ/AHQ (a), ARQ/ARQ (b and d) and AHQ/AFRQ (c).

There was one case in which PrP^{Sc} deposits were detectable with the PET blot only in the supratentorial (cerebral) brain structures and to a very small degree in the cerebellar cortex. The brainstem, including midbrain and spinal cord, were completely spared in this case, which was eventually considered to represent an early stage of atypical/Nor98 scrapie [23].

In Figure 7 the contrasts in PrP^{Sc} intensity existing in the gray and white matter between the two types of scrapie are demonstrated in a case of atypical/Nor98 scrapie and a classical scrapie case of the "basal ganglia stage." In classical scrapie it is the centromedial amygdaloid nuclei as well as the septal nuclei and basal ganglia that show substantially more PrP^{Sc} than the external capsule and the rostral commissure. In atypical/Nor98 scrapie, this principle turns out to be exactly the opposite, with the external capsule and the rostral commissure accumulating rather intense PrP^{Sc} deposits in contrast to the adjacent gray matter.

The lateral olfactory tract displayed PrP^{Sc} aggregates in both scrapie types with the respective PrP^{Sc} deposition patterns described above. Yet, the Islands of Calleja—clusters of neuronal granular cells in the olfactory tubercle—showed dense PrP^{Sc} deposits solely in classical scrapie cases and were completely devoid of PrP^{Sc} in atypical/Nor98 scrapie sheep. Regarding the hippocampus formation in classical scrapie cases, there was usually a more intense staining of the hippocampus and the fissura hippocampi compared to the dentate gyrus. In contrast to the atypical/Nor98 scrapie cases, there was no obvious accentuation of any layers. Atypical/Nor98 scrapie sheep showed a rather intense PrP^{Sc} staining of the granular layer of the dentate gyrus, the fissura hippocampi and the interconnective fibres between hippocampus and alveus (similar to the subcortical white matter) in comparison to the adjacent layers. The pyramid layer of the hippocampus appeared to be completely devoid of PrP^{Sc} deposits. The intensity of PrP^{Sc} staining in a single case was usually in agreement with the intensity of PrP^{Sc} deposits that could be found in the cerebral cortex of both scrapie types. As mentioned above, the complex PrP^{Sc} aggregates in classical scrapie were mainly confined to the deeper cortical layers (laminae V and VI) while reticular/synaptic PrP^{Sc} deposits in the cortices of atypical/Nor98 scrapie sheep were distributed more evenly, although an accentuation of laminae I and IV could be noted in some cases. Like in classical scrapie, differences regarding the distribution of PrP^{Sc} deposition could not be related to genotypes.

7.5 CONCLUSION

In summary, this study gives a basic description of PrP^{Sc} deposition patterns in classical as compared to atypical/Nor98 scrapie cases using the sensitive and specific PET blot method. We were able show a sequential appearance of PrP^{Sc} aggregates in the CNS of sheep with classical scrapie, but not in atypical/Nor98 scrapie. The four emerging stages of spread in classical scrapie were defined by the accumulation of PrP^{Sc} in certain neuroanatomical structures. These structures accumulated PrP^{Sc} aggregates in all animals belonging to this stage and the following stage/stages. Further conclusions drawn from this study regarding atypical/Nor98 scrapie might help in future to elucidate its origin and potentially related prion disease types like Creutzfeldt-Jakob disease type 1.

KEYWORDS

- **Chronic wasting disease**
- **Creutzfeldt-Jakob disease**
- **Goat**
- **Mouse**
- **Perineuronal**

COMPETING INTERESTS

The authors declare that they have no competing interests.

AUTHORS' CONTRIBUTIONS

Wiebke M Wemheuer carried out the PET blot studies, participated in immunohistochemistry, Western blot, tissue acquisition and the design of the study and drafted the manuscript. SLB participated in immunohistochemistry, Western blot, tissue acquisition and design of the study and co-edited the manuscript. Arne Wrede, Wilhelm E Wemheuer, Bertram Brenig, Bjørn Bratberg participated in tissue acquisition and diagnosing the cases. Walter J Schulz-Schaeffer conceived the study, participated in its design and coordination and co-edited the manuscript. All authors read and approved the final manuscript.

ACKNOWLEDGMENT

We would like to thank Tatjana Pfander, Nadine Rupprecht and Kerstin Brekerbohm for their skilful technical assistance. The work was supported by the VolkswagenStiftung (grants ZN 1294 and ZN 2168 to Walter J Schulz-Schaeffer).

REFERENCES

1. McGowan, J. P. Scrapie in sheep. *Scottish J Agric* **5**, 365–375 (1922).
2. Prusiner, S. B. Novel proteinaceous infectious particles cause scrapie. *Science* **216**, 136–144 (1982).
3. Oesch, B., Westaway, D., Walchli, M., McKinley, M. P., Kent, S. B., Aebersold R., Barry, R. A., Tempst, P., Teplow, D. B., Hood, L. E., Prusiner, S. B., and Weissmann, C. A cellular gene encodes scrapie PrP 27-30 protein. *Cell* **40**, 735–746 (1985).
4. Wang, F., Wang, X., Yuan, C. G., and Ma, J. Generating a prion with bacterially expressed recombinant prion protein. *Science* **327**, 1132–1135 (2010).
5. Bendheim, P. E., Brown, H. R., Rudelli, R. D., Scala, L. J., Goller, N. L., Wen, G. Y., Kascsak, R. J., Cashman, N. R., and Bolton, D. C. Nearly ubiquitous tissue distribution of the scrapie agent precursor protein. *Neurology* **42**, 149–156 (1992).
6. Bruce, M. E., Boyle, A., Cousens, S., McConnell, I., Foster, J., Goldmann, W., and Fraser, H. Strain characterization of natural sheep scrapie and comparison with BSE. *J Gen Virol* **83**, 695–704 (2002).
7. Hope, J., Wood, S. C., Birkett, C. R., Chong, A., Bruce, M. E., Cairns, D., Goldmann, W., Hunter, N., and Bostock, C. J. Molecular analysis of ovine prion protein identifies similarities between BSE and an experimental isolate of natural scrapie, CH1641. *J Gen Virol* **80**, 1–4 (1999).
8. Stack, M. J., Chaplin, M. J., and Clark, J. Differentiation of prion protein glycoforms from naturally occurring sheep scrapie, sheep-passaged scrapie strains (CH1641 and SSBP1), bovine

spongiform encephalopathy (BSE) cases and Romney and Cheviot breed sheep experimentally inoculated with BSE using two monoclonal antibodies. *Acta Neuropathol* **104**, 279–286 (2002).

9. Benestad, S. L., Sarradin, P., Thu, B., Schönheit, J., Tranulis, M. A., and Bratberg, B. Cases of scrapie with unusual features in Norway and designation of a new type, Nor98. *Vet Rec* **153**, 202–208 (2003).
10. EFSA. Opinion of the Scientific Panel on Biological Hazards on classification of atypical Transmissible Spongiform Encephalopathy (TSE) cases in Small Ruminants. *The EFSA Journal* 276, 1–30 (2005).
11. Benestad, S. L., Arsac, J. N., Goldmann, W., and Nöremark, M. Atypical/Nor98 scrapie. properties of the agent, genetics, and epidemiology. *Vet Res* **39**, 19 (2008).
12. Wemheuer, W. M., Benestad, S. L., Wrede A., Schulze-Sturm U., Wemheuer, W. E., Hahmann, U., Gawinecka, J., Schütz, E., Zerr, I., Brenig, B., Bratberg, B., Andreoletti, O., and Schulz-Schaeffer, W. J. Similarities between forms of sheep scrapie and Creutzfeldt-Jakob disease are encoded by distinct prion types. *Am J Pathol* **175**, 2566–2573 (2009).
13. Simon, S., Nugier, J., Morel, N., Boutal, H., Créminon, C., Benestad, S. L., Andréoletti, O., Lantier, F., Bilheude, J. M., Feyssaguet, M., Biacabe, AG., Baron, T., and Grassi, J. Rapid typing of transmissible spongiform encephalopathy strains with differential ELISA. *Emerg Infect Dis* **14**, 608–616 (2008).
14. Belt, P. B., Muileman, I. H., Schreuder, B. E., Bos-de Ruijter, J., Gielkens, A. L., and Smits, M. A. Identification of five allelic variants of the sheep PrP gene and their association with natural scrapie. *J Gen Virol* **76**, 509–517 (1995).
15. Hunter, N. PrP genetics in sheep and the applications for scrapie and BSE. *Trends Microbiol* **5**, 331–334 (1997).
16. Lühken, G., Buschmann, A., Groschup, M. H., and Erhardt, G. Prion protein allele A136 H154Q171 is associated with high susceptibility to scrapie in purebred and crossbred German Merinoland sheep. *Arch Virol* **149**, 1571–1580 (2004).
17. Lühken, G., Buschmann, A., Brandt, H., Eiden, M., Groschup, M. H., and Erhardt, G. Epidemiological and genetical differences between classical and atypical scrapie cases. *Vet Res* **38**, 65–80 (2007).
18. Moum, T., Olsaker, I., Hopp, P., Moldal, T., Valheim, M., Moum, T., and Benestad, S. L. Polymorphisms at codons 141 and 154 in the ovine prion protein gene are associated with scrapie Nor98 cases. *J Gen Virol* **86**, 231–235 (2005).
19. González, L., Martin, S., Begara-McGorum, I., Hunter, N., Houston, F., Simmons, M., and Jeffrey, M. Effects of agent strain and host genotype on PrP accumulation in the brain of sheep naturally and experimentally affected with scrapie. *J Comp Pathol* **126**, 17–29 (2002).
20. Miller, J. M., Jenny, A. L., Taylor, W. D., Marsh, R. F., Rubenstein, R., and Race, R. E. Immunohistochemical detection of prion protein in sheep with scrapie. *J Vet Diagn Invest* **5**, 309–316 (1993).
21. Moore, S. J., Simmons, M., Chaplin, M., and Spiropoulos, J. Neuroanatomical distribution of abnormal prion protein in naturally occurring atypical scrapie cases in Great Britain. *Acta Neuropathol* **116**, 547–559 (2008).
22. van Keulen, L. J., Schreuder, B. E., Meloen, R. H., Poelen-van den Berg, M., Mooij-Harkes, G., Vromans, M. E., and Langeveld, J. P. Immunohistochemical detection and localization of prion protein in brain tissue of sheep with natural scrapie. *Vet Pathol* **32**, 299–308 (1995).
23. Wemheuer, W. M., Benestad, S. L., Wrede, A., Wemheuer, W. E., Brenig, B., Bratberg, B., and Schulz-Schaeffer, W.J. Detection of classical and atypical/Nor98 scrapie by the paraffin-embedded tissue blot method. *Vet Rec* **164**, 677–681 (2009).
24. Andréoletti, O., Simon, S., Lacroux, C., Morel, N., Tabouret, G., Chabert, A., Lugan, S., Corbière, F., Ferre, P., Foucras, G., Laude, H., Eychenne, F., Grassi, J., and Schelcher, F. PrP^{Sc} accumulation in myocytes from sheep incubating natural scrapie. *Nat Med* **10**, 591–593 (2004).
25. Lacroux, C., Corbière, F., Tabouret, G., Lugan, S., Costes, P., Mathey, J., Delmas, J. M., Weisbecker, J. L., Foucras, G., Cassard, H., Elsen, J. M., Schelcher, F., and Andreoletti, O. Dynamics and genetics of PrP^{Sc} placental accumulation in sheep. *J Gen Virol* **88**, 1056–1061 (2007).

26. Ligios, C., Sigurdson, C. J., Santucciu, C., Carcassola, G., Manco, G., Basagni, M., Maestrale, C., Cancedda, M. G., Madau, L., and Aguzzi, A. PrP^{Sc} in mammary glands of sheep affected by scrapie and mastitis. *Nat Med* **11**, 1137–1138 (2005).
27. Thomzig, A., Schulz-Schaeffer, W., Wrede, A., Wemheuer, W., Brenig, B., Kratzel, C., Lemmer, K., and Beekes, M. Accumulation of pathological prion protein PrP^{Sc} in the skin of animals with experimental and natural scrapie. *PLoS Pathog* **3**, e66 (2007).
28. Ersdal, C., Ulvund, M. J., Espenes, A., Benestad, S. L., Sarradin, P., and Landsverk, T. Mapping PrP^{Sc} propagation in experimental and natural scrapie in sheep with different PrP genotypes. *Vet Pathol* **42**, 258–274 (2005).
29. van Keulen, L. J., Schreuder, B. E., Vromans M. E., Langeveld J. P., and Smits, M. A. Pathogenesis of natural scrapie in sheep. *Arch Virol Suppl* 57–71 (2000).
30. McBride, P. A., Schulz-Schaeffer, W. J., Donaldson, M., Bruce, M., Diringer, H., Kretzschmar, H. A., and Beekes, M. Early spread of scrapie from the gastrointestinal tract to the central nervous system involves autonomic fibers of the splanchnic and vagus nerves. *J Virol* **75**, 9320–9327 (2001).
31. van Keulen, L. J., Vromans, M. E., Dolstra, C. H., Bossers, A., and Van Zijderveld, F. G. Pathogenesis of bovine spongiform encephalopathy in sheep. *Arch Virol* **153**, 445–453 (2008).
32. Ryder, S. J., Dexter, G. E., Heasman, L., Warner, R., and Moore, S. J. Accumulation and dissemination of prion protein in experimental sheep scrapie in the natural host. *BMC Vet Res* **5**, 9 (2009).
33. Simmons, M. M., Konold, T., Simmons, H. A., Spencer, Y. I., Lockey, R., Spiropoulos, J., Everitt, S., and Clifford, D. Experimental transmission of atypical scrapie to sheep. *BMC Vet Res* **3**, 20 (2007).
34. Schütz, E., Scharfenstein, M., and Brenig, B. Genotyping of ovine prion protein gene (PRNP) variants by PCR with melting curve analysis. *Clin Chem* **52**, 1426–1429 (2006).
35. Schulz-Schaeffer, W. J., Fatzer, R., Vandevelde, M., and Kretzschmar, H. A. Detection of PrP(Sc) in subclinical BSE with the paraffin-embedded tissue (PET) blot. *Arch Virol Suppl* 173–180 (2000).
36. González, L., Martin, S., and Jeffrey, M. Distinct profiles of PrP(d) immunoreactivity in the brain of scrapie- and BSE-infected sheep. implications for differential cell targeting and PrP processing. *J Gen Virol* **84**, 1339–1350 (2003).
37. Kitamoto, T., and Tateishi, J. Human prion diseases with variant prion protein. *Philos Trans R Soc Lond B Biol Sci* **343**, 391–398 (1994).
38. Schulz-Schaeffer, W. J., Tschöke, S., Kranefuss, N., Dröse, W., Hause-Reitner, D., Giese, A., Groschup, M. H., and Kretzschmar, H. A. The paraffin-embedded tissue blot detects PrP(Sc) early in the incubation time in prion diseases. *Am J Pathol* **156**, 51–56 (2000).
39. Lezmi, S., Bencsik, A., and Baron, T. PET-blot Analysis Contributes to BSE Strain Recognition in C57Bl/6 Mice. *J Histochem Cytochem* **54**, 1087–1094 (2006).
40. Peden, A. H., Ritchie, D. L., Head, M. W., and Ironside, J. W. Detection and localization of PrP^{Sc} in the skeletal muscle of patients with variant, iatrogenic, and sporadic forms of Creutzfeldt-Jakob disease. *Am J Pathol* **168**, 927–935 (2006).
41. Thomzig, A., Kratzel, C., Lenz, G., Krüger, D., and Beekes, M. Widespread PrP^{Sc} accumulation in muscles of hamsters orally infected with scrapie. *EMBO Rep* **4**, 530–533 (2003).
42. Reder, A. T., Mednick, A. S., Brown, P., Spire, J. P., Van Cauter, E., Wollmann, R. L., Cervenakova, L., Goldfarb, L. G., Garay, A., Ovsiew, F., Gajdusek, D. C., and Roos, R. P. Clinical and genetic studies of fatal familial insomnia. *Neurology* **45**, 1068–1075 (1995).
43. Schulz-Schaeffer, W., McBride, P. A., Beekes, M., and Kretzschmar, H. A. Spread of PrP^{Sc} in orally infected animals during incubation time of prion disease. *XIV International Conference of Neuropathology* **663** (2000).
44. Jeffrey, M., Goodsir, C. M., Holliman, A., Higgins, R. J., Bruce, M. E., McBride, P. A., and Fraser, J. R. Determination of the frequency and distribution of vascular and parenchymal amyloid with polyclonal and N-terminal-specific PrP antibodies in scrapie-affected sheep and mice. *Vet Rec* **142**, 534–537 (1998).

45. Delattre, J. Y., Krol, G., Thaler, H. T., and Posner, J. B. Distribution of brain metastases. *Arch Neurol* **45**, 741–744 (1988).
46. Sisó, S., Jeffrey, M., and González, L. Neuroinvasion in sheep transmissible spongiform encephalopathies. The role of the haematogenous route. *Neuropathol Appl Neurobiol* **35**, 232–246 (2009).
47. Rüb, U., Schultz, C., Del, T. K., Gierga, K., Reifenberger, G., de Vos, R. A., Seifried, C., Braak, H., and Auburger, G. Anatomically based guidelines for systematic investigation of the central somatosensory system and their application to a spinocerebellar ataxia type 2 (SCA2) patient. *Neuropathol Appl Neurobiol* **29**, 418–433 (2003).
48. Nentwig, A., Oevermann, A., Heim, D., Botteron, C., Zellweger, K., Drogemuller, C., Zurbriggen, A., and Seuberlich, T. Diversity in Neuroanatomical Distribution of Abnormal Prion Protein in Atypical Scrapie. *PLoS Pathog* **3**, e82 (2007).
49. Parchi, P., Giese, A., Capellari, S., Brown, P., Schulz-Schaeffer, W., Windl, O., Zerr, I., Budka, H., Kopp, N., Piccardo, P., Poser, S., Rojiani, A., Streichemberger, N., Julien, J., Vital, C., Ghetti, B., Gambetti, P., and Kretzschmar, H. Classification of sporadic Creutzfeldt-Jakob disease based on molecular and phenotypic analysis of 300 subjects. *Ann Neurol* **46**, 224–233 (1999).
50. Buschmann, A., Biacabe, A. G., Ziegler, U., Bencsik, A., Madec, J. Y., Erhardt, G., Lühken, G., Baron, T., and Groschup, M. H. Atypical scrapie cases in Germany and France are identified by discrepant reaction patterns in BSE rapid tests. *J Virol Methods* **117**, 27–36 (2004).
51. Buschmann, A., Lühken, G., Schultz, J., Erhardt, G., and Groschup, M. H. Neuronal accumulation of abnormal prion protein in sheep carrying a scrapie-resistant genotype (PrPARR/ARR). *J Gen Virol* **85**, 2727–2733 (2004).
52. Gavier-Widen, D., Nöremark, M., Benestad, S., Simmons, M., Renstrom, L., Bratberg, B., Elvander, M., and Segerstad C. H. Recognition of the Nor98 variant of scrapie in the Swedish sheep population. *J Vet Diagn Invest* **16**, 562–567 (2004).
53. Onnasch, H., Gunn, H. M., Bradshaw, B. J., Benestad S. L., and Bassett, H. F. Two Irish cases of scrapie resembling Nor98. *Vet Rec* **155**, 636–637 (2004).
54. Orge, L., Galo, A., Machado, C., Lima, C., Ochoa, C., Silva, J., Ramos, M., and Simas, J. P. Identification of putative atypical scrapie in sheep in Portugal. *J Gen Virol* **85**, 3487–3491 (2004).
55. Heye, N., and Cervos-Navarro, J. Focal involvement and lateralization in Creutzfeldt-Jakob disease. correlation of clinical, electroencephalographic and neuropathological findings. *Eur Neurol* **32**, 289–292 (1992).
56. Hopp, P., Omer, M. K., and Heier, B. T. A case-control study of scrapie Nor98 in Norwegian sheep flocks. *J Gen Virol* **87**, 3729–3736 (2006).

8 Experimental H-Type Bovine Spongiform Encephalopathy

Hiroyuki Okada, Yoshifumi Iwamaru, Morikazu Imamura, Kentaro Masujin, Yuichi Matsuura, Yoshihisa Shimizu, Kazuo Kasai, Shirou Mohri, Takashi Yokoyama, and Stefanie Czub

CONTENTS

8.1 INTRODUCTION

Atypical bovine spongiform encephalopathy (BSE) has recently been identified in Europe, North America, and Japan. It is classified as H-type and L-type BSE according to the molecular mass of the disease-associated prion protein (PrP^{Sc}). To investigate the topographical distribution and deposition patterns of immunolabeled PrP^{Sc}, H-type BSE isolate was inoculated intracerebrally into cattle. The H-type BSE was successfully transmitted to 3 calves, with incubation periods between 500 and 600 days. Moderate to severe spongiform changes were detected in the cerebral and cerebellar cortices, basal ganglia, thalamus, and brainstem. The H-type BSE was characterized by the presence of PrP-immunopositive amyloid plaques in the white matter of the cerebrum, basal ganglia, and thalamus. Moreover, intraglial-type immunolabeled PrP^{Sc} was prominent throughout the brain. Stellate-type immunolabeled PrP^{Sc} was conspicuous in the gray matter of the cerebral cortex, basal ganglia, and thalamus, but not in the brainstem. In addition, PrP^{Sc} accumulation was detected in the peripheral nervous tissues, such as trigeminal ganglia, dorsal root ganglia, optic nerve, retina, and neurohypophysis. Cattle are susceptible to H-type BSE with a shorter incubation period, showing distinct and distinguishable phenotypes of PrP^{Sc} accumulation.

Bovine spongiform encephalopathy (BSE), which belongs to a group of diseases called transmissible spongiform encephalopathies (TSE), is a fatal neurodegenerative disorder of cattle. BSE was first identified in the United Kingdom in 1986 [1], then spread to European as well as North American countries and Japan, and has affected more than 190,000 cattle in the world. The infectious agent responsible for TSE is the disease-associated prion protein (PrP^{Sc}), which is thought to be a post-translationally modified form of the host-encoded membrane glycoprotein (PrP^{C}) [2]. According to the protein-only hypothesis, PrP^{Sc} is the principal component of the infectious agent.

On the basis of uniform pathology and biochemical profile of the protease-resistant prion protein (PrPres) among BSE-affected cattle, it is assumed that BSE in cattle is caused by only one prion strain. Since 2003, variants of BSE (named atypical BSE) have been detected in Japan, Europe, and North America and classified in at least two groups, namely, H-type and L-type BSE; according to the molecular mass of PrPres, compared with those of the classical BSE (named C-type BSE) [3]. The H-type BSE was first identified in France [4], and L-type BSE; called bovine amyloidotic spongiform encephalopathy (BASE), was first detected in Italy [5]. It is accepted that C-type BSE is caused by the consumption of BSE-contaminated feed, whereas the origins of H-type and L-type BSE remain enigmatic. Hypotheses for the origin of atypical BSE include (1) infection of cattle with different BSE agents, (2) infection of cattle with a non-bovine source or unrecognized forms of infectious TSE agents, (3) genetic mutations in the prion protein gene, and (4) spontaneous or so-called sporadic forms of TSE in cattle, limited to old age, like the sporadic form of human Creutzfeldt-Jakob disease (CJD) [6-10]. However, only one genetic mutation has been found in an H-type BSE case [11]. Sequence analysis of the open reading frame (ORF) of the prion protein gene (*PRNP*) has not revealed any mutations in atypical BSE cases in France [4], Italy [5], and Canada [12]. Therefore, it seems unrealistic to suggest a genetic origin of atypical BSE [13]. The transmissibility of atypical H-type and L-type BSE to mice [13-18] and cattle [19-22] has been confirmed, and these forms clearly differ from C-type BSE regarding incubation periods, PrPres profiles, protease susceptibility, and spatial distribution patterns of histopathological lesions and immunolabeled PrP^{Sc} [3, 6, 16, 20, 22]. Interestingly, C-type [23] and H-type [14, 15] BSE isolates were transmissible to wild-type mice already in the first passage, whereas L-type BSE agent failed to transmit in the first passage but was successfully transmitted to wild-type mice in the second passage [17].

Unfortunately, a detailed and all-encompassing analysis of neuropathology and topographical distribution of immunolabeled PrP^{Sc} in H-type BSE-affected cattle could not be performed, since only the obex region is routinely sampled for BSE surveillance testing and the remaining brain as well as the carcasses are not available in most countries [3, 10, 12, 13, 24-27]. Recently, clinical signs and biochemical properties of experimental German H-type BSE cases have been reported [20]. The primary objective of this study was to investigate the transmissibility of H-type BSE; using a field isolate detected in the active surveillance program in Canada [12]. The secondary objective was to extend the knowledge of the topographical distribution and deposition patterns of immunolabeled PrP^{Sc} in H-type BSE.

8.2 MATERIALS AND METHODS

8.2.1 Animal Inoculation

All animal experiments were approved by the Animal Ethical Committee and the Animal Care and Use Committee of National Institute of Animal Health. The Canadian H-type BSE case was of a Charolais cross displaying signs of recumbency prior to euthanasia [12]. By confirmatory immunohistochemistry, the staining pattern was characterized by a predominant reaction in the neuropil (including glial cells) and a relatively low level of intraneuronal, particulate, and stellate immunolabeling in the obex region. The molecular features of PrP^{Sc} in cattle were described for the Canadian H-type BSE [12] and C-type BSE [21]. Brain homogenates were prepared in nine volumes of phosphate-buffered saline (PBS; pH 7.4), using a multibead shocker (Yasui Kikai Co., Osaka, Japan). Two female and one neutered 3- to 4-month-old Holstein calves were challenged intracerebrally with 1 mL of the 10% brain homogenate prepared from the H-type BSE case. Intracerebral transmission of C-type BSE has previously been reported in cattle [21]. In brief, the inoculum was injected into the midbrain via an 18-gauge 7-cm-long disposable needle (NIPRO; Osaka, Japan), following which the needle was withdrawn from the brain. Two sham-inoculated Holstein calves served as controls, they were euthanized at the age of 27 months.

8.2.2 Tissue Processing for Histology

The left brain, including the brainstem and cerebellum, of the H-type BSE case was fixed in 10% neutral buffered formalin (pH 7.4), while the contralateral side was frozen at -80 °C for Western blot analysis of PrP^{Sc}. Tissues of C-type BSE for this study were derived from previously reported experimental cases [21]. Coronal slices of the formalin-fixed sample from each animal were cut serially, treated with 98% formic acid for 60 min at room temperature (RT) [28], embedded in paraffin wax, sectioned at 4-μm thickness, stained with hematoxylin & eosin (HE), and used for immunohistochemistry.

The lesion profile was determined in the HE-stained sections by scoring the vacuolar changes in 17 different brain areas [29]. The selected brain sections were stained with phenol Congo red [30]. In brief, Congo red dye dissolved at 0.2 g in 100 mL of distilled water was mixed with 9 g of NaCl and subsequently with an equal volume of 100% ethanol. This mixture was allowed to stand on ice for 10 min. After filtration, phenol (Nacalai Tesque, Kyoto, Japan) was added at 5 g in 100 mL of the supernatant, and the pH was adjusted to around 3.0 by adding glacial acetic acid. The sections were stained in this solution for 1 hr. After hematoxylin counterstaining, the sections were examined under a polarizing microscope.

8.2.3 PrP^{Sc} Immunolabeling

Dewaxed sections were treated with 3% hydrogen peroxide for 10 min, followed by incubation with 10 μg/mL proteinase K (PK) at RT for 10 min. Thereafter, the sections were subjected to an antigen retrieval protocol by alkaline hydrolysis at 60 °C for 10 min in 150 mM sodium hydroxide [31]. They were incubated with 10% normal goat serum for 10 min and then with the following anti-PrP primary antibodies for 60 min. SAF32, SAF54, SAF84, 12F10, F89/160.1.5, F99/97.6.1, T1 [32], 44B1 [33], and 43C5 [33] as the 9 monoclonal antibodies (mAbs) and B103 [34] and T4 [35] as the two rabbit polyclonal antibodies (pAbs) (Table 1). The working concentrations of the mAbs and pAbs were 1 μg/mL and 5 μg/mL, respectively. Immunolabeling was performed with an anti-mouse or anti-rabbit universal immunoperoxidase polymer (Nichirei Histofine Simple Stain MAX PO (M) or (R), Nichirei, Tokyo, Japan) for 30 min, and the reaction was visualized using 3, 3′-diaminobenzidine tetrachloride as the chromogen for 7 min. Finally, the sections were slightly counterstained with Mayer's hematoxylin. To observe the topographical distribution of PrP^{Sc} in the brain with the naked eye, the immunolabeled sections were photographed and viewed with Microsoft PowerPoint.

8.2.4 Western Blotting

The CNS tissues were homogenized in a buffer containing 100 mM NaCl and 50 mM *Tris*-HCl (pH 7.6). The homogenate was mixed with an equal volume of detergent buffer containing 4% (w/v) Zwittergent 3-14 (Merck, Darmstadt, Germany), 1% (w/v) Sarkosyl, 100 mM NaCl, and 50 mM *Tris*-HCl (pH 7.6) and incubated with 0.25-mg collagenase, followed by incubation with PK (final concentration, 40 μg/mL) at 37 °C for 30 min. The PK digestion was terminated using 2 mM Pefabloc. The sample was then mixed with 2-butanol:methanol (5:1) and centrifuged at 20,000 *g* for 10 min.

PrP^{Sc} was extracted from the peripheral nervous, extranervous, and lymphoid tissues by phosphotungstic acid precipitation as described previously [36]. The extracted samples were mixed with a gel-loading buffer containing 2% (w/v) sodium dodecyl sulfate (SDS) and boiled for 5 min before electrophoresis. The samples were then separated by 12% SDS-polyacrylamide gel electrophoresis (PAGE) and electrically blotted onto a polyvinylidene fluoride (PVDF) membrane (Millipore, Billerica, MA, USA). The blotted membrane was incubated with anti-PrP mAbs 6H4 (Prionics, Schlieren, Switzerland) and SAF84 at RT for 60 min. Signals were developed with a chemiluminescent substrate (SuperSignal, Pierce Biotechnology, Rockford, IL, USA). After PK

TABLE 1 Characteristics of the 11 antibodies and the epitope location of the bovine PrP.

Antibodies		Epitope		Clonality	Immunogen	Source**	Source**
		Location	Type*				
N-terminal region	**SAF32**	**62–92**	**L**	**Monoclonal**	**SAF from infected hamster brain**	**SPI-Bio (Montigny-le-Bretonneux, France)**	
	B103	103–121	L	Polyclonal	Cattle recPrP	FUJIREBIO (Tokyo, Japan) [34]	FUJIREBIO (Tokyo, Japan) [34]
	F89/160.1.5	148–155	L	Monoclonal	Cattle recPrP	VMRD (Pullman, WA; USA)	VMRD (Pullman, WA; USA)
	T1	149–153	L	Monoclonal	Mouse recPrP	Dr Tagawa	
	12F10	154–163	L	Monoclonal	Hamster recPrP	SPI-Bio	SPI-Bio SPI-Bio
Core region	SAF54	168–172	L	Monoclonal	SAF from infected hamster brain	SPI-Bio	
	44B1	168–242	DC	Monoclonal	Mouse recPrP	Dr Horiuchi [33]	Dr Horiuchi [33]
	SAF84	175–180	L	Monoclonal	SAF from infected hamster brain	SPI-Bio	SPI-Bio
	43C5	175–181	L	Monoclonal	Mouse recPrP	Dr Horiuchi [33]	
C-terminal region	T4	221–239	L	Polyclonal	Cattle recPrP	Dr Sata [35]	Dr Sata [35]
	F99/97.6.1	229–235	L	Monoclonal	Cattle recPrP	VMRD	VMRD VMRD

SAF; scrapie-associated fibrils, recPrP; recombinant prion protein.
*L; linear, DC; discontinuous.
**Manufacturer and reference number.

treatment, some samples were deglycosylated with *N*-glycosidase F (PNGase F; New England Biolabs, Beverly, MA; USA), according to the manufacturer's instructions.

8.2.5 Polymerase Chain Reaction (PCR) Amplification and DNA Sequencing

In each case, genomic DNA was isolated and purified from the liver using a GenElute mammalian genomic DNA purification kit (Sigma, St. Louis, MO; USA) according to the manufacturer's instructions. The ORF (792 bp) of *PRNP* was amplified using the primers 5′-ATGGTGAAAAGCCACATAG-3′ and 5′-CTATCCTACTATGAGA-AAAATG-3′. The purified PCR products of the bovine *PRNP* were directly sequenced using ABI 3100-*Avant* sequencer (Applied Biosystems, Foster City, CA; USA), with the abovementioned PCR primers. The nucleotide sequences obtained for the 3 challenged calves were aligned using the GENETYX software (GENETYX Co., Tokyo, Japan), with the following GenBank accession numbers. AY367641, AY367642, and AY367643.

8.3 DISCUSSION

This study demonstrated successful intraspecies transmission of H-type BSE characterized by a shorter incubation period as compared with C-type BSE [19]. To the best of our knowledge, thus far, neuropathological and immunohistochemical data for H-type BSE have only been reported from the medulla oblongata at the obex in German, United States, and Swedish field cases [10, 13, 24]. This is related to the fact that only the obex region is sampled for BSE rapid tests and other brain regions are often unavailable due to marked autolysis, limitations in collection infrastructure, or freezing artifacts [10, 13, 24, 25]. This is the first presentation of H-type lesion profiles involving the whole CNS and additional nervous tissues, although of experimentally infected animals.

Incubation periods in the cattle challenged with the Canadian H-type BSE (mean period, 18 months) were two months longer than those reported in cattle challenged with German H-type BSE [20]. This difference in incubation periods has several potential explanations, which include differences in agents tested, inoculum titers, and breeding conditions. Infectivity titer issues might be resolved by comparing second-passage infection experiment results.

The spongy changes were generally present in the gray matter throughout the brain and spinal cord, but were more conspicuous in the cerebral cortices, thalamus, hypothalamus, and midbrain. In most brain areas, vacuoles were generally detected in the neuropil and only occasionally in the neurons. The spatial distribution pattern of spongiform changes and immunolabeled PrP^{Sc} in the brain of an H-type BSE-infected Zebu, analyzed with N-terminal-specific mAb P4 and C-terminal-specific mAb F99/97.6.1, was similar to that in C-type BSE cases [38]. In natural and experimental C-type BSE cases, spongiform lesions are consistently distributed throughout the brain, but overall, the lesions in the thalamus and brainstem including the midbrain and medulla oblongata at the obex are more severe than those in the cerebral cortices [29, 39]. The results of the present study indicate that the vacuolar lesion score of the H-type BSE-challenged cattle was higher than that of C-type BSE-affected cattle [19, 29, 40, 41]. Moreover, the topographical distribution of PrP^{Sc} in the brain of BSE-infected sheep is similar irrespective of the different challenge routes such as intracerebral, intravascular, or intraperitoneal route [42], suggesting common patterns of neuroinvasion and CNS spread [43]. On the contrary, the minor differences detected in the distribution of PrP^{Sc} in the brain between deer that are orally and intracerebrally infected with BSE may be due to differences in the routes of infection [44].

The immunolabeling patterns of PrP^{Sc} in the cattle affected with H-type BSE were characterized by the presence of both PrP^{Sc}-positive plaques and intraglial- and stellate-type PrP^{Sc} accumulations in the brain. Severe intraneuronal- and intraglial-type PrP^{Sc} accumulations as well as plaque-like PrP^{Sc} aggregates with the absence of stellate-type PrP^{Sc} deposition have been reported in the obex region of H-type BSE-affected animals [10, 13]. These immunohistochemical features were detected in the obex region and coincided with those observed in the present study. However, neither amyloid plaques nor stellate-type PrP^{Sc} depositions have been reported in H-type BSE-affected cattle, most likely due to their limitation to the medulla oblongata at the obex [8, 10, 13, 24].

Two different types of plaques were found in this study: unicentric and multicentric PrP plaques. Most of these plaques were uniformly immunopositive for PrP; with a dense non-Congophilic core. The plaques that had a pale central core with a Congophilic reaction were less frequent. It has been suggested that Congophilic plaques may correspond with the late stage of plaque formation, whereas non-Congophilic plaques coincide with the early stage of CJD and Gerstmann-Sträussler-Scheinker syndrome [45]. The two types of PrP^{Sc}-positive plaques—unicentric and multicentric—have been described in L-type BSE [5, 19, 46]. The results indicate that the presence of PrP^{Sc} plaques in the forebrain but not in the brainstem is one of the neuropathological features in cattle affected with atypical BSE. In addition, glial-type PrP^{Sc} deposition in the white matter throughout the brain seems to be a characteristic feature of H-type BSE in cattle, as supported by identical findings in German and Swedish H-type BSE field cases [10, 13].

The extracellular PrP^{Sc} was immunolabeled with N-terminal-, core-, and C-terminal-specific antibodies, but intracellular PrP^{Sc} did not show immunoreactivity to the N-terminal-specific anti-PrP antibodies [47, 48]. Intracellular PrP^{Sc} has markedly diminished immunoreactivity to *N*-terminal-specific anti-PrP antibodies [47]. However, *N*-terminal-specific mAb P4, which recognizes an epitope at bovine PrP residues 101-107, showed intraneuronal PrP^{Sc} immunolabeling in sheep affected with C-type BSE [47] and in Zebu affected with H-type BSE [38]. These results indicate that the epitope region for either mAb P4 or core-specific anti-PrP antibodies is located upstream of an intracellular truncation site [38, 48]. The differences in intracellular PrP^{Sc} truncation sites between sheep scrapie and ovine BSE [47] as well as between C-type BSE and H-type BSE [38] most probably depend on the strain and the tissues and cells [47]. The intensity and patterns of PrP^{Sc} immunolabeling varied with the different anti-PrP antibodies used, and the difference in the PrP^{Sc} immunohistochemical labeling results might be related to the application of different technical protocols, especially antigen retrieval methods [49-51].

The Western blot profiles of PrPres for the H-type BSE-challenged cattle and the Canadian H-type BSE-infected brain homogenate used as inoculum were indistinguishable. Results of previous studies prove that H-type BSE isolates have distinct biological and biochemical properties compared with C-type and L-type BSE isolates [3, 52, 53]. The PrPres in H-type BSE, as detected by mAb SAF84 recognizing the C-terminus of PrP, was thought to be composed of two fragments with molecular masses of 19 kDa and 10–12 kDa, possessing a different cleavage site in the N-terminal region with PK digestion [53]. The higher molecular mass of the unglycosylated PrPres molecules, which included an additional 10–12 kDa fragment, in the Canadian H-type BSE case was maintained in the challenged animals. These unique molecular features of PrP in H-type BSE are also well preserved in transgenic and wild type mice [16, 53]. In addition, a distinct 10–12 kDa fragment detected with C-terminal-specific antibodies in H-type BSE might be associated with the presence of PrP plaques [53].

Although, PrP^{C} glycosylation seems to play a critical role in the maintenance of strain-dependent prion neurotropism [54, 55], a recent study has demonstrated that PrP^{Sc} glycosylation is not required for the maintenance of strain-specific neurotropisms [56]. Strain-dependent prion neurotropism is currently unknown, but several possibilities have been indicated [56]. Moreover, a local difference in the PrP^{Sc} replication rate may be attributed to a high degree of neurotropism in H-type BSE similar to that observed in C-type BSE [57].

Since 2003, sporadic and discontinuous occurrence of atypical BSE has been detected in Europe, North America, and Japan. Although, till date, the origin and frequency of atypical BSE is unknown, a high prevalence is found in older cattle over the age of eight years. This is the result of the active surveillance programs using rapid screening tests, with the exception of a Zebu case [38]. It has been reported that H-type BSE can be the result of a naturally occurring, heritable variant caused by glutamic acid/lysine polymorphism at codon 211 of the bovine *PRNP* gene (E211K) [11, 58]. However, our cases, although experimentally challenged via the intracranial route, and the original Canadian H-type BSE field case [11, 58] developed the disease without the novel mutation E211K within *PRNP*. Therefore, atypical BSE seemed to be sporadic rather than inherited with a higher risk in fallen stock than in healthy slaughtered cattle [8, 13, 25], suggesting that young adult cattle affected with atypical BSE might be dormant carriers. Further studies are required to determine the epidemiological significance and origin of atypical BSE.

The present study demonstrated successful intraspecies transmission of H-type BSE to cattle and the distribution and immunolabeling patterns of PrP^{Sc} in the brain of the H-type BSE-challenged cattle. The TSE agent virulence can be minimally defined by oral transmission of different TSE agents (C-type, L-type, and H-type BSE agents) [59]. Oral transmission studies with H-type BSE-infected cattle have been initiated and are underway to provide information regarding the extent of similarity in the immunohistochemical and molecular features before and after transmission. In addition, the present data will support risk assessments in some peripheral tissues derived from cattle affected with H-type BSE.

8.4 RESULTS

8.4.1 Clinical Signs

The three challenged calves developed initial signs of clinical disease approximately 12 months post challenge, which included disturbance, anxiety, and occasionally low head carriage. After 3–4 months of the onset of the clinical disease, the animals showed loss of body condition. Around 7–10 days prior to euthanasia, the animals developed ataxia of the forelimbs and hindlimbs and myoclonus and were unable to rise. The cattle were euthanized at 507 (case 1, code 7749), 574 (case 2, code 9458), and 598 (case 3, code 0728) days post challenge (mean ± standard deviation, 559.7 ± 47.2 days). The clinical signs were similar in all the three H-type BSE-challenged animals. The animals did not show any change in temperament, such as nervousness or aggression.

8.4.2 Neuropathology

The scores for the distribution and severity of vacuolation in the brain were similar among the three challenged calves. The vacuolar changes were generally observed in all the brain areas. In general, the vacuoles varied in size. The highest mean lesion scores appeared in the thalamic nuclei and neuropil of the central gray matter of the midbrain, and the lowest scores were found in the caudal cerebral and cerebellar cortices. In the vestibular and pontine nuclei, spongy changes were not as prominent as in the other brainstem nuclei. In the spinal cord of the animals with clinical disease, mild vacuolation was present in the neuropil of the gray matter. The detailed vacuolar lesion profile is shown in Figure 1. Lesion scores for C-type BSE in cattle have been described [21].

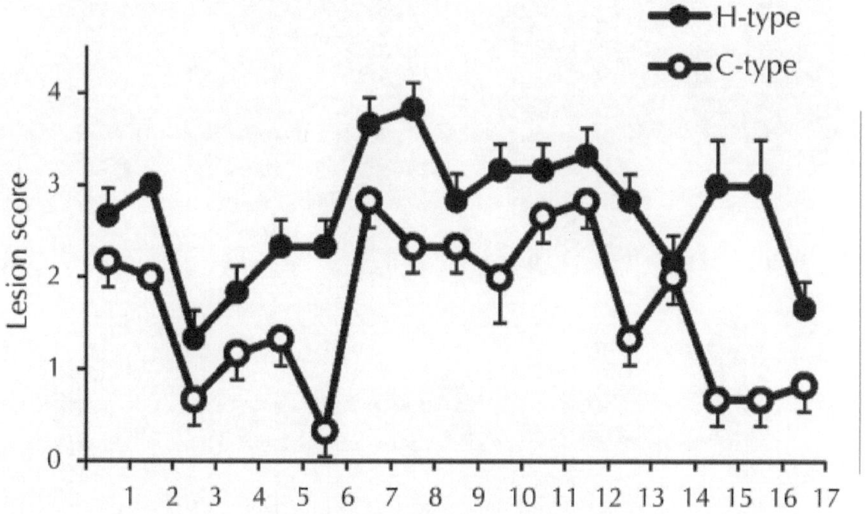

FIGURE 1 Lesion profile of H-type BSE-challenged cattle in 17 brain areas.

The different brain areas indicated are as follows. 1, nucleus of the solitary tract; 2, nucleus of the spinal tract of the trigeminal nerve; 3, hypoglossal nucleus; 4, vestibular nuclear complex; 5, cochlear nucleus; 6, cerebellar vermis; 7, central gray matter; 8, superior colliculus; 9, medial geniculate nucleus; 10, hypothalamus; 11, dorsomedial nucleus of the thalamus; 12, ventral intermediate nucleus of the thalamus; 13, frontal cortex; 14, septal nucleus; 15, caudate nucleus; 16, putamen; 17, claustrum. The lesion scores for C-type BSE are taken from a previous study [21].

8.4.3 PrP^{Sc} Immunohistochemistry

Initially, immunohistochemistry was performed with the C-terminus PrP specific antibody F99/97.6.1. Large amounts of PrP^{Sc} were deposited diffusely in the cerebral cortex, basal ganglia, thalamus, hypothalamus, brainstem, and spinal cord of all three challenged animals (Figure 2). The most conspicuous type of PrP^{Sc} deposition was fine or coarse particulate-type deposition in the neuropil of the gray matter throughout the brain and spinal cord of all the animals. Linear, perineuronal, and intraneuronal types of PrP^{Sc} staining, usually detected in C-type BSE-affected cattle, were observed in the cerebral cortex, basal ganglia, thalamus, and brainstem of the H-type BSE-challenged cattle (Figure 3). The deposition pattern of PrP^{Sc} was characterized by the presence of stellate, intraglial, and plaque forms in the brain of the H-type BSE-challenged cattle (Figure 3). The stellate-type PrP^{Sc} deposition was predominantly identified in the cerebral cortex, basal ganglia, thalamus, hypothalamus, and hippocampus and often in the cerebellar cortex, but was not visible in the brainstem and spinal cord. Intraglial-type PrP^{Sc} deposition was very consistent throughout the white matter of the central nervous system (CNS) and spinal cord.

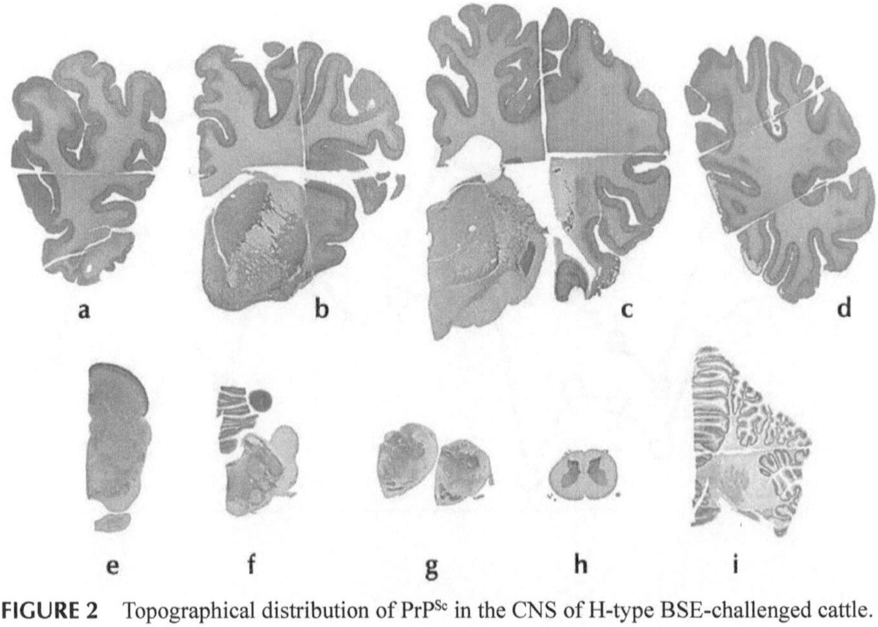

FIGURE 2 Topographical distribution of PrP^{Sc} in the CNS of H-type BSE-challenged cattle.

An H-type BSE-challenged case (case 1, code 7749) displays prominent immunolabeling in the cerebral and cerebellar cortices, basal ganglia, thalamus, brainstem, and spinal gray matter, but relatively sparse immunolabeling in the hypothalamus. The nine different areas indicated are as follows. a, frontal cortex; b, septal nucleus; c, temporal and parietal cortices and thalamus; d, occipital cortex; e, midbrain; f, pons; g, medulla oblongata at the obex; h, spinal cord; i, cerebellum. The sections show immunohistochemical labeling with mAb F99/97.6.1 and hematoxylin counterstaining.

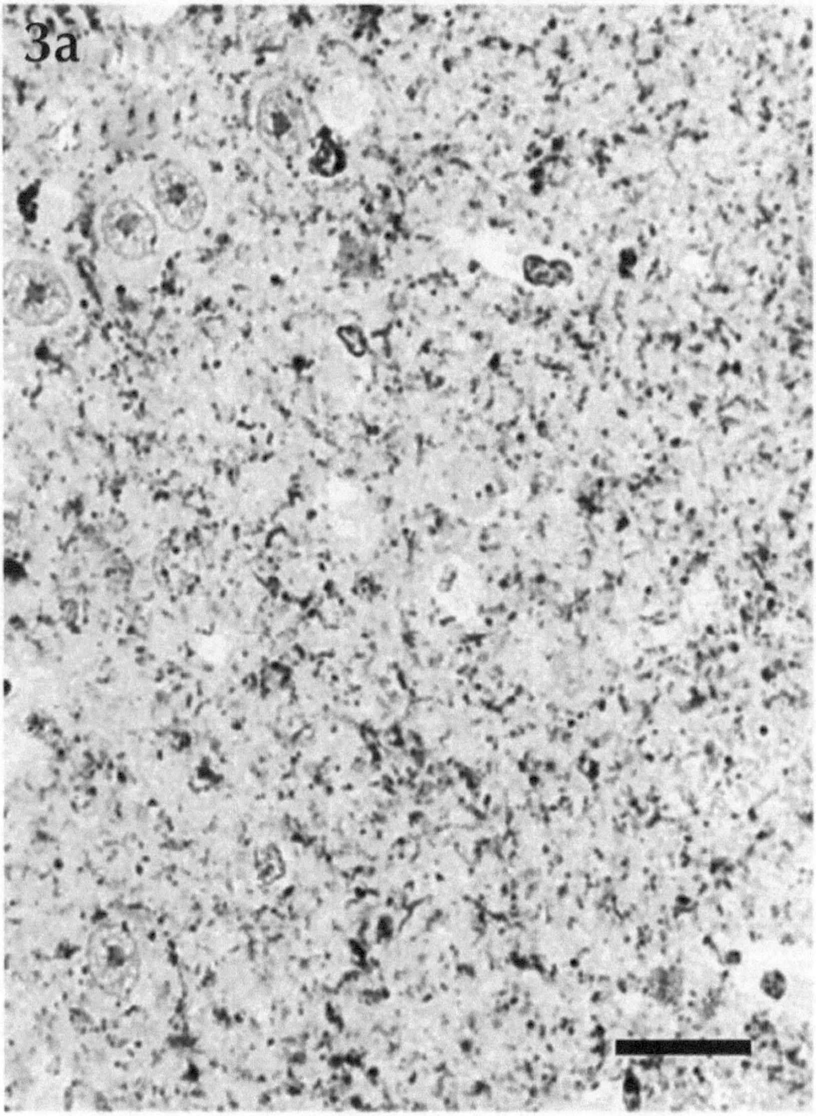

FIGURE 3 *(Continued)*

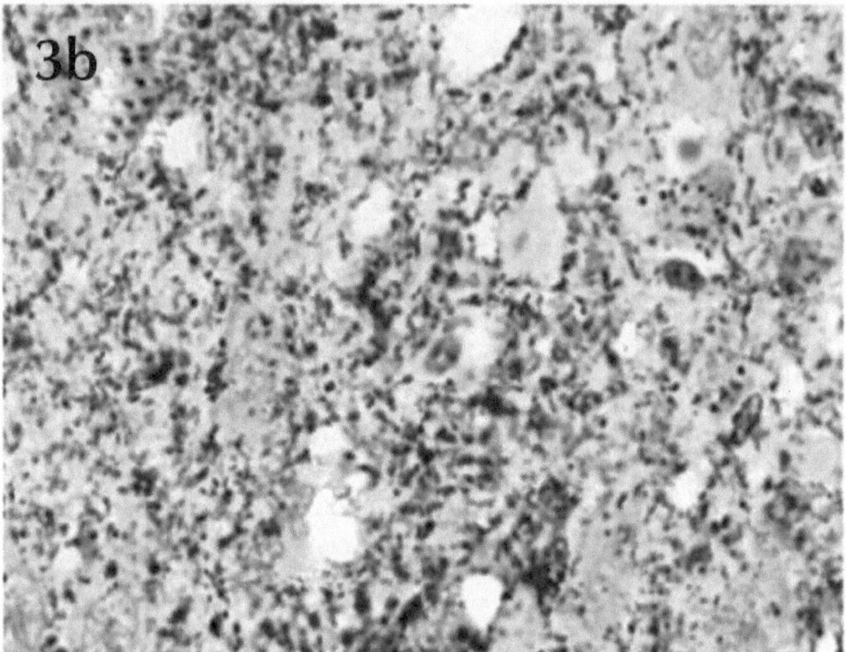

FIGURE 3 *(Continued)*

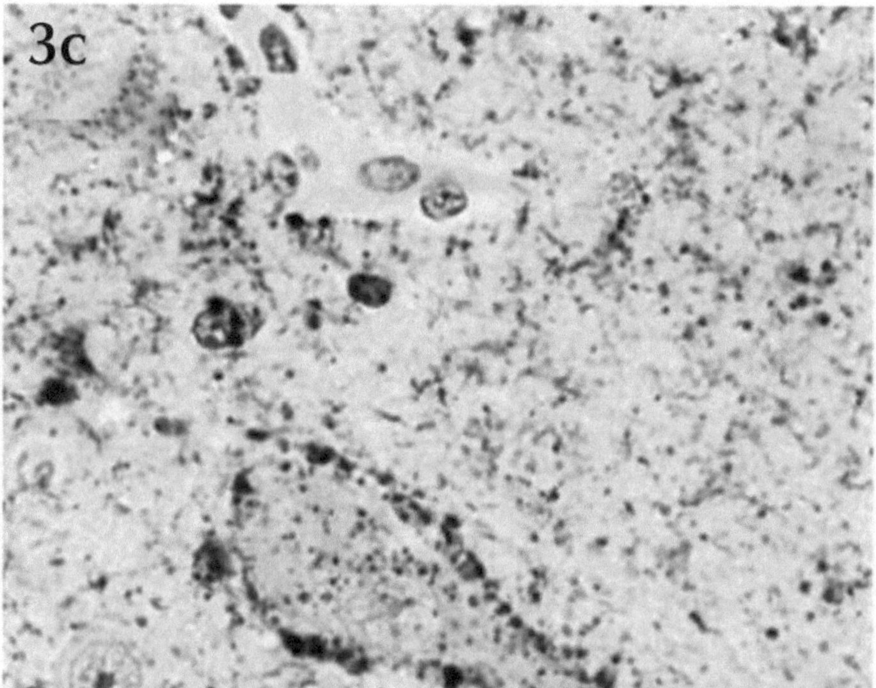

FIGURE 3 *(Continued)*

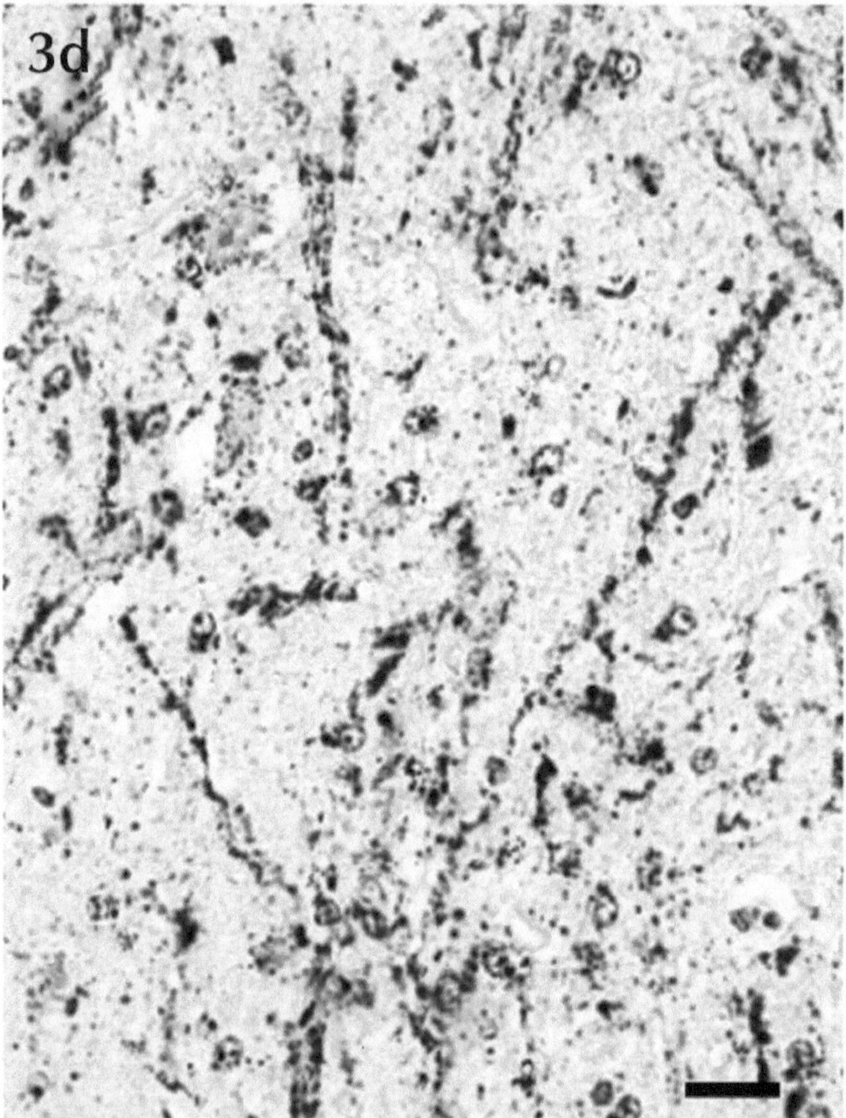

FIGURE 3 *(Continued)*

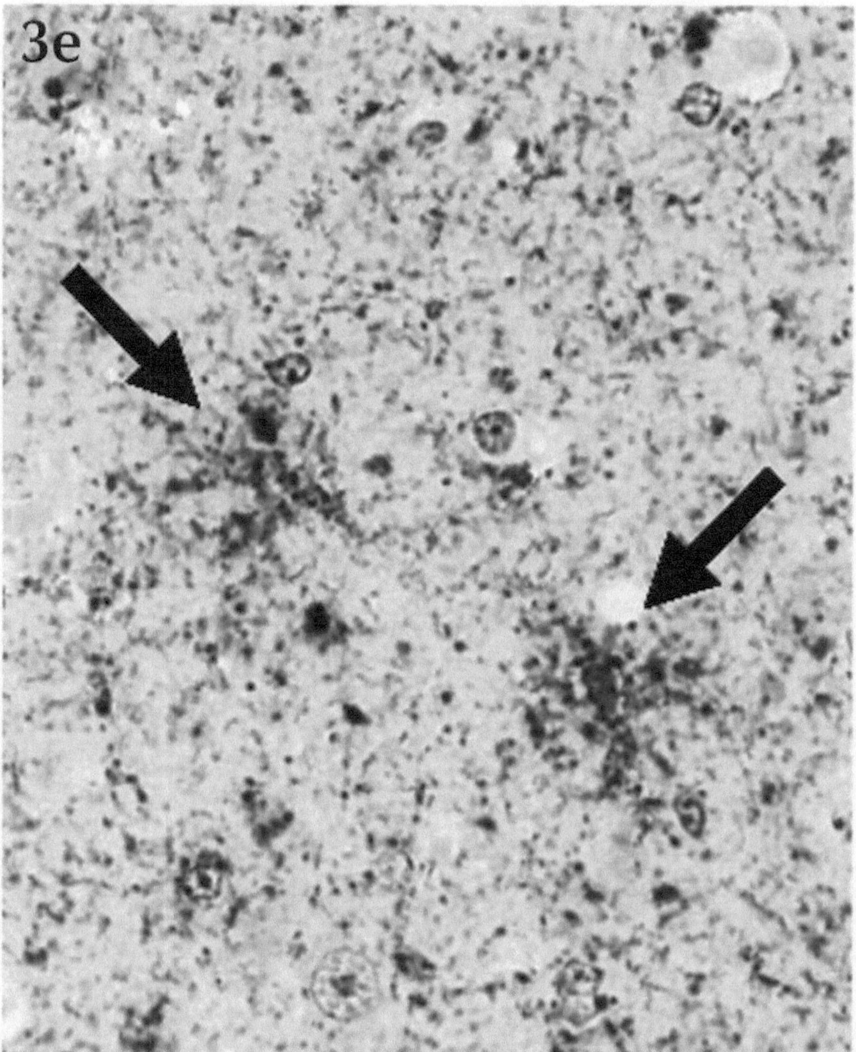

FIGURE 3 *(Continued)*

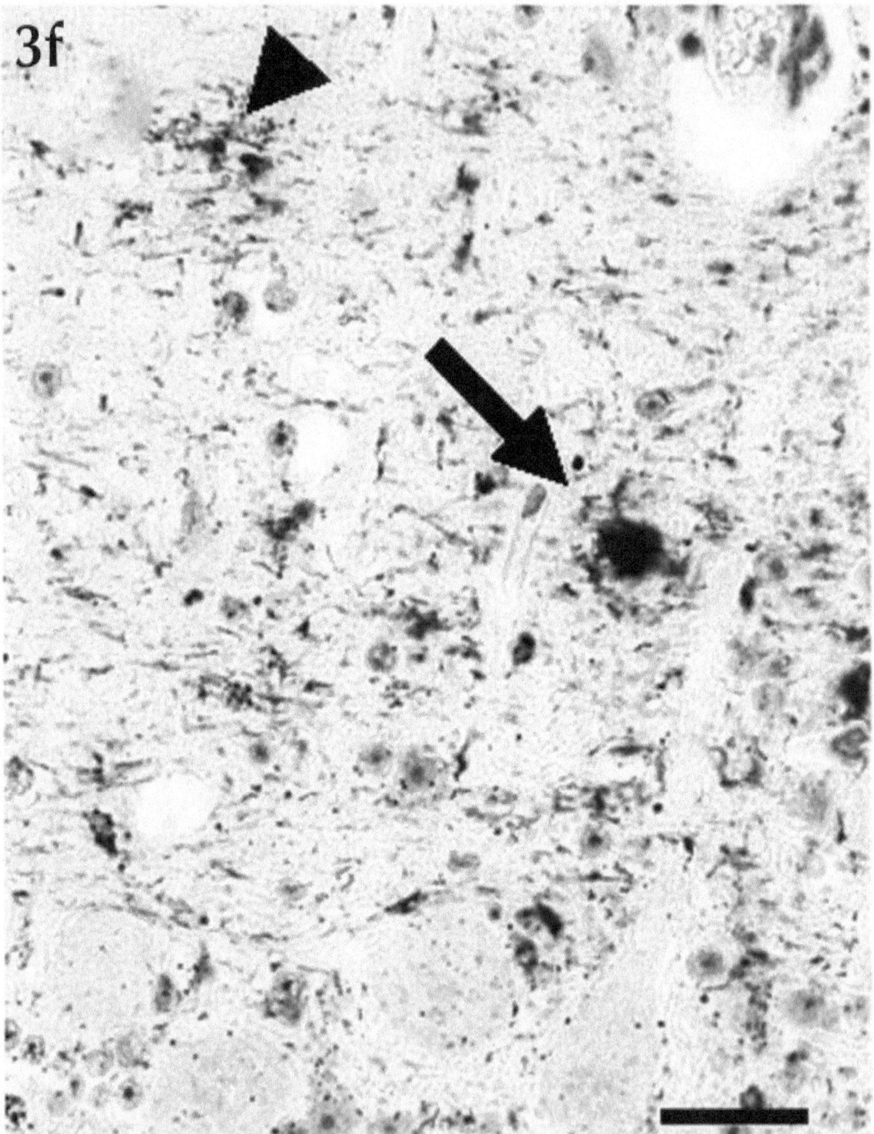

FIGURE 3 *(Continued)*

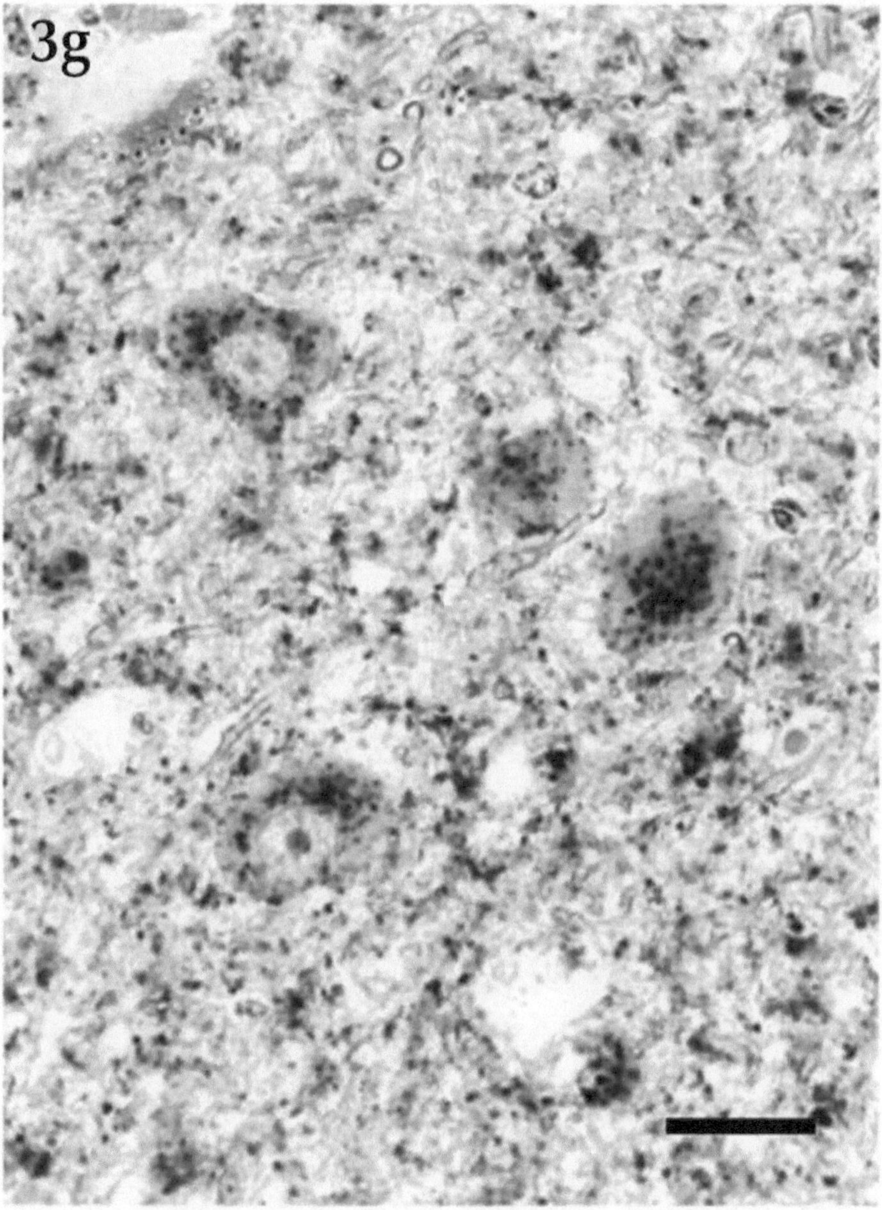

FIGURE 3 *(Continued)*

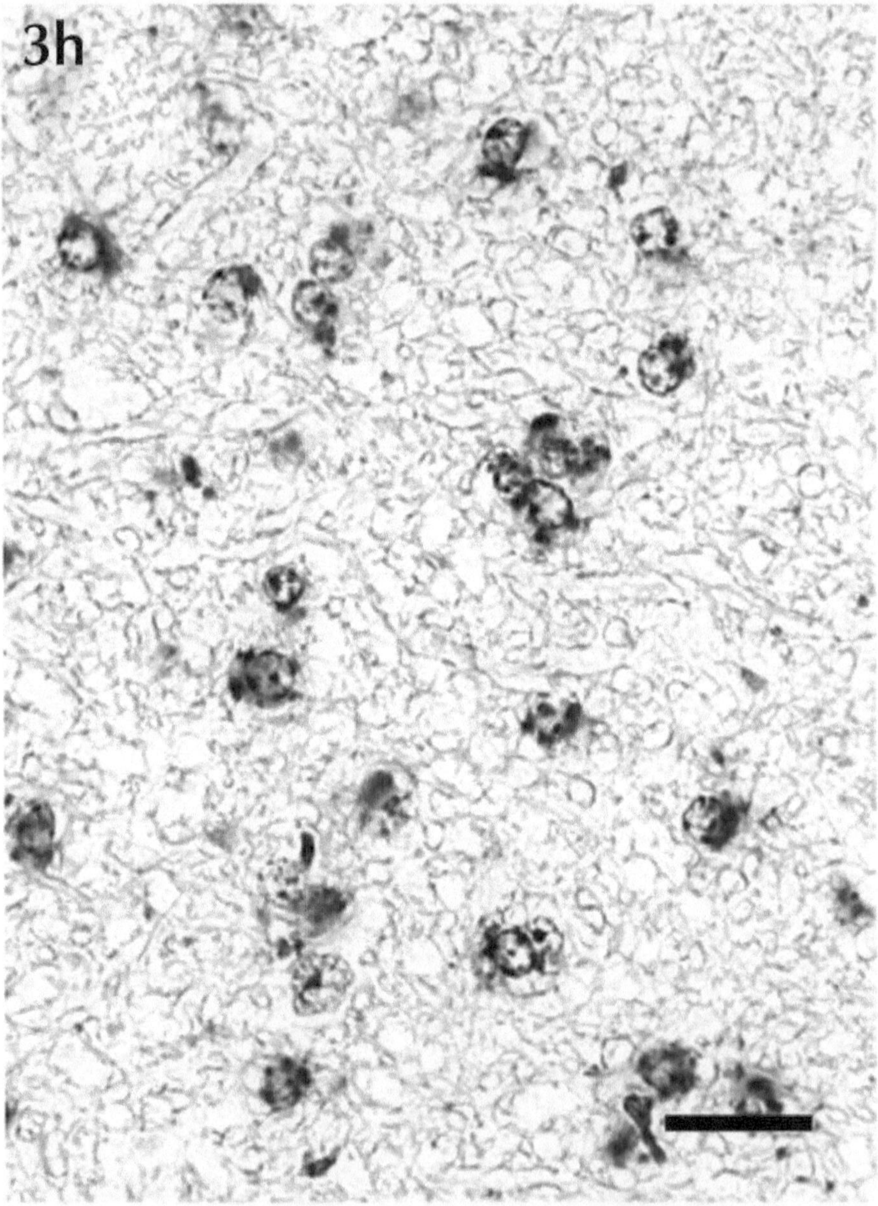

FIGURE 3 PrP^{Sc} accumulation patterns in the brain of H-type BSE-challenged cattle.

Fine particulate-type PrP^{Sc} accumulation in the neuropil of the frontal cortex (a). Coarse particulate-type (b), perineuronal-type (c), and linear-type (d) PrP^{Sc} deposition in the thalamus. Stellate-type deposition in the caudate nucleus (e, arrows). Plaque-like (f, arrow) and stellate-type (f, arrowhead) deposition in the cerebral cortex.

Intraneuronal-type PrP^{Sc} accumulation in the olivary nucleus (g). Intraglial-type PrP^{Sc} deposition in the cerebellar medulla (h). The images show immunohistochemical labeling with mAb F99/97.6.1 and hematoxylin counterstaining. Scale bars = 20 µm.

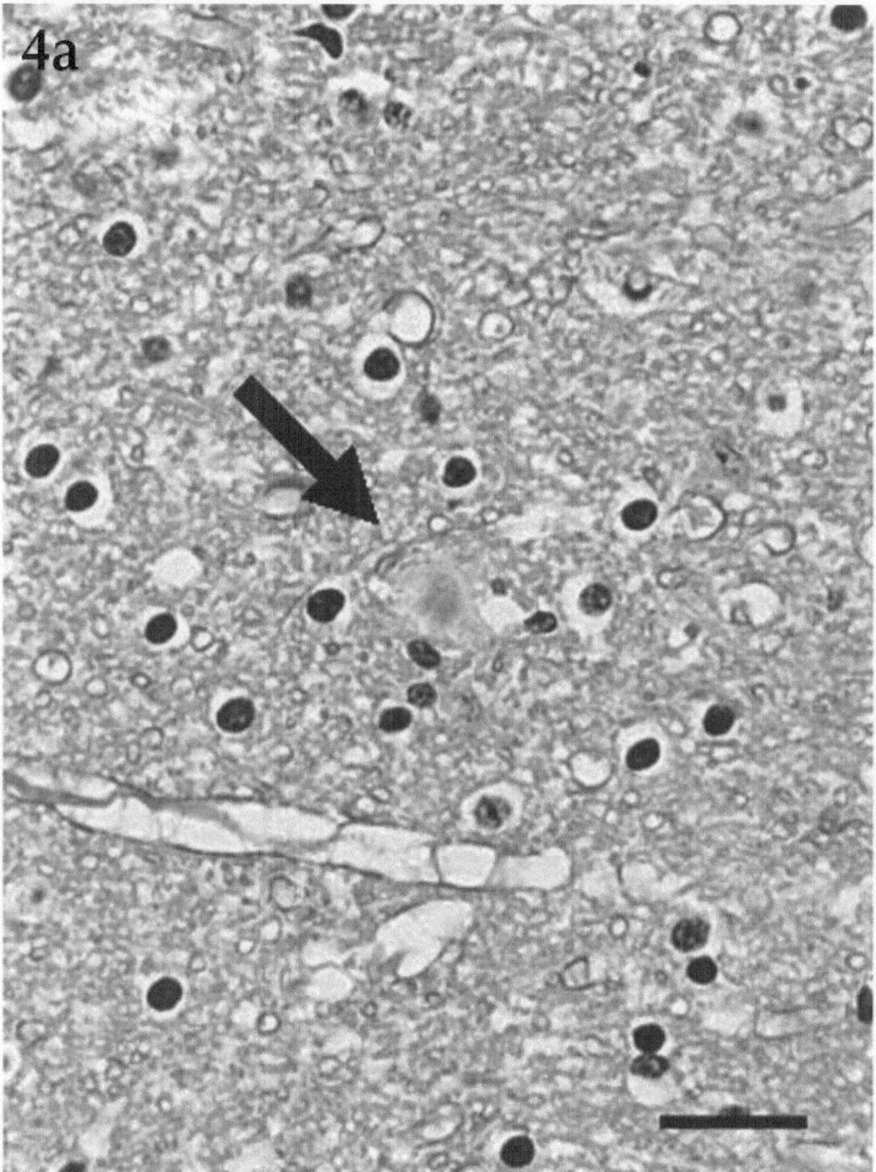

FIGURE 4 *(Continued)*

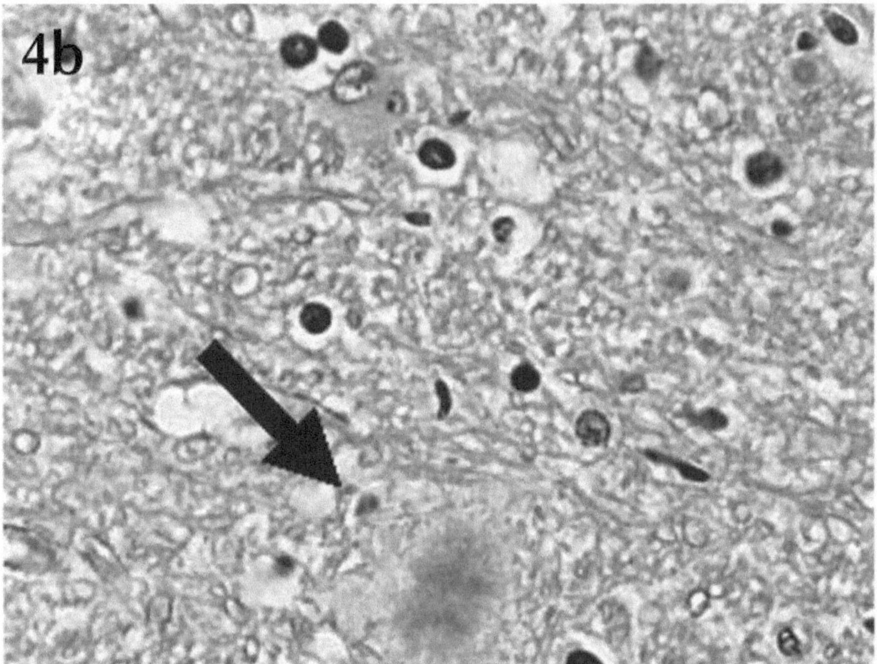

FIGURE 4 *(Continued)*

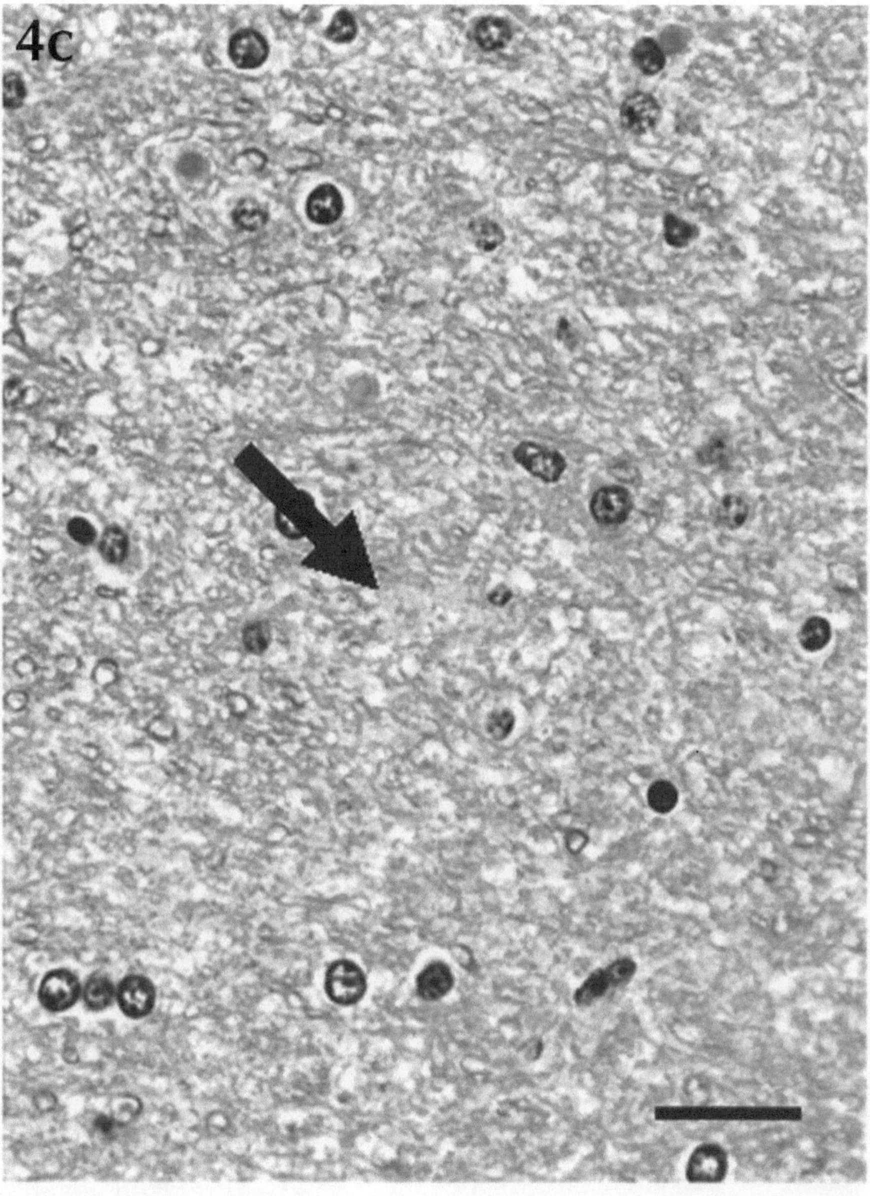

FIGURE 4 *(Continued)*

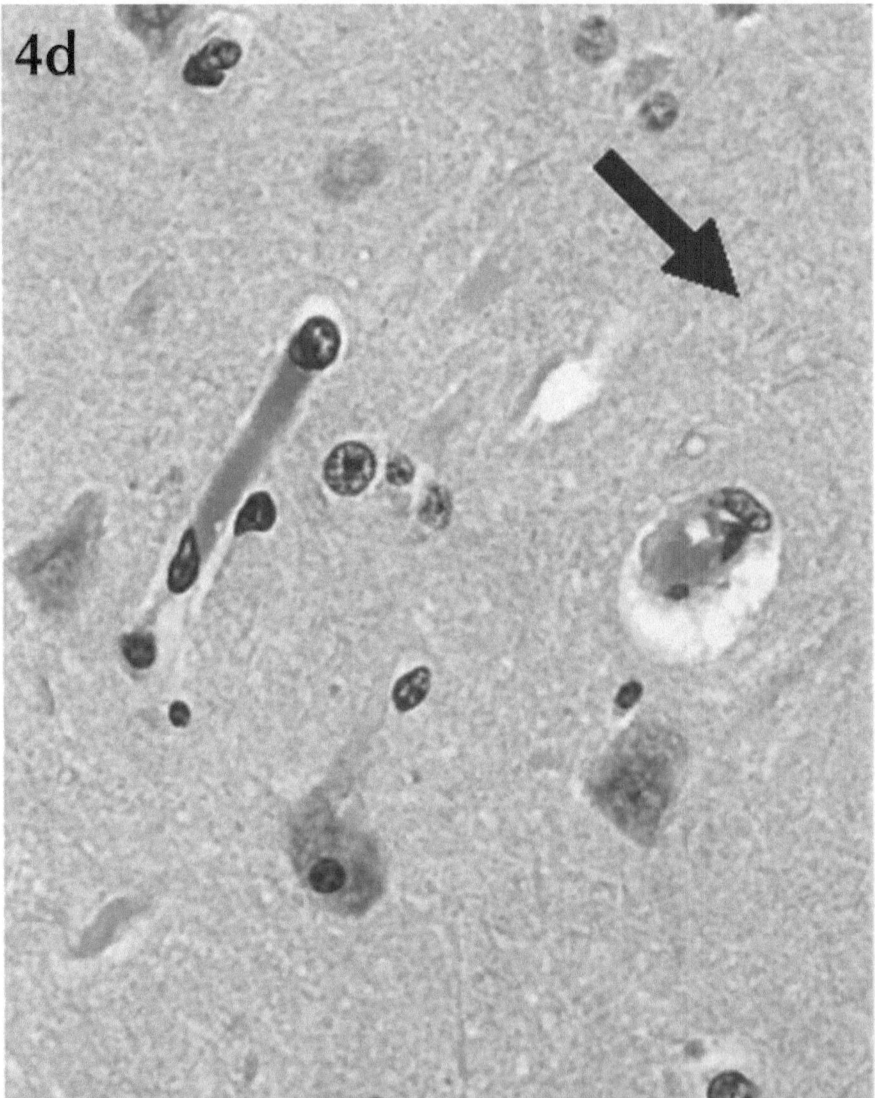

FIGURE 4 *(Continued)*

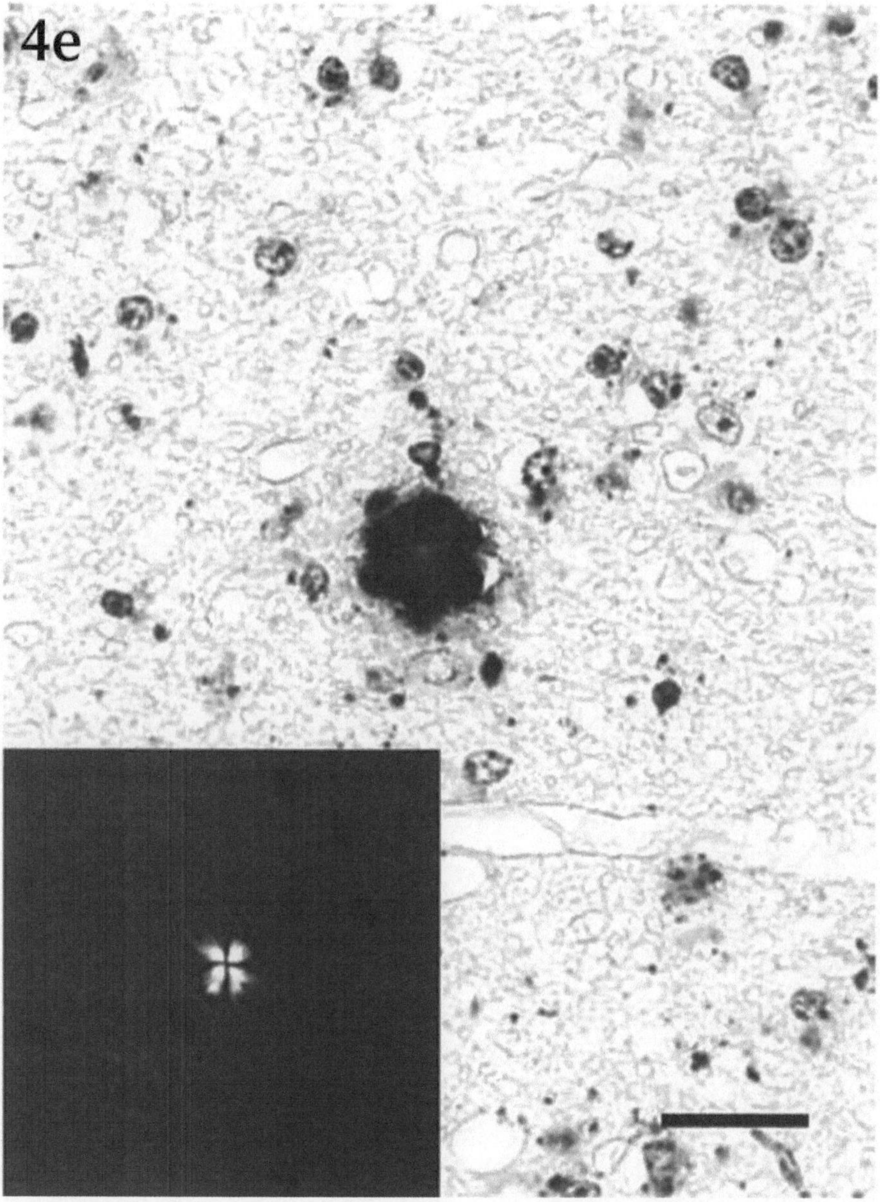

FIGURE 4 *(Continued)*

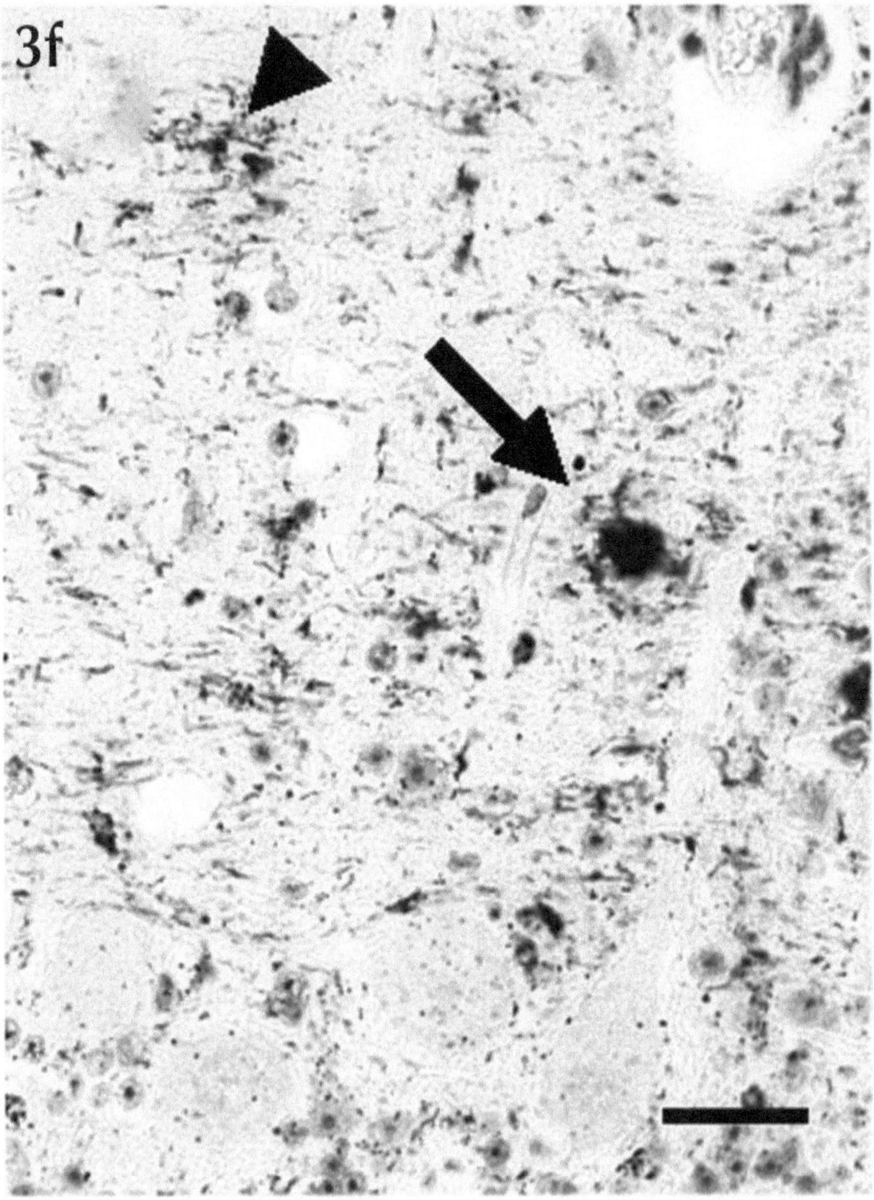

FIGURE 4 *(Continued)*

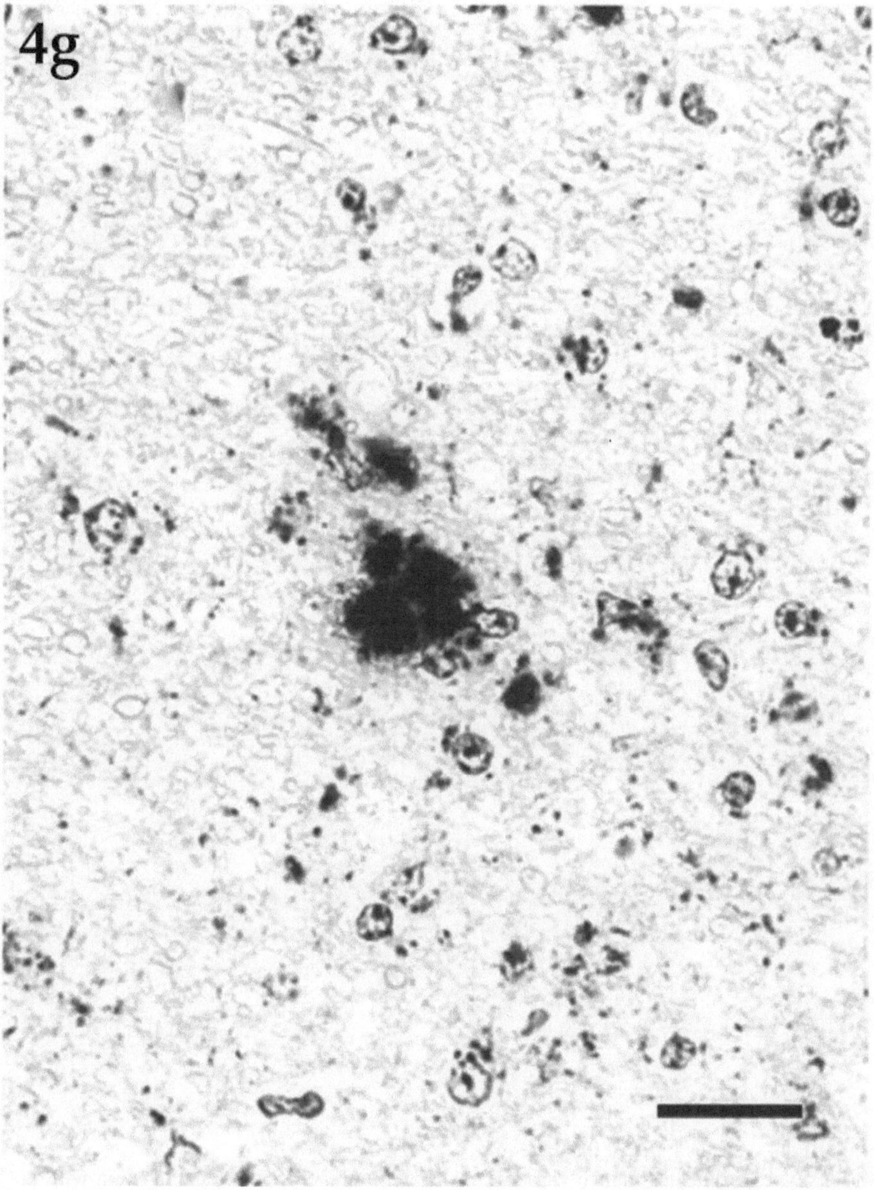

FIGURE 4 *(Continued)*

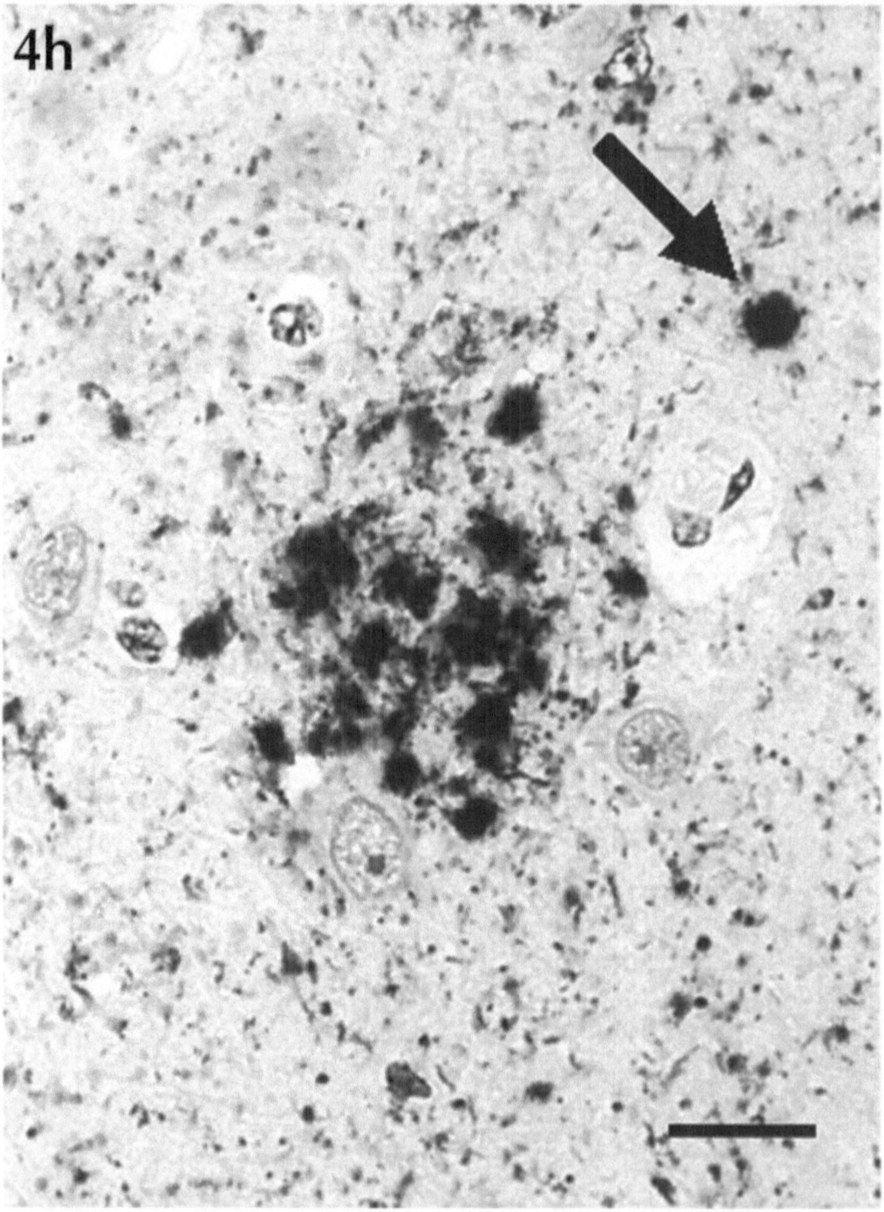

FIGURE 4 Various types of plaques.
Various types of plaques stained with HE (a–d) and immunohistochemical labeling with mAb F99/97.6.1 (e–h). Unicentric (a and e) and multicentric (b–d and f–h) cores of plaques (arrows) in the white matter of the thalamus (a–c and e–g) and in the deep cortical area of the frontal lobe (d and h). The inset in the lower left corner of (e) shows an amyloid plaque detected by Congo red staining in the serial section, it shows birefringence under polarized light. Scale bars = 20 μm.

PrP^{Sc}-positive plaques were scattered throughout the cerebral white matter (Figure 4). Some PrP^{Sc}-positive were also detected in the deep cortical layer of the cerebrum, internal capsule of the corpus striatum and thalamic white matter, and cerebellum, but were absent from the brainstem and spinal cord. On the basis of the morphology of the cores, as revealed after application of immunohistochemistry, the plaques were classified into unicentric and multicentric types (Figure 4). Unicentric plaques had a single core and were up to 25 μm in diameter. Multicentric plaques were composed of multiple smaller cores clustered together or a central core surrounded by even smaller plaques or aggregates, varied in shape and could extend up to 40 μm in diameter. Both types of plaques were subdivided into two subtypes, that is, those with a dense compact core and others with a pale central core. The dense compact core plaques were less than 20 μm in diameter and smaller than the plaques with a pale central core, they were uniformly immunolabeled and were difficult to detect in HE or Congo red-stained sections. Furthermore, plaques with a pale central core were generally larger than that with a dense compact core and stained pale basophilic or amphophilic with HE and positively with Congo red under polarized light. The periphery of these plaques looked like a halo unstained with HE and Congo red but well immunolabeled with PrP-specific antibodies. In addition, the granular form of plaque-like deposits, was rarely detected in the deep cerebral cortex, basal ganglia, and thalamic nuclei and not detected in the white matter. These deposits were not stained with Congo red and were composed of aggregates about 5 μm in diameter.

8.4.4 Variability of PrP^{Sc} Immunolabeling with Antibodies

The H-type cases were further investigated by immunohistochemistry also using a set of other PrP specific antibodies covering the different regions of PrP (Table 1). Results of PrP^{Sc} immunolabeling with each antibody are summarized in Table 2. The immunolabeling intensity of each type of PrP^{Sc} varied with the different primary antibodies. The strongest immunolabeling for both extracellular and intracellular PrP^{Sc} was evident with mAb F99/97.6.1 and pAb T4, which recognized the C-terminal region of PrP, whereas mAb SAF32 and pAb B103, which recognized the N-terminal region of PrP, showed no or weak immunolabeling. In general, the core-specific antibodies used in this study produced varying degrees of immunolabeling intensity for each extracellular-type PrP^{Sc} deposit. Intraneuronal-type PrP^{Sc} deposits in H-type BSE showed weaker immunolabeling with the core-specific antibodies than those in C-type BSE. In addition, the core-specific antibodies did not show any immunolabeling for intraglial-type PrP^{Sc} deposits.

8.4.5 PrP^{Sc} Deposition in Additional Structures

The positive PrP^{Sc} immunolabeling was detected in the trigeminal and dorsal root ganglia, neurohypophysis, retina, and optic nerve. In the neurohypophysis, fine granular PrP^{Sc} depositions were detected in the unmyelinated nerve fibers, and intracytoplasmic immunolabeling was detected in the pituicytes. In the retina, intense granular immunolabeling was observed in the ganglion cell layer as well as the inner and outer plexiform layers. In the optic nerve, intraglial immunolabeling was prominent. The PrP^{Sc} was occasionally found in satellite cells and ganglion cells of the trigeminal and dorsal

TABLE 2 Comparison of immunolabeling intensities of different PrP^{Sc} types between C-type and H-type BSE...

	Extracellular PrP^{Sc} type								Intracellular PrP^{Sc} type						Intracellular PrP^{Sc} type			
Antibody	**Particulate**		**Perinuclear**		**Linear**		**Plaque-like**		**Stellate**		**Intraneu-ronal**		**Intraglial**		**Intraneuronal**		**Intraglial**	
	C-type	**H-type**	**C-type**	**H-type**	**C-type**	**H-type**	**C-type**	**H-type**	**C-type**	**H-type**	**C-type**	**H-type**	**C-type**	**H-type**	**C-type**	**H-type**	**C-type**	**H-type**
SAF32	+	–	±	–	–	–	+	±	+	–	±	–	±	–	±	–	±	–
B103	2+	+	+	±	2+	±	+	+	2+	–	+	–	+	–	+	–	+	–
F89/160.1.5	2+	2+	2+	2+	2+	2+	3+	3+	3+	2+	2+	+	±	–	2+	+	±	–
T1	2+	+	2+	±	2+	+	2+	3+	2+	2+	2+	+	+	–	2+	+	+	–
12F10	3+	2+	3+	+	3+	+	3+	2+	3+	2+	3+	+	2+	–	3+	+	2+	–
SAF54	2+	±	2+	–	2+	–	2+	+	3+	+	2+	+	+	–	2+	+	+	–
44B1	3+	2+	3+	2+	3+	2+	3+	3+	3+	2+	3+	2+	+	+	3+	2+	+	+
SAF84	3+	2+	3+	+	3+	3+	3+	3+	3+	3+	3+	2+	2+	–	3+	2+	2+	–
43C5	3+	2+	3+	+	3+	+	3+	2+	3+	2+	3+	2+	2+	+	3+	2+	2+	+
T4	3+	2+	3+	2+	3+	2+	3+	3+	3+	3+	3+	3+	+	3+	3+	3+	+	3+
F99/97.6.1	3+	3+	3+	3+	3+	3+	3+	3+	3+	3+	3+	3+	2+	3+	3+	3+	2+	3+

-, none, ±, negligible, +, weak, 2+, distinct, 3+, strong

root ganglia. No PrP^{Sc} immunolabeling was detected in the lymphoid tissues, including the spleen, tonsils, Peyer's patches, and lymph nodes, in any of the animals.

8.4.6 Western Blot Analysis

PrPres was detected by Western blot analysis using mAbs 6H4 and SAF84 in all the animals challenged with H-type BSE (Figure 5), whereas the control animals were negative for PrPres. The Western blot analysis with 6H4 showed that the diglycosylated, monoglycosylated, and unglycosylated fragments of PK-treated PrPres derived from H-type BSE-challenged cattle were more than those of PrPres derived from the C-type BSE-affected cattle. On the contrary, the glycoform profiles of PrPres derived from both the H-type BSE- and C-type BSE-challenged cattle were similar. With mAb SAF84, a multiple banding pattern was detected from the H-type BSE sample. After deglycosylation with PNGase F treatment, in addition to ~19 kDa PrPres fragment, a 10–12 kDa PrPres fragment was detected from the H-type BSE sample. This additional 10–12 kDa fragment was not recognized with 6H4. The molecular size and glycoform patterns of PrPres were similar and conserved in the H-type BSE-challenged cattle.

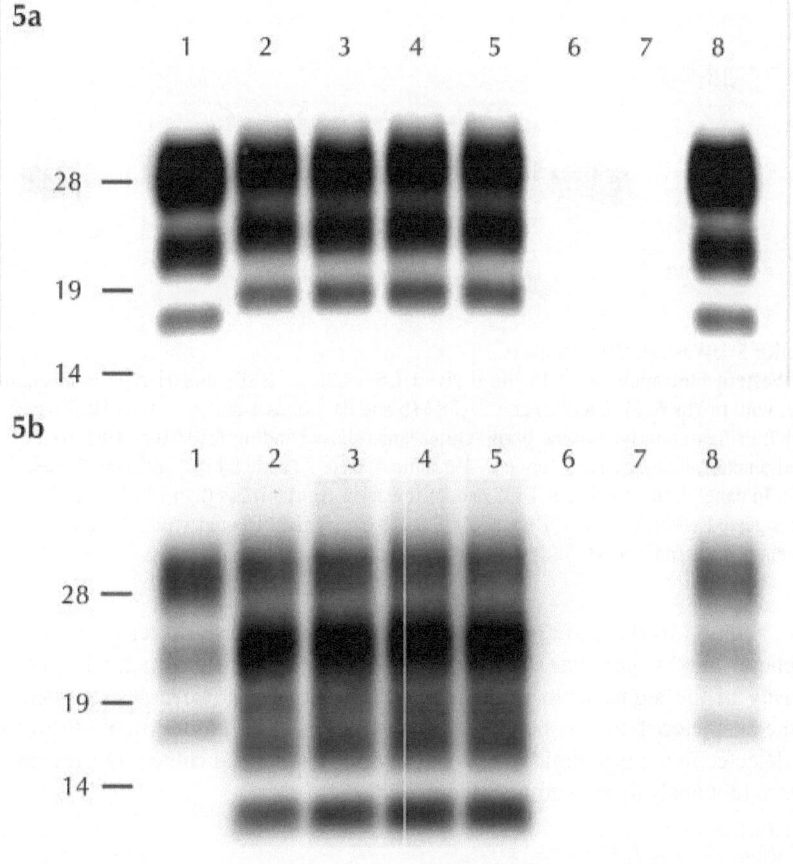

FIGURE 5 *(Continued)*

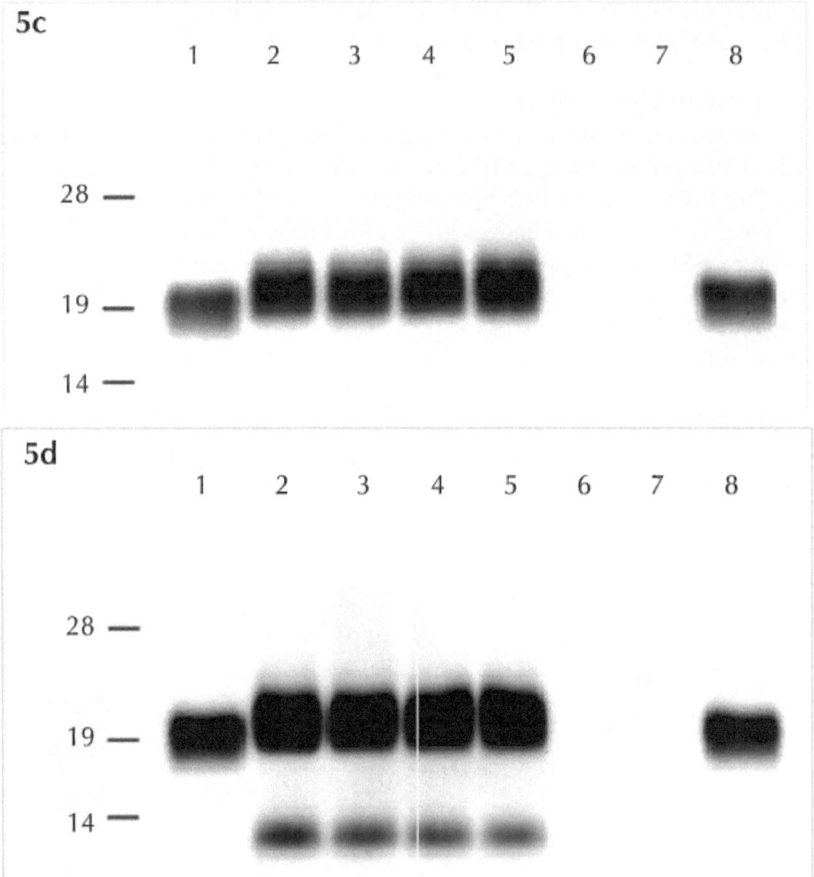

FIGURE 5 Western Blot Analysis.
The Western blot analysis of PrPres derived from C-type BSE- and H-type BSE-challenged cattle, with mAbs 6H4 (a and c) and SAF84 (b and d). Lanes 1 and 8, C-type BSE tissue, lanes 6 and 7, PrPres-negative bovine brain. Other lanes show banding for H-type BSE tissue. lane 2, Canadian case, lane 3, case 1 (code 7749), lane 4, case 2 (code 9458), and lane 5, case 3 (code 0728). In panel d, the band near 14 kDa is of lower molecular mass than 14 kDa. All the samples were digested with 50-µg/mL PK at 37 °C for 1 h, and then treated with PNGase F (c and d). Molecular mass markers (kDa) are shown on the left.

In addition to the brain and spinal cord, PrPres was also detected in most of the peripheral nerves, ganglia, optic nerve, retina, hypophysis, and adrenal gland. The intensity of the signal from most of these tissues was, however, barely detectable, but the characteristic triple banding was always detected (Figure 6). No PrPres signal was detected in the lymphoid tissues of the three challenged calves. The results of the western blot analysis are summarized in Table 3.

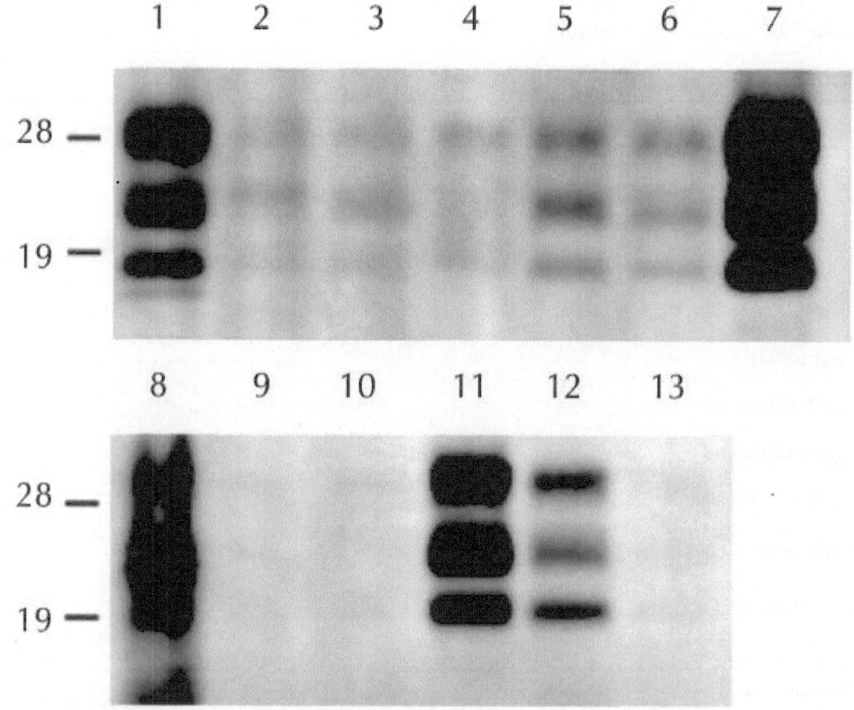

FIGURE 6 Western blot analysis of PrPres in nervous tissue.

Western blot analysis of PrPres in the nervous tissue sampled from the H-type BSE-challenged cattle (case 1, code 7749). 1, cranial cervical ganglion; 2, vagal nerve (cervical division); 3, vagal nerve (thoracic division); 4, phrenic nerve; 5, sympathetic trunk (thoracic division); 6, sympathetic trunk (lumbar division); 7, trigeminal ganglion; 8, adrenal gland (medulla); 9, facial nerve; 10, hypoglossal nerve; 11, celiac-mesenteric ganglion complex; 12, cauda equina; 13, suprascapular nerve. Each lane was loaded with 100 mg of tissue equivalent. Western blots were probed with mAb 6H4 to detect PrPres. Molecular mass standards (kDa) are indicated on the left.

TABLE 3 Immunohistochemical and Western blot analyses of PrP^{Sc} in tissue samples obtained from cattle intracerebrally challenged with H-type BSE.

Tissue	Case code							
	7749		9458		0728		0728	
	IHC	WB	IHC	WB	IHC	WB	IHC	WB
Central nervous system								
Cerebral cortex	+	+	+	+	+	+	+	+

TABLE 3 *(Continued)*

Tissue	Case code							
	7749		9458		0728		0728	
	IHC	WB	IHC	WB	IHC	WB	IHC	WB
Obex	+	+	+	+	+	+	+	+
Cerebrum	+	+	+	+	+	+	+	+
Spinal cord	+	+	+	+	+	+	+	+
Peripheral nervous system								
Cauda equina	+	+	+	+	+	+	+	+
Dorsal root ganglia	+	+	+	+	+	+	+	+
Trigeminal ganglia	+	+	+	+	+	+	+	+
Cranial cervical ganglia	–	+	–	+	–	+	–	+
Stellate ganglia	–	+	–	+	–	+	–	+
Sympathetic trunk	–	+	–	+	–	+	–	+
Celiac-mesenteric ganglion complex	–	+	–	+	–	+	–	+
Vagus nerve	–	+	–	+	–	+	–	+
Facial nerve	–	+	–	+	–	–	–	–
Hypoglossal nerve	–	+	–	+	–	–	–	–
Phrenic nerve	–	+	–	+	–	+	–	+
Accessory nerve	–	–	–	+	–	+	–	+
Suprascapular nerve	–	+	–	+	–	+	–	+
Brachial plexus	–	+	–	+	–	+	–	+
Median nerve	–	+	–	+	–	+	–	+
Radial nerve	–	+	–	+	–	+	–	+
Sciatic nerve	–	+	–	+	–	+	–	+
Tibial nerve	–	+	–	+	–	+	–	+
Optic nerve	+	+	+	+	+	+	+	+

TABLE 3 *(Continued)*

Tissue	Case code							
	7749		9458		0728		0728	
	IHC	WB	IHC	WB	IHC	WB	IHC	WB
Retina	+	+	+	+	+	+	+	+
Pituitary gland	+	+	+	+	+	+	+	+
Adrenal gland	–	+	–	+	–	+	–	+
Lymphoid tissues								
Spleen	–	–	–	–	–	–	–	–
Tonsils (palatine, pharyngeal, lingual)	–	–	–	–	–	–	–	–
Thymus	–	–	–	–	–	–	–	–
Parotid lymph node	–	–	–	–	–	–	–	–
Mandibular lymph node	–	–	–	–	–	–	–	–
Lateral retropharyngeal lymph node	–	–	–	–	–	–	–	–
Superficial cervical lymph node	–	–	–	–	–	–	–	–
Brachiocephalic lymph node	–	–	–	–	–	–	–	–
Axillary lymph node	–	–	–	–	–	–	–	–
Superficial inguinal lymph node	–	–	–	–	–	–	–	–
Subiliac lymph node	–	–	–	–	–	–	–	–
Popliteal lymph node	–	–	–	–	–	–	–	–
Hepatic lymph node	–	–	–	–	–	–	–	–
Internal iliac lymph node	–	–	–	–	–	–	–	–
External iliac lymph node	–	–	–	–	–	–	–	–
Mesenteric lymph node	–	–	–	–	–	–	–	–

IHC; immunohistochemistry, WB; western blotting, +, positive for PrP^{Sc}, -, negative for PrP^{Sc}.

8.4.7 Analysis of *PRNP*

The mature PrP sequences (amino acids 25-242) of the *PRNP* ORF in the three challenged animals were compared with the representative bovine *PRNP* sequence (GenBank accession number. AJ298878) [37]. The *PRNP* sequence for case 2 (code 9458) was the same as the reference sequence, and that for case 1 (code 7749) was also normal with a synonymous polymorphism at codon 78 (G or A, no amino acid substitution). The two animals had 6 copies of the octarepeat region on both the *PRNP* alleles. The *PRNP* sequence for case 3 (code 0728) was also normal, and both alleles contained five copies of the octarepeat region.

KEYWORDS

- **Bovine spongiform encephalopathy**
- **Multicentric PrP plaques**
- **Unicentric PrP plaques**

COMPETING INTERESTS

The authors declare that they have no competing interests.

AUTHORS' CONTRIBUTIONS

Conception and design of experiments. Takashi Yokoyama and Hiroyuki Okada. Conduction of experiments. Hiroyuki Okada; Yoshifumi Iwamaru, Morikazu Imamura, Kentaro Masujin, and Yuichi Matsuura. Intracerebral inoculation of H-type BSE isolate and collection of samples from H-type BSE-infected cattle. Yoshifumi Iwamaru, Hiroyuki Okada, Morikazu Imamura, Kentaro Masujin, Yoshihisa Shimizu, and Kazuo Kasai. Manuscript draft preparation and data analysis. Hiroyuki Okada, Yoshifumi Iwamaru, Morikazu Imamura, and Yuichi Matsuura. Participation in scientific discussion of the results. Stefanie Czub. Study supervision. Shirou Mohri. All the authors have read and approved the final manuscript.

ACKNOWLEDGMENT

We thank Dr Yuichi Tagawa (National Institute of Animal Health) for providing the anti-prion mAb T1 and Dr Motoshi Horiuchi (Graduate School of Veterinary Medicine, Hokkaido University) for providing mAbs 44B1 and 43C5. Expert technical assistance was provided by Mutsumi Sakurai, Miyo Kakizaki, Junko Endo, Noriko Amagai, Tomoko Murata, Naomi Furuya, Naoko Tabeta, Nobuko Kato, and the animal caretakers. This work was supported by grants from the BSE and other Prion Disease Control Projects of the Ministry of Agriculture, Forestry and Fisheries of Japan and from the Canadian Food Inspection Agency.

REFERENCES

1. Wells, G. A., Scott, A. C., Johnson, C. T., Gunning, R.F., Hancock, R.D., Jeffrey, M., Dawson, M., and Bradley, R. A novel progressive spongiform encephalopathy in cattle. *Vet Rec* **121**, 419420 (1987).

2. Prusiner, S.B. Molecular biology of prion diseases. *Science* **252**, 1515–1522 (1991).
3. Jacobs, J. G., Langeveld, J. P., Biacabe, A. G., Acutis, P. L., Polak, M. P., Gavier-Widen, D., Buschmann, A., Caramelli, M., Casalone, C., Mazza, M., Groschup, M., Erkens, J. H., Davidse, A., van Zijderveld, F. G., and Baron, T. Molecular discrimination of atypical bovine spongiform encephalopathy strains from a geographical region spanning a wide area in Europe. *J Clin Microbiol* **45**, 1821–1829 (2007).
4. Biacabe, A. G., Laplanche, J. L., Ryder, S., and Baron, T. Distinct molecular phenotypes in bovine prion diseases. *EMBO Rep* **5**, 110–115 (2004).
5. Casalone, C., Zanusso, G., Acutis, P., Ferrari, S., Capucci, L., Tagliavini, F., Monaco, S., and Caramelli, M. Identification of a second bovine amyloidotic spongiform encephalopathy. molecular similarities with sporadic Creutzfeldt-Jakob disease. *Proc Natl Acad Sci U. S. A* **101**, 3065–3070 (2004).
6. Baron, T., Biacabe, A. G., Arsac, J. N., Benestad, S., and Groschup, M. H. Atypical transmissible spongiform encephalopathies (TSEs) in ruminants. *Vaccine* **25**, 5625–5630 (2007).
7. Biacabe, A. G., Morignat, E., Vulin, J., Calavas, D., Baron, T. G. Atypical bovine spongiform encephalopathies, France, 2001–2007. *Emerg Infect Dis* **14**, 298–300 (2008).
8. Brown, P., McShane, L. M., Zanusso, G., and Detwile, L. On the question of sporadic or atypical bovine spongiform encephalopathy and Creutzfeldt-Jakob disease. *Emerg Infect Dis* **12**, 1816–1821 (2006).
9. Clawson, M. L., Richt, J. A., Baron, T., Biacabe, A. G., Czub, S., Heaton, M. P., Smith, T. P., and Laegreid, W. W. Association of a bovine prion gene haplotype with atypical BSE. *PLoS One* **3**, e1830 (2008).
10. Gavier-Widén, D., Nöremark, M., Langeveld, J. P., Stack, M., Biacabe, A. G., Vulin, J., Chaplin, M., Richt, J. A., Jacobs, J., Acín, C., Monleón, E., Renström, L., Klingeborn, B., and Baron, T. G. Bovine spongiform encephalopathy in Sweden. an H-type variant. *J Vet Diagn Invest* **20**, 2–10 (2008).
11. Richt, J. A. and Hall, S. M. BSE case associated with prion protein gene mutation. *PLoS Pathog* **4**, e1000156 (2008).
12. Dudas, S., Yang, J., Graham, C., Czub, M., McAllister, T. A., Coulthart, M. B., and Czub, S. Molecular, biochemical and genetic characteristics of BSE in Canada. *PLoS One* **5**, e10638 (2010).
13. Buschmann, A., Gretzschel, A., Biacabe, A. G., Schiebel, K., Corona, C., Hoffmann, C., Eiden, M., Baron, T., Casalone, C., and Groschup, M. H. Atypical BSE in Germany—proof of transmissibility and biochemical characterization. *Vet Microbiol* **117**, 103–116 (2006).
14. Baron, T. G., Biacabe, A. G., Bencsik, A., and Langeveld, J. P. Transmission of new bovine prion to mice. *Emerg Infect Dis* **12**, 1125–1128 (2006).
15. Baron, T., Vulin, J., Biacabe, A. G., Lakhdar, L., Verchere, J., Torres, J. M., and Bencsik, A. Emergence of classical BSE strain properties during serial passages of H-BSE in wild-type mice. *PLoS One* **6**, e15839 (2011).
16. Béringue, V., Bencsik, A., Le Dur, A., Reine, F., Laï, T. L., Chenais, N., Tilly, G., Biacabé, A. G., Baron, T., Vilotte, J. L., and Laude H. Isolation from cattle of a prion strain distinct from that causing bovine spongiform encephalopathy. *PLoS Pathog* **2**, e112 (2006).
17. Capobianco, R., Casalone, C., Suardi, S., Mangieri, M., Miccolo, C., Limido, L., Catania, M., Rossi, G., Di Fede, G., Giaccone, G., Bruzzone, M. G., Minati, L., Corona, C., Acutis, P., Gelmetti, D., Lombardi, G., Groschup, M. H., Buschmann, A., Zanusso, G., Monaco, S., Caramelli, M., and Tagliavini, F. Conversion of the BASE prion strain into the BSE strain. the origin of BSE? *PLoS Pathog* **3**, e31 (2007).
18. Okada, H., Masujin, K., Imamaru, Y., Imamura, M., Matsuura, Y., Mohri, S., Czub, S., and Yokoyama, T. Experimental transmission of H-type Bovine Spongiform Encephalopathy to bovinized transgenic mice. *Vet Pathol* (2011).
19. Lombardi, G., Casalone, C., D'Angelo, A., Gelmetti, D., Torcoli, G., Barbieri, I., Corona, C., Fasoli, E., Farinazzo, A., Fiorini, M., Gelati, M., Iulini, B., Tagliavini, F., Ferrari, S., Caramelli, M., Monaco, S., Capucci, L., and Zanusso, G. Intraspecies transmission of BASE induces clinical dullness and amyotrophic changes. *PLoS Pathog* **4**, e1000075 (2008).

20. Balkema-Buschmann, A., Ziegler, U., McIntyre, L., Keller, M., Hoffmann, C., Rogers, R., Hills, B., and Groschup, M. H. Experimental challenge of cattle with German atypical bovine spongiform encephalopathy (BSE) isolates. *J Toxicol Environ Health A* **74**, 103–109 (2011).
21. Fukuda, S., Iwamaru, Y., Imamura, M., Masujin, K., Shimizu, Y., Matsuura, Y., Shu, Y., Kurachi, M., Kasai, K., Murayama, Y., Onoe, S., Hagiwara, K., Sata, T., Mohri, S., Yokoyama, T., and Okada, H. Intraspecies transmission of L-type-like Bovine Spongiform Encephalopathy detected in Japan. *Microbiol Immunol* **53**, 704–707 (2009).
22. Dobly, A., Langeveld, J., van Keulen, L., Rodeghiero, C., Durand, S., Geeroms, R., Van Muylem, P., De Sloovere, J., Vanopdenbosch, E., and Roels, S. No H- and L-type cases in Belgium in cattle diagnosed with bovine spongiform encephalopathy (1999–2008) aging seven years and older. *BMC Vet Res* **6**, 26 (2010).
23. Fraser, H., Bruce, M. E., Chree, A., McConnell, I., and Wells, G. A. Transmission of bovine spongiform encephalopathy and scrapie to mice. *J Gen Virol* **73**, 1891–1897 (1992).
24. Richt, J. A., Kunkle, R. A., Alt, D., Nicholson, E. M., Hamir, A. N., Czub, S., Kluge, J., Davis, A. J., and Hall, S. M. Identification and characterization of two bovine spongiform encephalopathy cases diagnosed in the United States. *J Vet Diagn Invest* **19**, 142–154 (2007).
25. Stack, M., Focosi-Snyman, R., Cawthraw, S., Davis, L., Jenkins, R., Thorne, L., Chaplin, M., Everitt, S., Saunders, G., and Terry, L. Two unusual bovine Spongiform encephalopathy cases detected in Great Britain. *Zoonoses Public Health* **56**, 376–383 (2009).
26. Polak, M. P., Zmudzinski, J. F., Jacobs, J. G., and Langeveld, J.P. Atypical status of bovine spongiform encephalopathy in Poland. a molecular typing study. *Arch Virol* **153**, 69–79 (2008).
27. Polak, M. P. and Zmudzinski, J. F. Distribution of a pathological form of prion protein in the brainstem and cerebellum in classical and atypical cases of bovine spongiform encephalopathy. *Vet J* (2011).
28. Taylor, D. M., Brown, J. M., Fernie, K., and McConnell, I. The effect of formic acid on BSE and scrapie infectivity in fixed and unfixed brain-tissue. *Vet Microbiol* **58**, 167–174 (1997).
29. Simmons, M. M., Harris, P., Jeffrey, M., Meek, S. C., Blamire, I. W., and Wells, G. A. BSE in Great Britain. consistency of the neurohistopathological findings in two random annual samples of clinically suspect cases. *Vet Rec* **138**, 175–177 (1996).
30. Sai, S., Hayama, M., and Hotchi, M. A new amyloid stain by phenol Congo red. *Pathol Clin Med* **4**, 1229–1232 (1986). (in Japanese).
31. Okada, H., Sato, Y., Sata, T., Sakurai, M., Endo, J., Yokoyama, T., and Mohri, S. Antigen retrieval using sodium hydroxide for prion immunohistochemistry in bovine spongiform encephalopathy and scrapie. *J Comp Pathol* **144**, 251–256 (2011).
32. Shimizu, Y., Kaku-Ushiki, Y., Iwamaru, Y., Muramoto, T., Kitamoto, T., Yokoyama, T., Mohri, S., and Tagawa, Y. A novel anti-prion protein monoclonal antibody and its single-chain fragment variable derivative with ability to inhibit abnormal prion protein accumulation in cultured cells. *Microbiol Immunol* **54**, 112–121 (2010).
33. Kim, C. L., Karino, A., Ishiguro, N., Shinagawa, M., Sato, M., and Horiuchi, M. Cell-surface retention of PrP^{C} by anti-PrP antibody prevents protease-resistant PrP formation. *J Gen Virol* **85**, 3473–3482 (2004).
34. Horiuchi, M., Yamazaki, N., Ikeda, T., Ishiguro, N., and Shinagawa, M. A cellular form of prion protein (PrP^{C}) exists in many non-neuronal tissues of sheep. *J Gen Virol* **76**, 2583–2587 (1995).
35. Takahashi, H., Takahashi, R. H., Hasegawa, H., Horiuchi, M., Shinagawa, M., Yokoyama, T., Kimura, K., Haritani, M., Kurata, T., and Nagashima, K. Characterization of antibodies raised against bovine-PrP-peptides. *J Neurovirol* **5**, 300–307 (1999).
36. Shimada, K., Hayashi, H. K., Ookubo, Y., Iwamaru, Y., Imamura, M., Takata, M., Schmerr, M. J., Shinagawa, M., and Yokoyama, T. Rapid PrP(Sc) detection in lymphoid tissue and application to scrapie surveillance of fallen stock in Japan. variable PrP(Sc) accumulation in palatal tonsil in natural scrapie. *Microbiol Immunol* **49**, 801–804 (2005).
37. Hills, D., Comincini, S., Schlaepfer, J., Dolf, G., Ferretti, L., and Williams, J. L. Complete genomic sequence of the bovine prion gene (PRNP) and polymorphism in its promoter region. *Anim Genet* **32**, 231–232 (2001).

38. Seuberlich, T., Botteron, C., Wenker, C., Café-Marçal, V. A., Oevermann, A., Haase, B., Leeb, T., Heim, D., and Zurbriggen, A. Spongiform encephalopathy in a miniature zebu. *Emerg Infect Dis* **12**, 1950–1953 (2006).
39. Wells, G. A., Wilesmith, J. W., and McGill, I. S. Bovine spongiform encephalopathy. a neuropathological perspective. *Brain Pathol* **1**, 69–78 (1991).
40. Vidal, E., Marquez, M., Tortosa, R., Costa, C., Serafin, A., and Pumarola, M. Immunohistochemical approach to the pathogenesis of bovine spongiform encephalopathy in its early stages. *J Virol Methods* **134**, 15–29 (2006).
41. Breslin, P., McElroy, M., Bassett, H., and Markey, B. Vacuolar lesion profile of BSE in the Republic of Ireland. *Vet Rec* **159**, 889–890 (2006).
42. González, L., Martin, S., Houston, F. E., Hunter, N., Reid, H. W., Bellworthy, S. J., and Jeffrey, M. Phenotype of disease-associated PrP accumulation in the brain of bovine spongiform encephalopathy experimentally infected sheep. *J Gen Virol* **86**, 827–838 (2005).
43. Sisó, S., González, L., and Jeffrey, M. Neuroinvasion in prion diseases. the roles of ascending neural infection and blood dissemination. *Interdiscip Perspect Infect Dis* **2010**, 747892 (2010).
44. Martin, S., Jeffrey, M., Gonzalez, L., Siso, S., Reid, H. W., Steele, P., Dagleish, M. P., Stack, M. J., and Chaplin, M. J., Balachandran A. Immunohistochemical and biochemical characteristics of BSE and CWD in experimentally infected European red deer (Cervus elaphus elaphus). *BMC Vet Res* **5**, 26 (2009).
45. Miyazono, M., Iwaki, T., Kitamoto, T., Kaneko, Y., Doh-ura, K., and Tateishi, J. A comparative immunohistochemical study of Kuru and senile plaques with a special reference to glial reactions at various stages of amyloid plaque formation. *Am J Pathol* **139**, 589–598 (1991).
46. Hagiwara, K., Yamakawa, Y., Sato, Y., Nakamura, Y., Tobiume, M., Shinagawa, M., and Sata, T. Accumulation of mono-glycosylated form-rich, plaque-forming PrP^{Sc} in the second atypical bovine spongiform encephalopathy case in Japan. *Jpn J Infect Dis* **60**, 305–308 (2007).
47. Jeffrey, M., Martin, S., and Gonzalez, L. Cell-associated variants of disease-specific prion protein immunolabelling are found in different sources of sheep transmissible spongiform encephalopathy. *J Gen Virol* **84**, 1033–1045 (2003).
48. Jeffrey, M., Martin, S., Gonzalez, L., Ryder, S. J., Bellworthy, S. J., and Jackman, R. Differential diagnosis of infections with the bovine spongiform encephalopathy (BSE) and scrapie agents in sheep. *J Comp Pathol* **125**, 271–284 (2001).
49. Bencsik, A. A., Debeer, S. O., and Baron, T. G. An alternative pretreatment procedure in animal transmissible spongiform encephalopathies diagnosis using PrP^{Sc} immunohistochemistry. *J Histochem Cytochem* **53**, 1199–1202 (2005).
50. Furuoka, H., Yabuzoe, A., Horiuchi, M., Tagawa, Y., Yokoyama, T., Yamakawa, Y., Shinagawa, M., and Sata, T. Species-specificity of a panel of prion protein antibodies for the immunohistochemical study of animal and human prion diseases. *J Comp Pathol* **136**, 9–17 (2007).
51. Van Everbroeck, B., Pals, P., Martin, J. J., and Cras, P. Antigen retrieval in prion protein immunohistochemistry. *J Histochem Cytochem* **47**, 1465–1470 (1999).
52. Baron, T., Bencsik, A., Biacabe, A. G., Morignat, E., and Bessen, R. A. Phenotypic similarity of transmissible mink encephalopathy in cattle and L-type bovine spongiform encephalopathy in a mouse model. *Emerg Infect Dis* **13**, 1887–1894 (2007).
53. Biacabe, A. G., Jacobs, J. G., Bencsik, A., Langeveld, J. P., and Baron, T. G. H-type bovine spongiform encephalopathy. complex molecular features and similarities with human prion diseases. *Prion* **1**, 61–68 (2007).
54. Lehmann, S. and Harris, D. A. Blockade of glycosylation promotes acquisition of scrapie-like properties by the prion protein in cultured cells. *J Biol Chem* **272**, 21479–21487 (1997).
55. Priola, S. A. and Lawson, V. A. Glycosylation influences cross-species formation of protease-resistant prion protein. *Embo J* **20**, 6692–6699 (2001).
56. Piro, J. R., Harris, B. T., Nishina, K., Soto, C., Morales, R., Rees, J. R., and Supattapone, S. Prion protein glycosylation is not required for strain-specific neurotropism. *J Virol* **83**, 5321–5328 (2009).

57. Stumpf, M. P. and Krakauer, D. C. Mapping the parameters of prion-induced neuropathology. *Proc Natl Acad Sci U. S. A* **97**, 10573–10577 (2000).
58. Nicholson, E. M., Brunelle, B. W., Richt, J. A., Kehrli, M. E Jr, and Greenlee, J. J. Identification of a heritable polymorphism in bovine PRNP associated with genetic transmissible spongiform encephalopathy. evidence of heritable BSE. *PLoS One* **3**, e2912 (2008).
59. Manuelidis, L. Transmissible encephalopathy agents. virulence, geography and clockwork. *Virulence* **1**, 101–104 (2010).

9 A Novel Form of Human Disease

Ana B. Rodríguez-Martínez, Joseba M. Garrido, Juan J. Zarranz, Jose M. Arteagoitia, Marian M. de Pancorbo, Begoca Atarés, Miren J. Bilbao, Isidro Ferrer, and Ramyn A. Juste

CONTENTS

9.1 INTRODUCTION

The sporadic CreutzfeldtJakob disease (sCJD) is a rare neurodegenerative disorder in humans included in the group of Transmissible Spongiform Encephalopathies or prion diseases. The vast majority of sCJD cases are molecularly classified according to the abnormal prion protein (PrP^{Sc}) conformations along with polymorphism of codon 129 of the prion protein gene (*PRNP*) gene. Recently, a novel human disease, termed "protease-sensitive prionopathy", has been described. This disease shows a distinct clinical and neuropathological phenotype and it is associated to an abnormal prion protein more sensitive to protease digestion.

We report the case of 75-year-old-man who developed a clinical course and presented pathologic lesions compatible with sCJD, and biochemical findings reminiscent of "protease-sensitive prionopathy". Neuropathological examinations revealed spongiform change mainly affecting the cerebral cortex, putamen/globus pallidus and thalamus, accompanied by mild astrocytosis and microgliosis, with slight involvement of the cerebellum. The confluent vacuoles were absent. Diffuse synaptic PrP deposits in these regions were largely removed following proteinase treatment. The PrP deposition, as revealed with 3F4 and 1E4 antibodies, was markedly sensitive to pre-treatment with proteinase K (PK). molecular analysis of PrP^{Sc} showed an abnormal prion protein more sensitive to PK digestion, with a five-band pattern of 28, 24, 21, 19, and 16 kDa, and three aglycosylated isoforms of 19, 16, and 6 kDa. This PrP^{Sc} was estimated to be 80% susceptible to digestion while the pathogenic prion protein associated with classical forms of sCJD were only 2% (type VV2) and 23% (type MM1) susceptible. No mutations in the *PRNP* gene were found and genotype for codon 129 was heterozygous methionine/valine.

A novel form of human disease with abnormal prion protein sensitive to protease and MV at codon 129 was described. Although clinical signs were compatible with sCJD, the molecular subtype with the abnormal prion protein isoforms showing enhanced protease sensitivity was reminiscent of the "protease-sensitive prionopathy". It remains to be established whether the differences found between the latter and this case are due to the polymorphism at codon 129. Different degrees of PK susceptibility were easily determined with the chemical polymer detection system which could help to detect proteinase-susceptible pathologic prion protein in diseases other than the classical ones.

9.2 HISTORY

The sCJD is a fatal neurodegenerative disorder which constitutes the most common form of human transmissible spongiform encephalopathy (TSE) occurring at a rate of 1–1.5 cases per million of the population per annum [1]. The clinical features may vary but classic sCJD cases present a rapidly progressive dementia accompanied by focal neurological signs that progress to akinetic mutism and death within 4–6 months

[2, 3]. Neuropathologic hallmarks are neuronal loss, spongiosis and reactive gliosis, which are variable in nature, severity and location [4, 5]. Two protease resistant PrP^{Sc} types have been described associated with sCJD, both presenting a three-band pattern. diglycosylated, monoglycosylated and aglycosylated. Type 1 is characterised by an aglycosylated isoform of 21 kDa whereas type 2 isoform is 19 kDa in size [6]. Few sCJD cases have been described with PrP^{Sc} conformations different from type 1 and type 2 after PK digestion [7-10]. Two of them were characterized by the absence of the diglycosylated isoform [7, 8], while a unique PrP^{Sc} resistant fragment of 6 kDa size was observed in another case [9]. More recently, a novel human disease, defined by the authors as "protease-sensitive prionopathy" (PSPr) has been described. It showed a distinct clinical and neuropathological phenotype and a more sensitive to PK digestion PrP^{Sc} [10]. Patients presented behavioral and psychiatric manifestations and longer duration of the disease and all of them were valine homozygous. Histopathologically, minimal spongiform degeneration with larger vacuoles than in typical sCJD as well as minimal astrogliosis were described. This lesion profile mainly affected the cerebral neocortex, basal ganglia and thalamus. Abnormal PrP was less resistant to PK digestion and it showed a ladder-like pattern on Western blot, with PrP fragments ranging from 29 to 6 kDa, all detected with Mab 1E4.

9.3 CASE PRESENTATION

9.3.1 Clinical findings

A 74-year-old man presented to his general practitioner in August 2006 complaining of memory loss and was then referred to the neurologist. He showed a rapid global cognitive decline associated with aggressiveness, bizarre behavior and language loss. This was accompanied by severe anomia, disinhibition and a score of 10/30 on MMSE. There were no focal signs, myoclonus or ataxia. The clinical deterioration was very rapid and by December 2006 he was in an akinetic-mutism-like syndrome with abnormal posturing. Two cranial magnetic resonance imaging (MRI), in October and December 2006, including T1, T2, FLAIR, and DWI sequences, showed moderate signs of brain atrophy but no increase in abnormal cortical or basal ganglia signal. The electroencephalogram (EEG) was non-diagnostic and protein 14-3-3 level in the cerebrospinal fluid (CSF) was normal. The patient died in March 2007. Family history of dementia included an 80-year-old brother diagnosed with probable Alzheimer disease.

9.3.2 Genetic findings

No mutations were found in the open reading frame after sequencing the *PRNP*. A heterozygosis methionine valine (MV) was observed in codon 129.

9.3.3 Neuropathology

Moderate-to-mild spongiform change was present in the neocortex, putamen/globus pallidus and thalamus, with the lesions being more evident in the putamen and frontal cortex. Confluent vacuoles were not found in any region. Except for a few focal vacuoles in the deeper molecular layer, the cerebellar cortex was otherwise unremarkable.

Neurons were largely preserved in the cerebral cortex and basal ganglia although focal astrogliosis was seldom observed. Mild-to-moderate microgliosis was present in the cerebral cortex and basal ganglia, and subcortical white matter, respectively. The immunostaining of PrP without PK pre-treatment showed strong staining characterized by fine punctate deposits (synaptic-like) and irregular granular, often confluent, deposits that could be categorized as diffuse synaptic. Perineuronal and cerebellar plaque-like deposits, kuru plaques and florid plaques were absent. Following PK treatment, the vast majority of staining disappeared, except a few granular PrP PK-resistant deposits. The cerebellum showed a discrete PrP synaptic-like pattern in the molecular and granular layers that vanished after PK pre-treatment. The sensitivity to PK pre-treatment was best visualized in consecutive sections with and without pre-treatment with PK (**Figure 1**). Parallel sections stained with the 3F4 antibody showed marked reduction of PrP immunoreactivity, as evaluated by densitometry, involving 70–80% of the total PrP in tissue sections. This was further confirmed by incubating tissue sections with the 1E4 antibody, and comparing the PrP immunohistochemical pattern of one sCJD MV1 case with the proband. As shown in **Figure 2**, 3F4 and 1E4 synaptic PrP immunoreactivity in the common MV1 case showed resistant PrP immunoreactivity. In contrast, 3F4 and 1E4 immunoreactivity was practically abolished after PK pre-treatment in the proband.

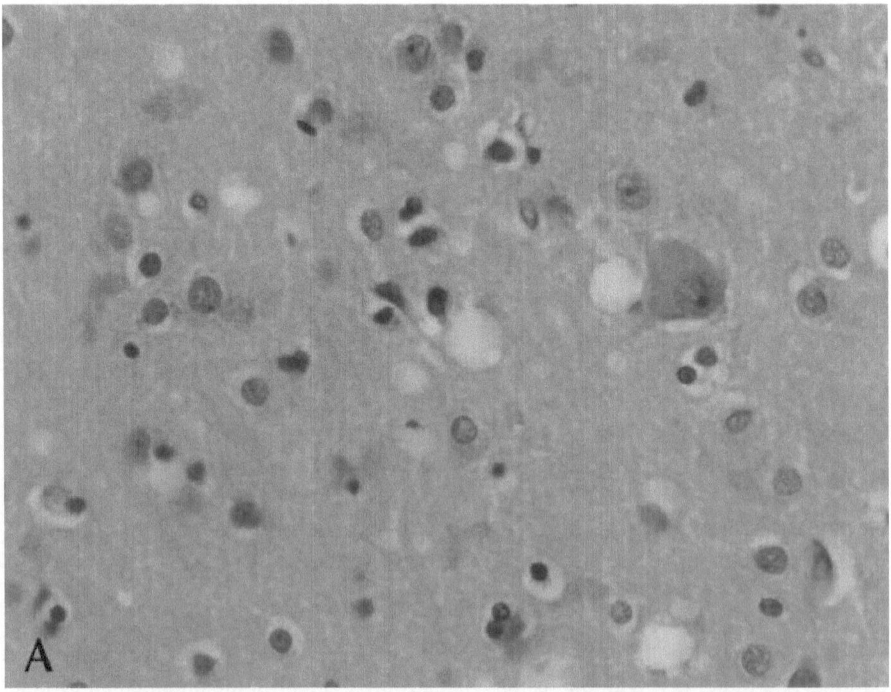

FIGURE 1 *(Continued)*

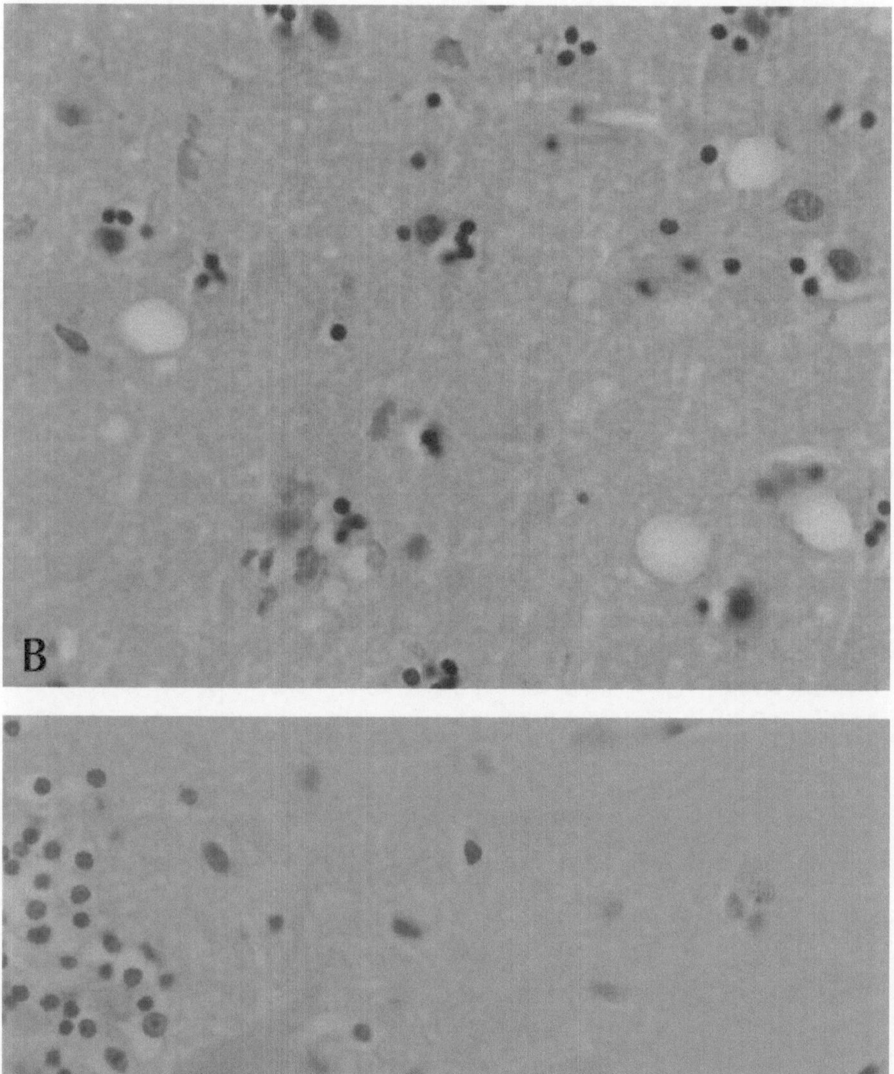

FIGURE 1 *(Continued)*

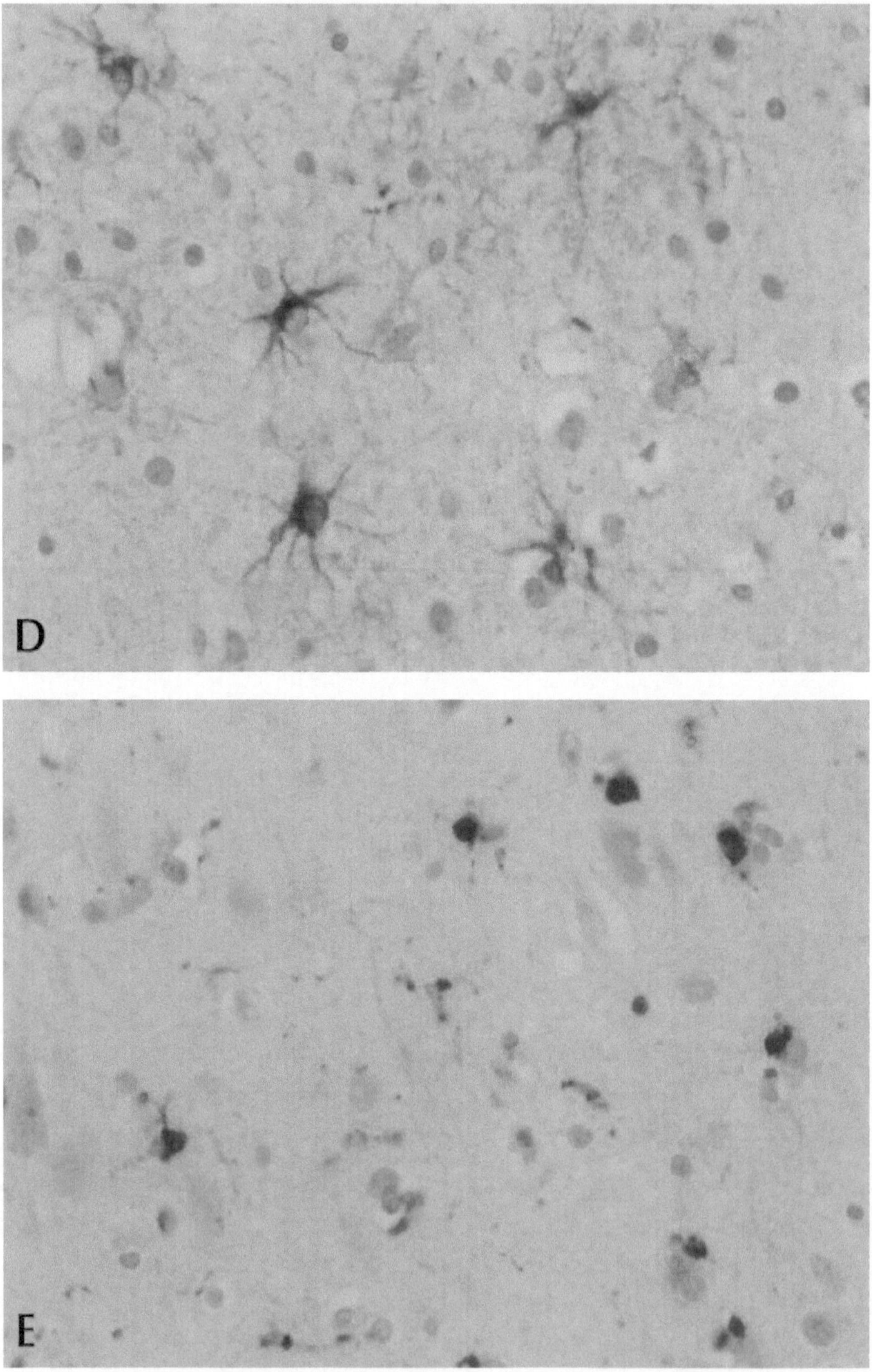

FIGURE 1 *(Continued)*

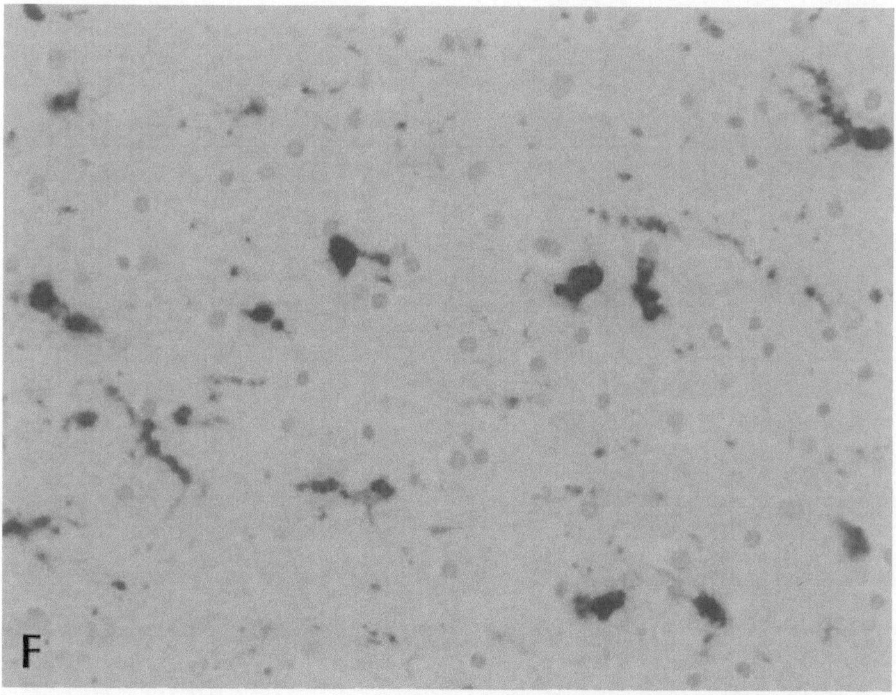

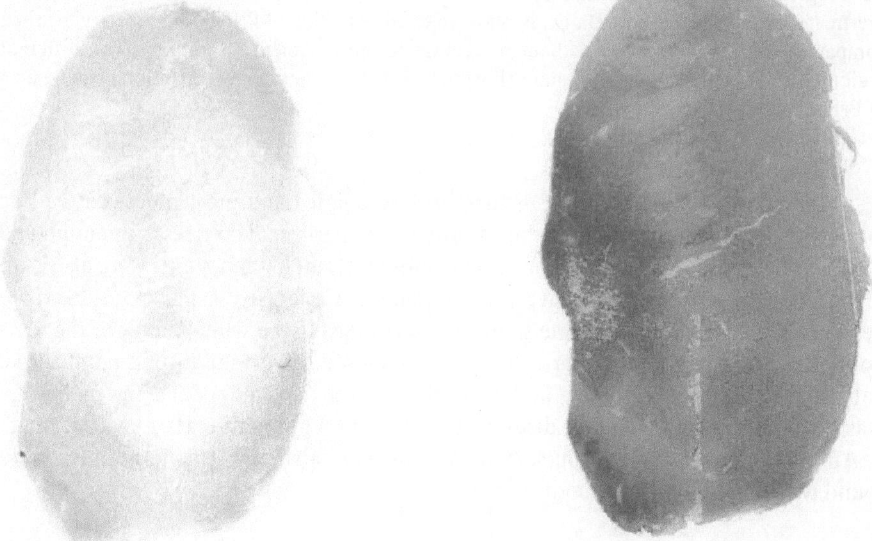

FIGURE 1 *(Continued)*

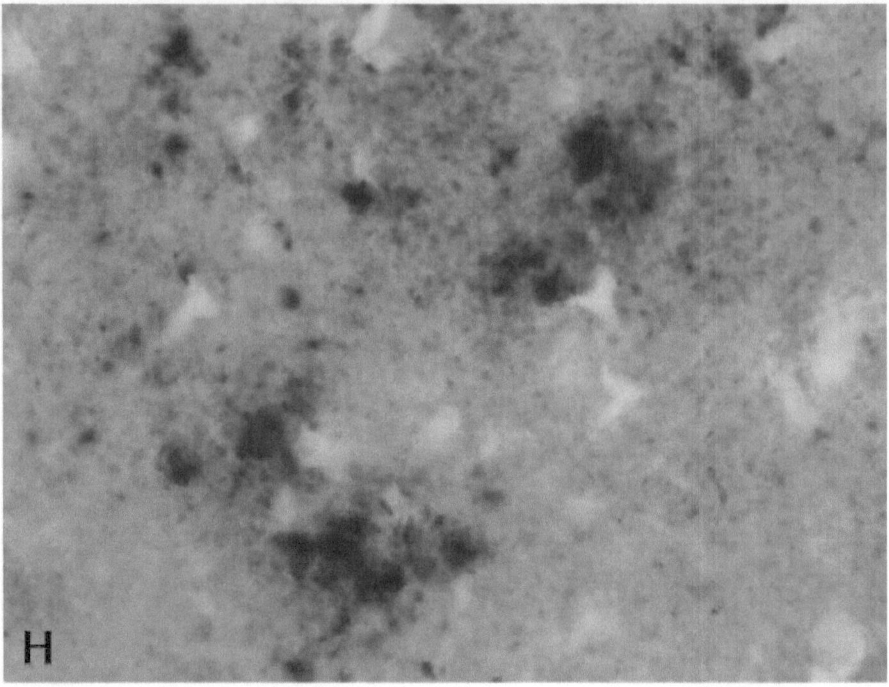

FIGURE 1 3F4 immunohistochemistry without and with proteinase K pre-treatment in the same regions of consecutive serial sections. The parallel (A., B., C., D., and E., F) cortical regions pre-treated with proteinase K (B., D., F) show marked reduction of PrP immunoreactivity when compared with serial sections without proteinase K pre-treatment (A., C., E). The different regions with variable amounts of total PrP were selected in order to have a comprehensive idea of PrP sensitivity.

In addition to these changes, neurofibrillary tangles and pre-tangles, as well as granules (grains), were present in the entorhinal and perirhinal cortices, subiculum and CA1 and CA3 regions of the hippocampus. A few pre-tangles and grains were also seen in the amygdala. These changes were accompanied by a few hyper-phosphorylated tau deposits in neurons of the dentate gyrus, coiled bodies in the white matter of the temporal lobe, and peri-ventricular astrocytes. Scattered αB-crystallin-immunoreactive ballooned neurons were present in the entorhinal cortex and amygdala. Tau pathology was consistent with Alzheimer disease stage III and argyrophilic grain disease stage 3. Amyloid plaques and α-synuclein inclusions were absent. No abnormalities were found with anti-TDP-43 antibodies.

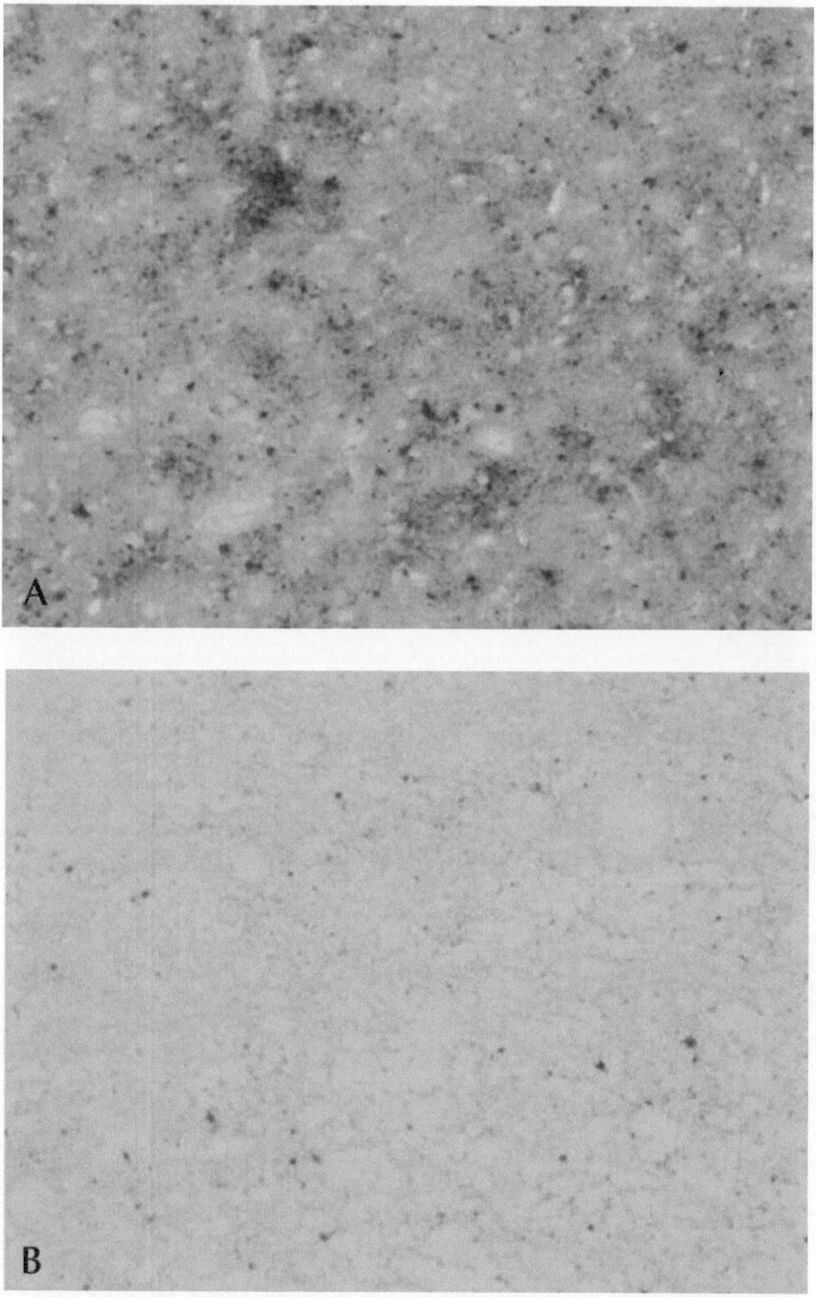

FIGURE 2 *(Continued)*

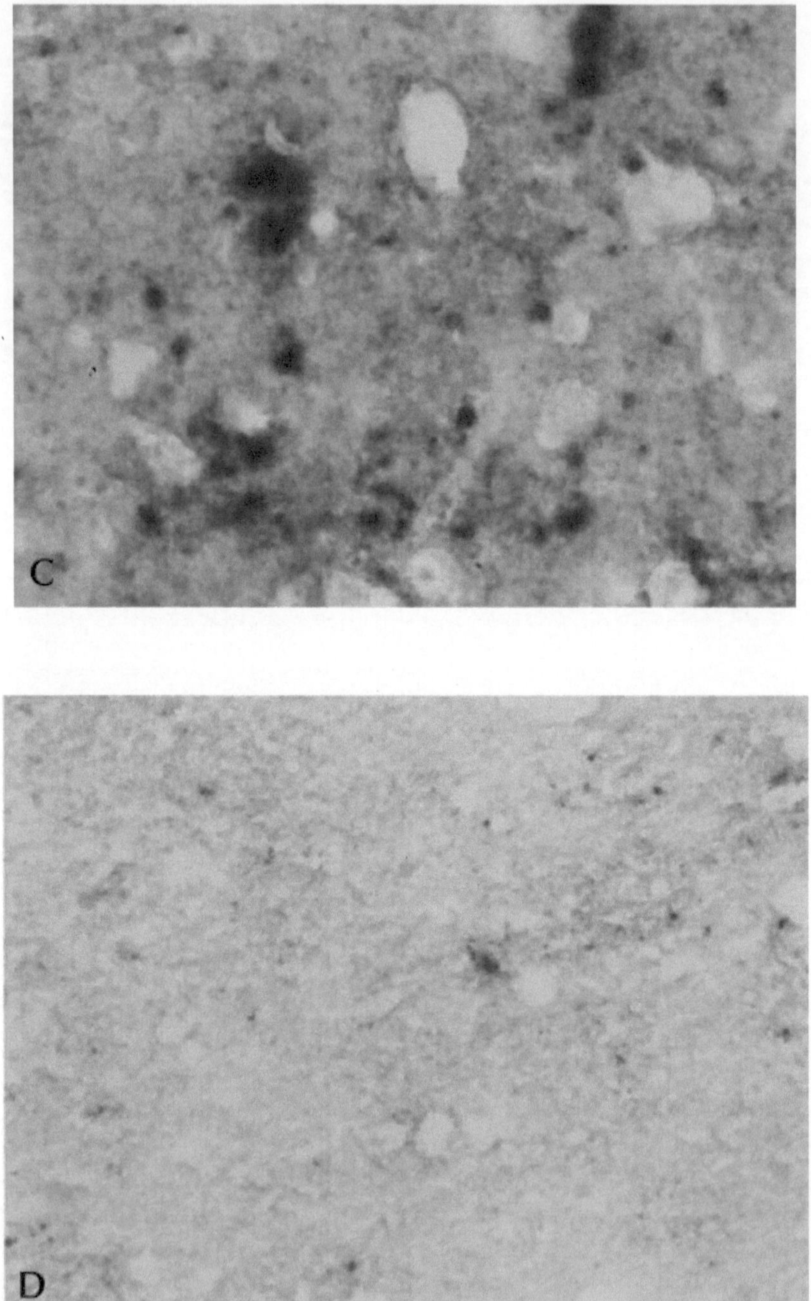

FIGURE 2 *(Continued)*

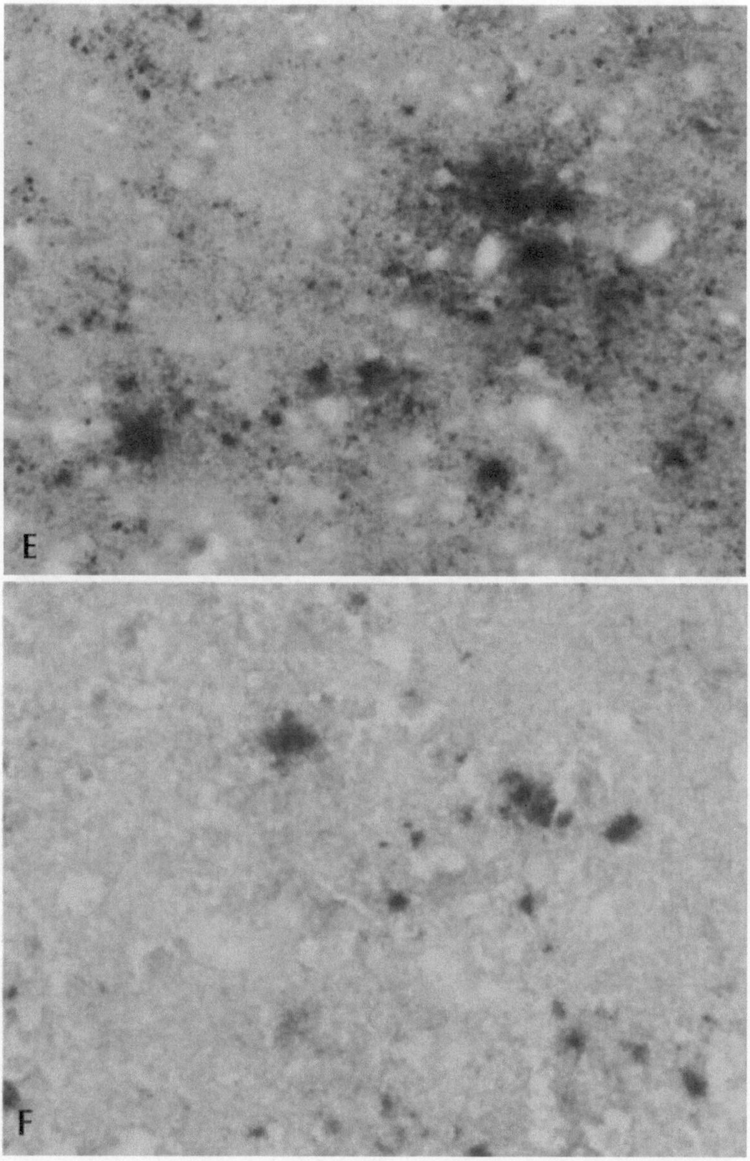

FIGURE 2 3F4 and 1E4 immunohistochemistry in MV1 sCJD and the proband. 3F4 (A., B., E., F) and 1E4 (C., D., G., H) immunohistochemistry without (A., C., E., G) and with (B., D., F., H) PK pre-treatment in the cerebellum of one case of sCJD MV1 with synaptic PrP deposition (A–D) and in the cerebellum of the proband (E–H) show different patterns, when PK-treatment was performed. In the MV1 sCJD case, synaptic PrP deposition, as revealed with 3F4 and 1E4 antibodies, is observed in the molecular and granular layers of the cerebellum. PrP immunoreactivity is largely resistant to the treatment with PK. Synaptic PrP immunoreactivity with the antibodies 3F4 and 1E4 is also found in the molecular and granular layer in the proband, however, immunostaining is lost following incubation with PK. Notice that formic acid treatment did not seem to modify PK susceptibility.

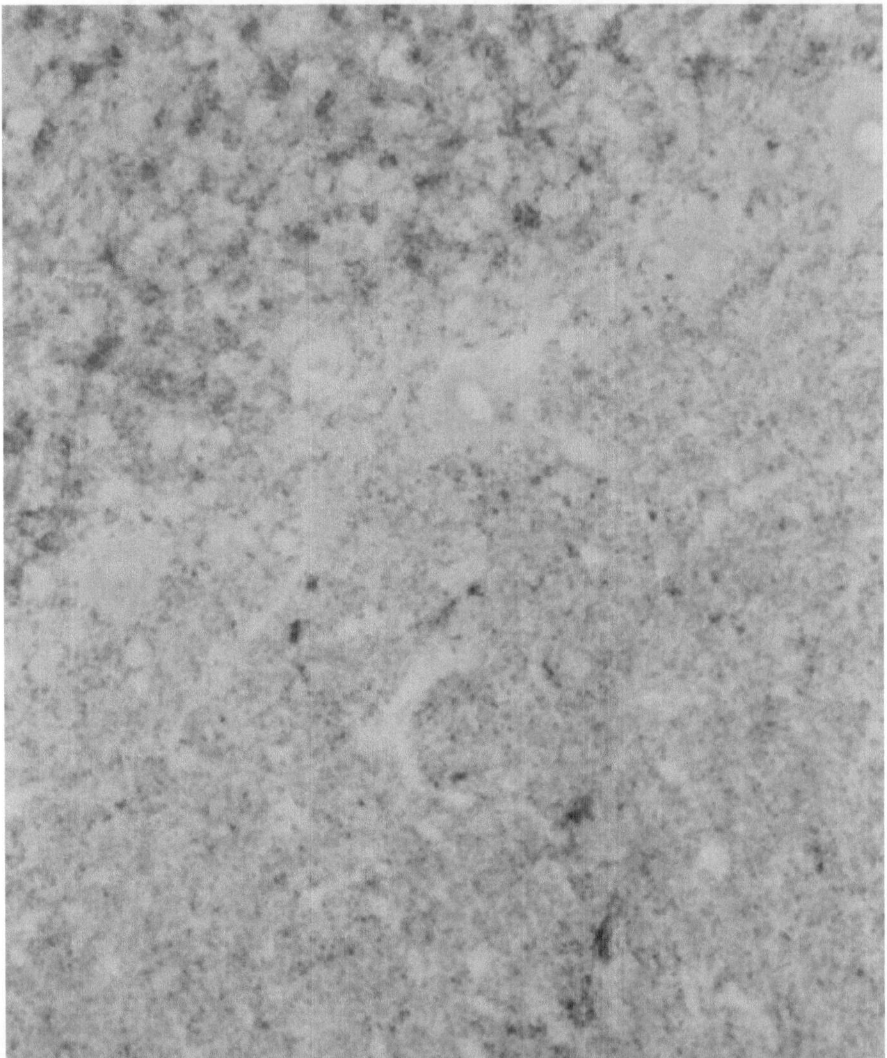

FIGURE 3 *(Continued)*

FIGURE 3 *(Continued)*

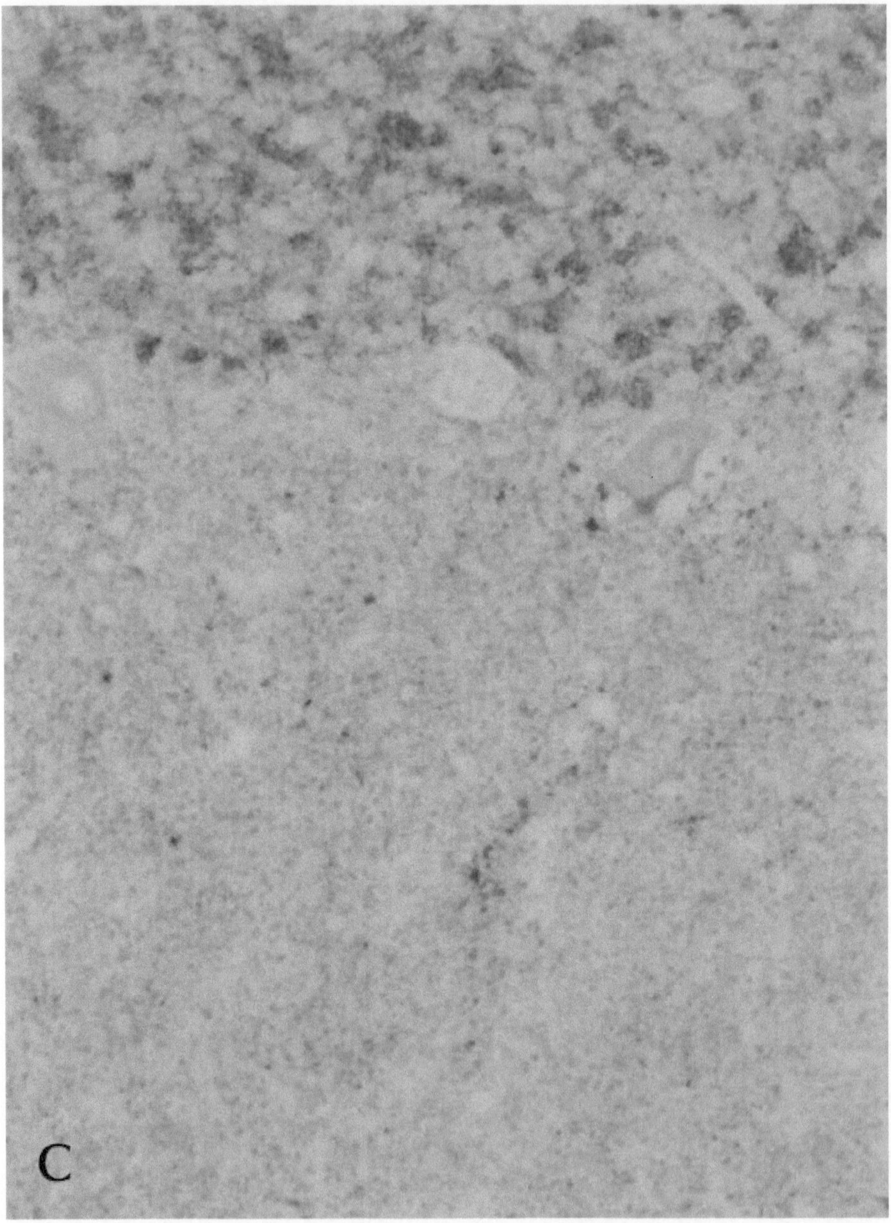

FIGURE 3 *(Continued)*

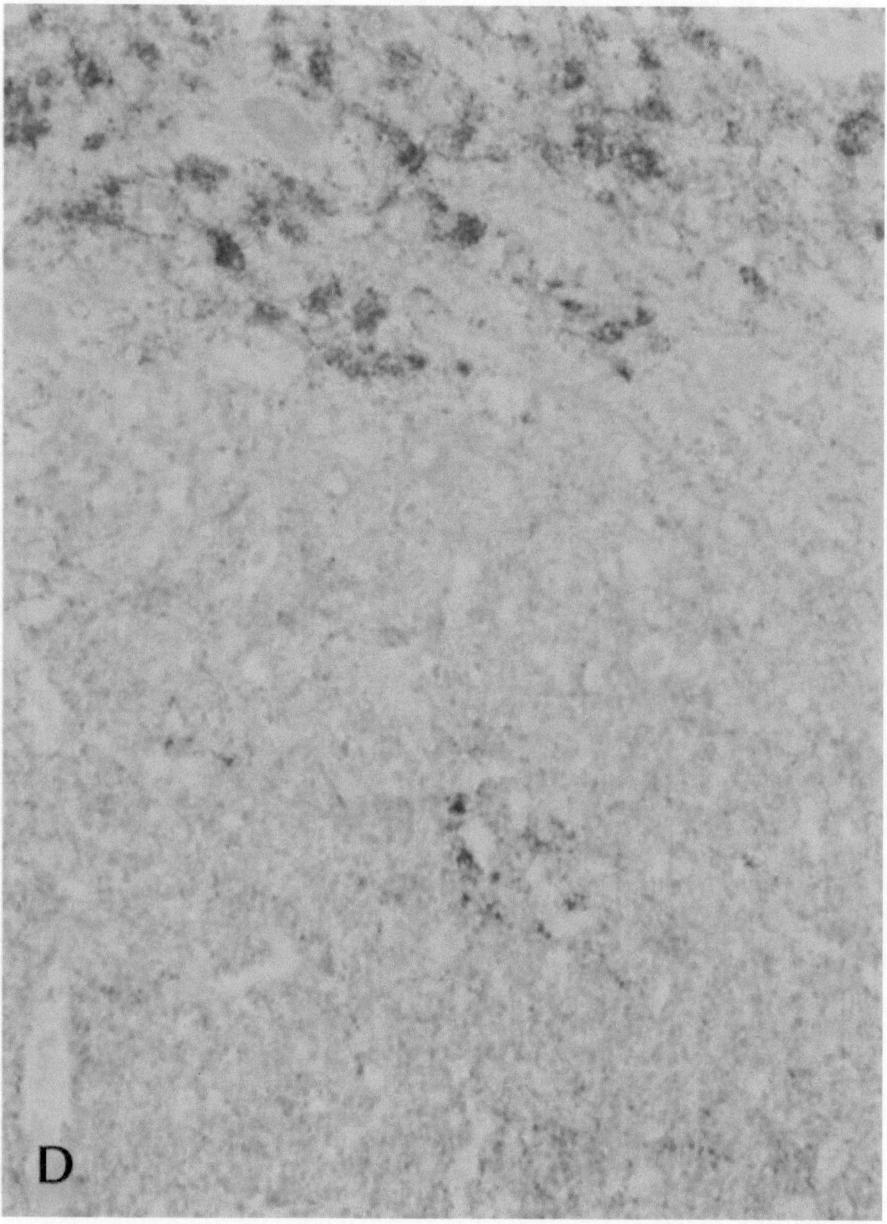

FIGURE 3 *(Continued)*

FIGURE 3 *(Continued)*

FIGURE 3 *(Continued)*

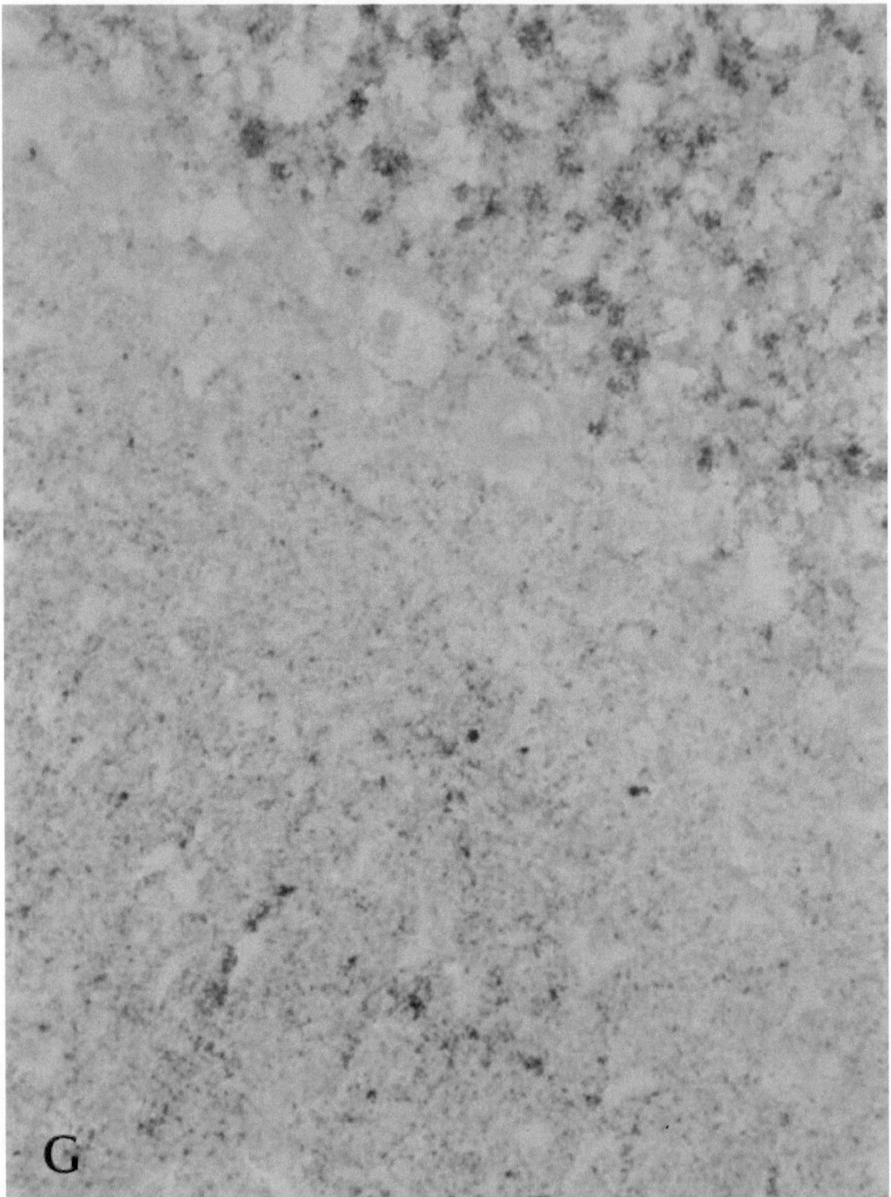

FIGURE 3 *(Continued)*

FIGURE 3 Main neuropathological findings. A. Mild spongiform change in the frontal cortex, B. Mild spongiform change in the putamen/globus pallidus characterised by predominance of large vacuoles, C. A few small vacuoles in the vicinity of a Purkinje cell, D. Focal astrocytosis in the cerebral cortex, E and F. Microgliosis with globular reactive microglia in the cerebral cortex and subcortical white matter, respectively, G. PrP immunostaining with and without PK pre-treatment in the putamen/globus pallidus. PrP immunoreactivity practically disappears in PK-treated section. H. PrP immunostaining in the cerebral cortex without PK pre-treatment showing PrP-positive punctate (synaptic-like) deposits and large granular confluent deposits forming coarse plaque-like accumulations, I. A few PrP-immunoreactive granular deposits are seen in sections after PK pre-treatment. A–C. Haematoxylin and eosin, D. GAFP immunohistochemistry, E and F. CD68 immunostaining, G-I. PrP immunohistochemistry.

9.3.4 Biochemical analysis

The standard PrP Western-blot procedure (10% brain homogenate and final PK concentration of 440 μg/ml) failed to detect PrP^{Sc}. Increasing the volume loaded into the gel from 5 to 10 μl yielded an extremely weak signal corresponding to 24 and 19 kDa under saturating film exposure times. The decreasing PK concentrations (440, 100, and 50 μg/ml) showed an increase in the PrP^{Sc} signal, which was suggestive of a PK-sensitive prion protein. Even then, only two bands of 24 and 19 kDa were visible. After increasing the brain homogenate percentage to 20%, the same two-band pattern was obtained.

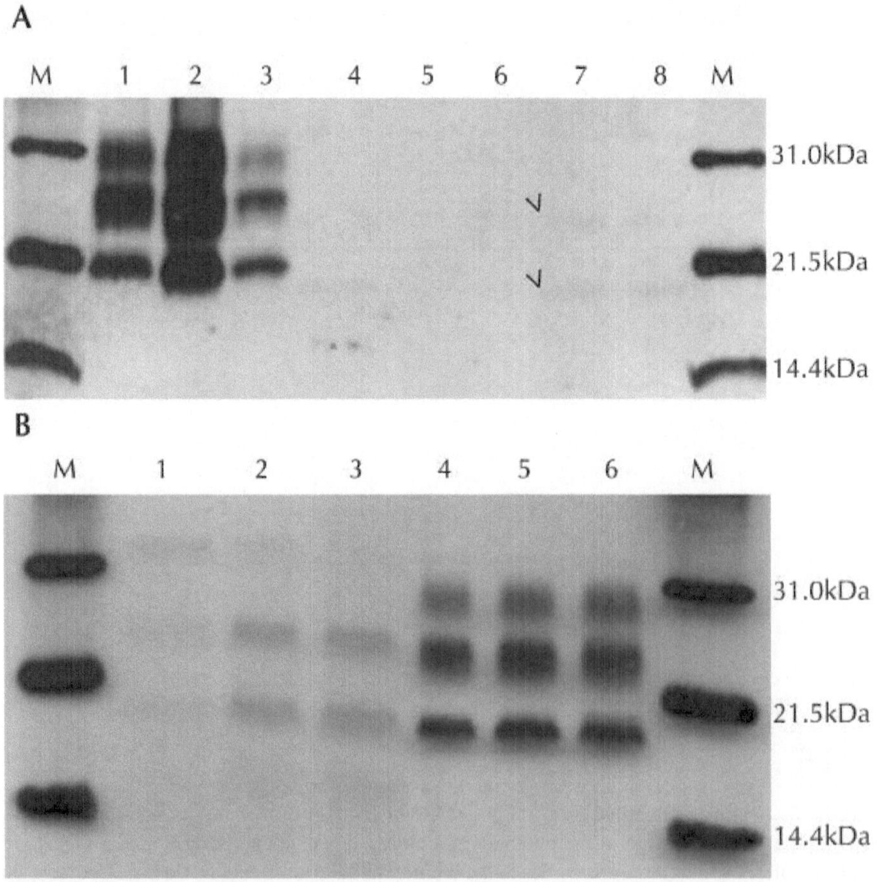

FIGURE 4 Immunoblot images under standard conditions. A. 10% brain homogenate digested with 440 μg/ml PK final concentration and incubated with Mab 3F4. M. Molecular weight marker. 1–3. sCJD MM1 thalamus, frontal and temporal cortex, respectively. 4–8. Occipital cortex, cerebellum, parietal cortex, frontal cortex and temporal cortex of the patient, respectively. Film exposure time 5 min. Arrowheads indicate band position. B. 10% brain homogenate digested with decreasing PK concentration. 440 μg/ml (1, 4), 100 μg/ml (2, 5) and 50 μg/ml (3, 6) and incubated with Mab 3F4. Temporal cortex of the patient (1–3) and occipital cortex of a sCJD VV2 case (4–6).

Using the TeSeE® kit, characterized by softer PK digestion conditions followed by steps of purification and concentration of the protein and staining with Sha31 Mab, the presence of two unexpected bands of 21 and 16 kDa was revealed. This band profile was observed in all the brain regions and it constituted a striking result, since their molecular weight was different from that previously detected with Mab 3F4 and 6H4. Performing a combination of digestion, purification and concentration of the sample according to TeSeE® kit recommendations, along with detection using 3F4 and 6H4, yielded a novel pattern. Not only the previous bands of 24, 21, 19, and 16 kDa were present in each of the samples, but also a very weak band of 28 kDa and a fragment of approximately 6 kDa size were observed in some brain regions. Furthermore, differences of signal intensity were obtained with 3F4 and 6H4 antibodies suggesting differential affinity for PrP^{Sc} which could be interpreted as a different protein conformation in which the 3F4-binding epitope was more exposed than the 6H4 one.

Deglycosylation analysis revealed three aglycosylated isoforms of 19, 16, and 6 kDa, which were more intense in the cortex (parietal, frontal, and temporal) and weaker in the occipital cortex and putamen/globus pallidus. In the thalamus region, two bands were detected, a more intense one of 19 kDa and a weaker one of 16 kDa. Finally, the cerebellum was the only region where a single aglycosylated band of 19 kDa was observed, similar to that found in sCJD type 2. However, we cannot rule out the possibility that this finding was the result of the presence of a small amount of PrP^{Sc} and an underrepresentation of the other bands, as observed in the thalamus, where a 16 kDa size band was only observed under longer film exposure times.

Evaluation of sensitivity to PK digestion was achieved by measuring the absorbance of PrP^{Sc} before and after treatment with PK using the IDEXX HerdChek BSE Test. This technology is based on selective PrP^{Sc} capture by a specific chemical polymer through polyionic interactions in the presence of PrP^{C} from a brain homogenate sample. Absorbance values decrease with serial dilutions, so it can be assumed that the quantity of PrP^{Sc} is directly proportional to the absorbance. The goal of this protocol was to perform relative quantification of PrP^{Sc} without treatment with proteases. We consider that the introduction of a digestion step could be useful to easily evaluate the relative resistance to PK digestion. The results showed that the absorbance values decreased after PK treatment in all the samples (Table 1). For multi-infarct encephalopathy (MIE) and sCJD VV2, the signal detection was reduced in a 4.32% and 2.02%, respectively, but the reductions were not statistically significant. By contrast, samples from the proband and sCJD MM1 showed a statistically significant ($p < 0.005$) reduction of the signal in a 79.82% and 22.68%, respectively. Absorbance values for MIE were below the cut-off, at the same level as negative controls. The remaining values were above the cut-off.

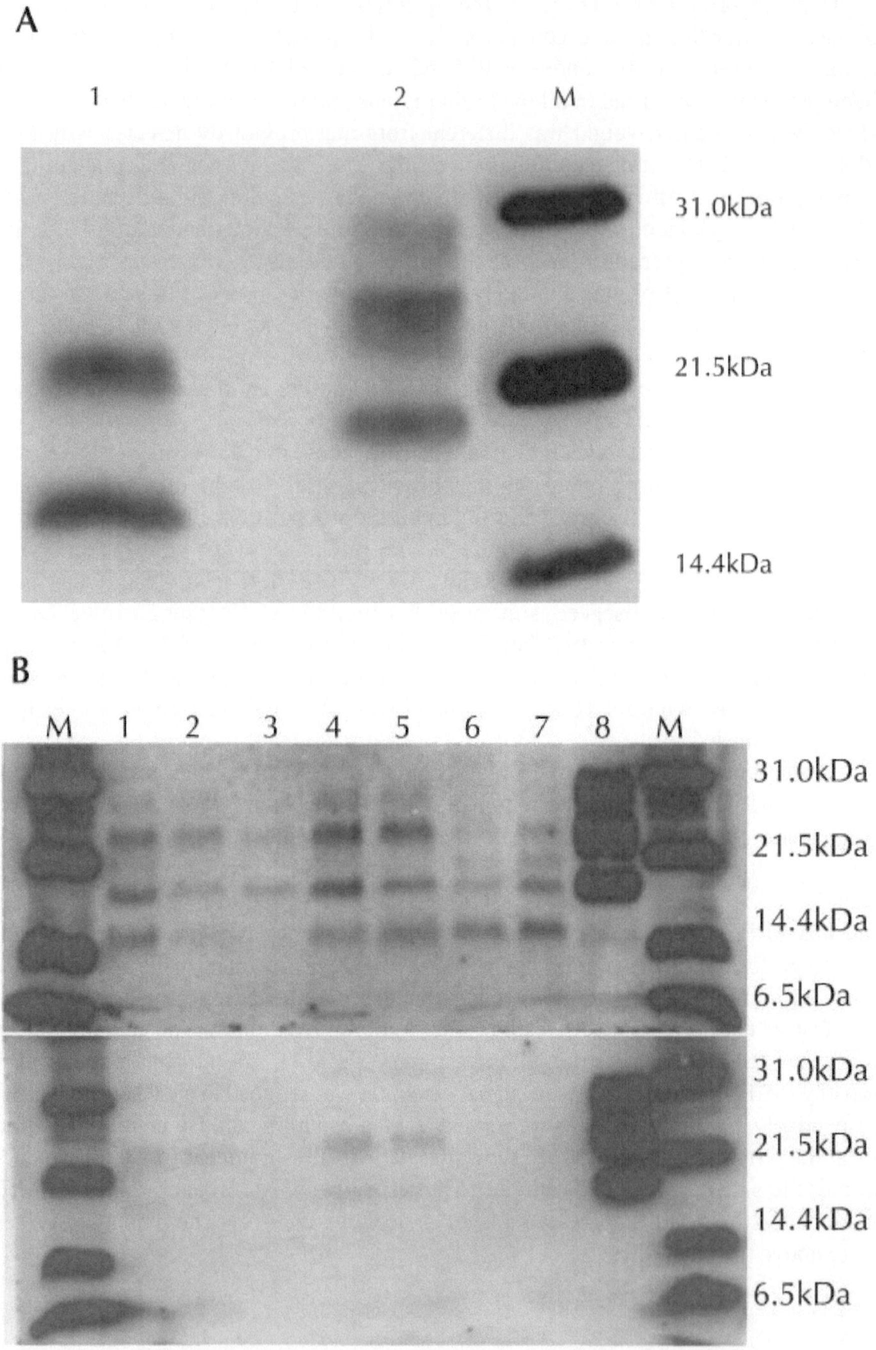

FIGURE 5 *(Continued)*

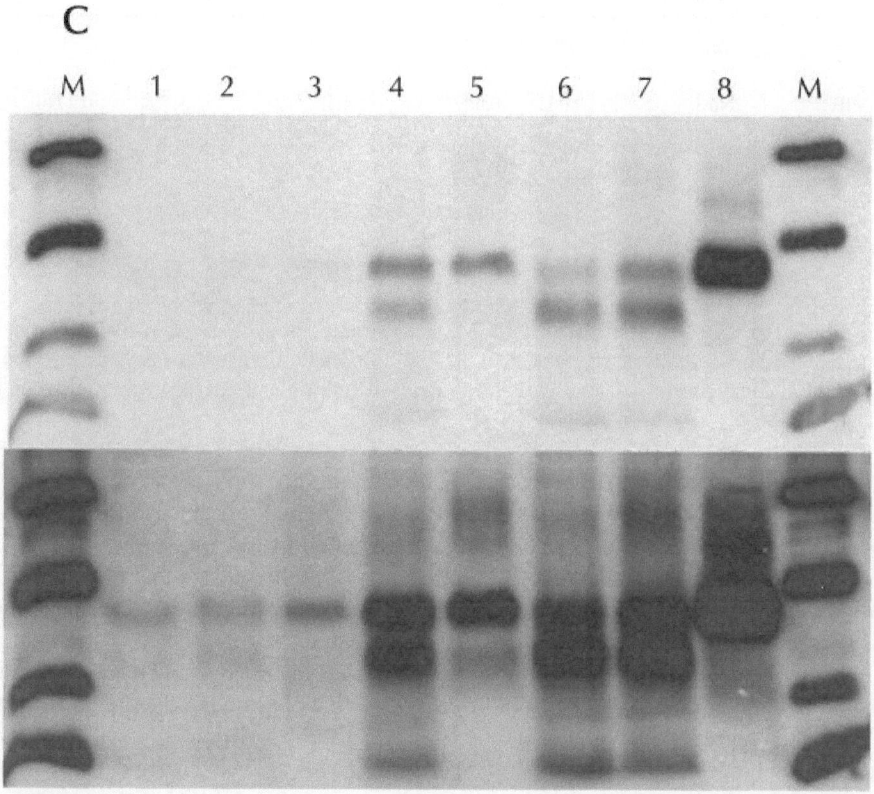

FIGURE 5 Immunoblot images under modified methodology. A. Immunoblot with TeSeE® Western kit of patient occipital cortex (1) and sCJD VV2 control (2) incubated with Mab Sha31. B. Immunoblot of TeSeE® digested and purified samples incubated with Mab 3F4 (upper) and 6H4 (lower). C. PNGase F digestion and detection with Mab 3F4. Film exposure time. 1 min (upper) and 3 min (lower). M. molecular weight marker. 1–8, respectively. occipital cortex, putamen/globus pallidus, cerebellum, parietal cortex, thalamus, frontal cortex, temporal cortex and sCJD VV2 control.

TABLE 1 Results of the evaluation of proteinase K digestion susceptibility.

Case	PK treatment	Mean	Std. deviation	Signal		
				Reduction (%)	P value	P value
Proband	No	1.189	0.698	**79.82**	**<0.0001**	**<0.0001**
	Yes	0.240	0.140			
sCJD MM1	No	3.737	0.206	**22.68**	**0.0041**	**0.0041**

TABLE 1 *(Continued)*

	Yes	2.890	1.076			
MIE	No	0.062	0.009	4.37	0.2740	0.2740
	Yes	0.059	0.004			
sCJD VV2	No	3.629	0.115	2.02	0.1133	0.1133
	Yes	3.556	0.152			

Proband. The case reported here, sCJD MM1. Sporadic Creutzfeldt-Jakob disease type 1 methionine homozygous at codon 129, sCJD VV2. Sporadic Creutzfeldt-Jakob disease type 2 valine homozygous at codon 129, MIE. A case of multi-infarct encephalopathy used as a negative control

These differentiated levels of signal reduction are indicative of three levels of resistance to PK digestion. high, intermediate and low. A high resistance to PK digestion would be represented by a low percentage of signal reduction, as observed for sCJD VV2. In this case, the reduction of 2% in the signal would indicate that PK digestion would only degrade a minimal fraction of PrP^{Sc}, thus suggesting high resistance of the abnormal prion protein. Intermediate resistance would be represented by a slightly higher percentage of signal reduction as observed in sCJD MM1, in which 22% would point to a higher degradable fraction of PrP^{Sc} than that observed in the previous case. This would represent a protein type only slightly sensitive to degradation with proteases, depending on the brain region. Further investigations are being carried out in order to elucidate whether this level of degradation is associated with MM1 protein type or a phenomenon specific to this subject. Finally, low resistance to PK digestion would be represented by a high percentage of signal reduction, for example the 79% observed in the proband, indicating that a high fraction of PrP^{Sc} is degradable. This suggests the existence of abnormal prion protein types extremely susceptible to protease digestion that might potentially be overlooked by detection methods based on the characteristic proteinase resistance of the pathologic prion protein.

9.4 METHODS

9.4.1 Clinical findings

The patient was subjected to standard clinical, electroencephalographic and MRI examinations.

9.4.2 Genetic findings

Analysis of *PRNP* was performed by standard sequencing methods.

9.4.3 Neuropathology

Only the brain was removed at autopsy for neuropathological and biochemical examination. Following the recommended safety guidelines, fresh samples from seven areas (occipital, frontal, parietal and temporal cortex, putamen/globus pallidus, thalamus and cerebellum) were processed for neuropathological analyses. Selected samples of the cerebral cortex, putamen/globus pallidus, thalamus, cerebellum and brain stem fixed in 4% formalin were treated with formic acid, and then post-fixed in formalin and embedded in paraffin. De-waxed sections were stained with haematoxylin and eosin and Klüver-Barrera, or processed for immunohistochemistry, following the En Vision+ system method, for glial fibrillary acidic protein (GFAP), CD68 for microglia, hyper-phosphorylated tau epitopes (antibody AT8), 3Rtau and 4Rtau, β-amyloid 1-40 and β-amyloid 1-42, α-synuclein, αB-crystallin, ubiquitin, TDP-43, and prion protein (antibodies 3F4 and 1E4) without and with PK pre-treatment. Densitometry of immunohistochemical sections not counterstained with haematoxylin was analysed by using modified Total Laboratory v2.01 software. Measurements were expressed as arbitrary units in parallel PrP immunostained sections without and with PK pre-treatment. The results were presented as a percentage of decreased immunoreactivity of PK-treated in comparison with PK-untreated sections.

9.4.4 Biochemical analysis

The eight brain regions corresponding to cortex (frontal, temporal, occipital, and parietal), cerebellum, caudate nucleus, thalamus, and putamen/globus pallidus were analysed. The 10% and 20% (w/v) brain homogenates were prepared in lysis buffer [21]. The homogenates were cleared by centrifugation at 2100 rpm and 4°C (Heraeus Biofuge Fresco) for 5 min. Supernatants were treated with different final concentrations of PK (440, 100, and 50 μg/ml) for 60 min at 37°C. The reaction was terminated by the addition of Pefabloc SC (Roche-Diagnostics) to a final concentration of 1 mM. An equal volume of 2× loading buffer (modified from [21] 125 mM Tris-HCL pH 7, 4% SDS, 20% glycerol, 0, 02% bromophenol blue, 200 mM DTT) was added and the samples were denatured at 96°C for 8 min before electrophoresis on 16% SDS-Tris-glycine gels (5% stacking) for 90 min at 150 V. The gels were electroblotted onto PVDF membrane (Immobilon-P, Millipore) and blocked as described elsewhere [21]. After a short wash in PBST 1×, membranes were incubated either with anti-PrP monoclonal antibody 3F4 (epitope 109-112. MKHM) (Sigma) (1:20.000) or 6H4 (epitope 144-152 DYEDRYYRE) (Prionics), diluted 1:5.000 for 1 h. Following a washing step in PBST 1× for 45 min, membranes were incubated with an alkaline phosphatase conjugated goat anti-mouse IgG antibody (Sigma) diluted 1:10.000, and secondary against 6H4 (Prionics, dilution 1:5000) respectively in PBST 1× for 1 h at room temperature or at 4°C overnight. After a washing step of 45 min in PBST 1× and equilibration in 200 mM Tris-HCl, 10 mM MgCl2, pH 9, 8 [22] for 5 min, membranes were developed in chemiluminescent substrate (CDP-STAR, Tropix) and visualized on X-Omat AR film (Kodak).

In addition, samples were examined by TeSeE® Western Blot (Bio-Rad) following the manufacturer's recommendations. Briefly, 20% brain homogenate was incubated with PK and detergent solution for 10 min at 37°C before addition of buffer B. After a short mixture, samples were centrifuged at 15,000g for 7 min. The pellet was solubilised in 1× loading buffer by incubating at 100°C for 5 min. Samples were then centrifuged at 15,000g for 15 min and supernatants were denatured at 100°C for 4 min before electrophoresis. Electrophoresis separation was performed as described. Proteins were transferred onto a PVDF membrane at 115 V for 60 min and 4°C. Following transfer, the membrane was soaked successively with PBS buffer, ethanol, and distilled water, and then saturated for 30 min with blocking solution. The membrane was incubated for 30 min at room temperature with monoclonal antibody Sha31 against epitope YEDRYYRE (145-152, huPrP), diluted 1:10 in PBST. The following a washing step with PBST, the membrane was incubated for 20 min with goat anti-mouse IgG antibody conjugated to horseradish peroxidase diluted 1:10 in PBST. Finally, membranes were developed in chemiluminescent substrate (Western Blotting detection system, ECL, Amersham) and visualised on film.

A combination of both protocols was also used. In such cases, samples were digested and purified according to TeSeE® Western Blot procedure and incubation with monoclonal antibodies 3F4 and 6H4 was performed as described earlier.

9.4.5 Deglycosilation analysis

In order to detect the non-glycosylated isoforms, samples (either PK digested or purified with TeSeE®) were subjected to PNGase F (New England Biolabs) digestion overnight at 37°C and PrP^{Sc} was recovered as described elsewhere [23].

9.4.6 Evaluation of sensitivity to PK digestion

For the evaluation of sensitivity to PK digestion the IDEXX HerdCheck BSE Test was performed according to the manufacturer's instructions [24] with modifications. Tissue samples from eight brain regions of four cases were analysed. These cases included a sCJD control MM1, sCJD control VV2, MIE as prion disease negative control and the proband. In brief, 0.25 g of tissue was homogenized in a tissue-disruption tube for two cycles of 23 s at 6500 rpm in a homogenizer. A fraction of brain homogenate was treated with 100 μg/ml PK (Sigma) for 1 hr at 37°C. The reaction was stopped by adding 1 mM Pefabloc SC (Roche). After this, 100 μl of homogenized samples (PK treated and untreated) were then diluted with 25 μl of working plate diluent, mixed by pipetting six times and transferred (100 μl) to the BSE antigen capture enzyme immunoassay plate. The plate was incubated at 34°C for 20 min at 200 rpm in a Thermo shaker PHMP-4 (Grant Instruments, Cambridge Ltd) in order to allow the disease-associated conformer (PrP^{Sc}) to bind to the immobilized ligand with high affinity. The plate was washed three times with 1X Wash 1 in a Biotek ELx50™Microplate washer to remove unbound materials, including PrP^{C}. The plate was then incubated with 100 μl of CC-conjugate for 25 min at 34°C and washed five times with 1X Wash 2. Finally, the plate was incubated with 100 μl horseradish peroxidase (HRPO) substrate for 15 min at 34°C in the dark prior to reading the optical density at 450 nm and 650 nm with a plate reader (Model SUNRISE, TECAN). Color development was related to the relative amounts of PrP^{Sc} captured by the ligand immobilized in the microtiter plate well. Negative controls, positive controls and samples were analysed in duplicate. Results were analysed with Magellan V6.3 software (Tecan Austria GmbH). Calculations of negative controls means (NCmean) were automatically made by the software according to the formula (NCmean = (A1 (A_{450} - A_{650}) + B1 (A_{450} - A_{650})) / 2 where A1 and B1 are the plate wells for negative control. The value corresponding to (A_{450} - A_{650}) was calculated and applied as a correction factor to the absorbance values of the samples.

The statistical significance of mean comparisons was checked with the Student's t-test for independent samples.

9.5 DISCUSSION

The present report describes a case of a novel human disease with abnormal prion protein sensitive to protease and MV heterozygosity at polymorphic codon 129 of the *PRNP* gene. A clinical picture of memory impairment as a first symptom, followed by a rapid evolution leading to an akinetic mutism in a 7-month course, was compatible with sCJD [11]. However, post-mortem examinations showed that neuropathological and biochemical findings did not neatly conform to any of the principal subtypes (MM/MV1, VV2, MV2K with kuru type amyloid plaques, MM/MV2C with predominant cortical pathology with confluent vacuoles and perivacuolar PrP staining, MM2T with prominent thalamic pathology and atrophy, and VV1) [2, 6, 12].

In reference to biochemical and molecular findings, two main striking biochemical features were observed. Sensitivity-to-protease, resulting in an extremely weak PrP^{Sc} signal in immunoblotting, and the multiband profile. Digestion with decreasing concentrations of PK revealed a more PK-sensitive protein than that observed in the control. Further data supporting this was obtained by means of a methodology based on the selective capture of PrP^{Sc} by a specific chemical polymer. We first tested this technique in controls in order to determine whether the anti-PrP specific antibody could detect human PrP^{Sc} strains. After demonstrating that the technique was applicable to human samples, we treated homogenates with PK prior to submitting the samples (PK treated and untreated) to subsequent analyses. Data obtained this way showed that treatment with proteases reduced the absorbance values proportionally to PrP^{Sc} capture. This suggested a degradation of a fraction of PrP^{Sc} molecules, which was minimal when the protein type was highly resistant, intermediate when it was slightly susceptible and variable depending on the brain region, and maximal when it was highly susceptible. Furthermore, treatment with milder PK digestion conditions and detection with Mab 3F4 showed a biochemical profile of five to six bands of 28, 24, 21, 19, 16, and 6 kDa. Differential affinity of antibodies 3F4 and 6H4 for PrP^{Sc} suggested a protein conformation on which epitope-recognizing Mab 3F4 was more exposed than that of Mab 6H4. This protein conformation appeared to be different from that of controls since it did not show this unequal affinity for these antibodies. The deglycosylation analysis revealed the presence of up to three aglycosylated isoforms of 19, 16, and 6 kDa, suggesting the coexistence of several PrP^{Sc} strains [13-15].

Recently, a novel human prion disease defined as a protease-sensitive prionopathy (PSPr) was described in 11 cases. It is characterized by a distinct clinical and neuropathological phenotype and by a PrP^{Sc} more sensitive to PK.; with a distinctive electrophoretic profile [10]. The case showed some features compatible with this novel disease [10] such as a family history of dementia, prominent neuropsychiatric symptoms early in the evolution and the absence of specific abnormalities in the ancillary test such as EEG.; 14-3-3 protein in CSF and MRI. In contrast, our case had a short clinical course of seven months (versus a median of 20 months in the Gambetti et al. series) [10]. Regarding abnormal prion protein, evaluation of the sensitivity to PK showed that 80% of the detectable abnormal PrP was PK sensitive, with a ladder-like pattern on Western blot as seen in other PSPr cases. However, it differed from them on the earlier age at onset (74 years vs. upper range value 71), the aforementioned clinical course, higher PrP^{Sc} isoforms (three aglycosylated isoforms of 19, 16, and 6 kDa, versus two of 20 and 6 kDa) and MV genotype for *PRNP* 129 polymorphism. Codon 129 appears to be the most reliable factor to explain these dissimilarities, as recently reported in other PSPr MM and MV cases [16, 17], and as observed in other human prion diseases [11, 18, 19]. However, the existence of other unknown factors cannot be discarded.

Regarding the neuropathological findings, the present case differs from common subtypes of sCJD.; including those cases presenting with combined molecular subtypes [12, 20]. Main involvement of the putamen/globus pallidus, thalamus and cerebral cortex with slight cerebellar involvement, together with a lower band of 19 kDa, is not common in pure MV cases [2, 12]. Cases with MV1 show typical synaptic pattern of PrP immunostaining and slight involvement of the cerebellum, whereas MV2 cases exhibit large numbers of kuru plaques [2, 12]. Large confluent vacuoles are common in MM2C but these also differ from the moderately large, non-confluent vacuoles observed in the present case. Finally, PrP sensitive to PK.; as revealed in immunohistochemical sections, and further validated by molecular studies, does not occur in common subtypes of sCJD [2, 12].

Neuropathologically, this form differs from MV2 by its lack of confluent vacuoles. It also differs from VV2 and MV2K in the absence of cerebellar plaque-like deposits and kuru plaques, respectively. Mixed forms MM/MV1+2C and MV2K+2C are also different for analogous reasons. Finally, MM2T.; characterized by thalamo-olivary atrophy, and MM2V, characterized by florid plaques, can be clearly distinguished from the present form. The present case bears similarities to MM/MV1 and VV1 although the band pattern of PrP is obviously different. The original report stressed the size of vacuoles as a distinctive feature, with the vacuoles being larger than those currently seen in MM/MV1 and VV1 cases. Small confluent vacuoles were present in the cerebral cortex in the proband. Sparser vacuoles, a bit larger in size, occurred in the putamen/globus pallidus and thalamus, and they differed from the confluent microspongiosis usually seen in MM/MV1 and VV1 cases. However, it is difficult to draw fair conclusions on this point based on our particular case.

In summary, although clinical signs pointed to sCJD, deposition of PrP sensitive to PK digestion and abnormal prion protein with a ladder-like pattern indicated that our case fitted better with a diagnosis of "protease-sensitive prionopathy". However, heterozygosis MV in codon 129 of the prion protein gene suggested that it might rather be a novel form of human disease with abnormal prion protein sensitive to protease. From a technical point of view, it should be noted that the use of milder digestion conditions could provide interesting information on the characteristics of "less frequent" PrP^{Sc} strains involved in human TSEs. Additionally, the application of new methods that allow the detection of PrP^{Sc} without PK digestion could be of great value in evaluating the level of resistance to PK of abnormal prion protein types and the specific relation between relative amounts of PrP^{Sc} and clinical and neuropathological phenotypes.

9.6 CONCLUSION

A novel form of human disease with abnormal prion protein sensitive to protease was described. Although clinical signs were compatible with sCJD.; the molecular subtype with the abnormal prion protein isoforms showing enhanced protease sensitivity and a ladder-like pattern was reminiscent of the "protease-sensitive prionopathy". Whether or not the genotypic difference from previously reported PSPr cases influences the clinical and neuropathological phenotype, as well as the prion protein conformation and its profile after digestion with PK, remains elusive. Nevertheless, this case established a significant difference with that form of disease. The introduction of modifications in the analysis and detection methodology, mainly focused on applying milder digestion conditions, is necessary in order to detect these proteinase-sensitive proteins. This could also be complemented by the use of analytical approaches that allow quantification of PrP^{Sc} before and after treatment with PK. In this manner, pathologic prion protein could be further characterized using a new perspective that would help to study the phenotypic variability of human prion diseases.

It should not be overlooked that the method presented herein opens a way to more easily detecting pathologic proteinase-susceptible prions associated with other neurodegenerative diseases.

KEYWORDS

- **Sporadic CreutzfeldtJakob disease**
- **Proteinase K**
- **Prion protein gene**
- **Immunohistochemical**

CONSENT

Written informed consent was obtained from the next of kin of the patient for publication of this case report. A copy of the written consent form is available for review from the editor-in-chief of this journal.

COMPETING INTERESTS

The authors declare that they have no competing interests.

AUTHORS' CONTRIBUTIONS

ABRM carried out the molecular analyses, and drafted the manuscript. ABRM.; JMG and RAJ designed the molecular analyses. JJZ helped to draft the neurological and neuropathological sections of the manuscript. JMA provided neurological examination data and helped to draft the manuscript. MMP carried out sequencing analysis. BAP and IFA performed neuropathological studies and IFA drafted the neuropathological section. MJB diagnosed the patient. RAJ coordinated all the information and completed the writing of the final manuscript. All authors read and approved the final manuscript.

ACKNOWLEDGEMENTS

This work received a Support to Health Research grant from the Planning and Arranging Director of the Department of Health of the Basque Government (grant #2006111037). We thank the DNA Bank of the University of the Basque Country for technical support. We also thank Dr. Natalia Elguezabal and Dr. Ana Hurtado for English revision.

REFERENCES

1. Ladogana, A., Puopolo, M., Croes, E. A., Budka, H., Jarius, C., Collins, S., Klug, G. M., Sutcliffe, T., Giulivi, A., Alperovitch, A., et al. Mortality from Creutzfeldt-Jakob disease and related disorders in Europe, Australia, and Canada. *Neurology* **64,** 1586–1591 (2005).
2. Parchi, P., Giese, A., Capellari, S., Brown, P., Schulz-Schaeffer, W., Windl, O., Zerr, I., Budka, H., Kopp, N., Piccardo, P., et al. Classification of sporadic Creutzfeldt-Jakob disease based on molecular and phenotypic analysis of 300 subjects. *Ann. Neurol.* **46,** 224–233 (1999).
3. Ironside, J. W., Ritchie, D. L., and Head, M. W. Phenotypic variability in human prion diseases. *Neuropathol. Appl. Neurobiol.* **31,** 565–579 (2005).
4. Budka, H. Histopathology and immunohistochemistry of human transmissible spongiform encephalopathies (TSEs). *Arch. Virol. Suppl.* **16,** 135–142 (2000).
5. Ironside, J. W., Review. Creutzfeldt-Jakob disease. *Brain Pathol.* **6,** 379–388 (1996).
6. Parchi, P., Castellani, R., Capellari, S., Ghetti, B., Young, K., Chen, S. G., Farlow, M., Dickson, D. W., Sima, A. A., Trojanowski, J. Q., et al. Molecular basis of phenotypic variability in sporadic Creutzfeldt-Jakob disease. *Ann. Neurol.* **39,** 767–778 (1996).
7. Giaccone, G., Di Fede, G., Mangieri, M., Limido, L., Capobianco, R., Suardi, S., Grisoli, M., Binelli, S., Fociani, P., Bugiani, O., et al. A novel phenotype of sporadic Creutzfeldt-Jakob disease. *J. Neurol. Neurosurg. Psychiatry* **78,** 1379–1382 (2007).
8. Zanusso, G., Polo, A., Farinazzo, A., Nonno, R., Cardone, F., Di Bari, M., Ferrari, S., Principe, S., Gelati, M., Fasoli, E., et al. Novel prion protein conformation and glycotype in Creutzfeldt-Jakob disease. *Arch. Neurol.* **64,** 595–599 (2007).
9. Krebs, B., Bader, B., Klehmet, J., Grasbon-Frodl, E., Oertel, W. H., Zerr, I., Stricker, S., Zschenderlein, R., and Kretzschmar, H. A. A novel subtype of Creutzfeldt-Jakob disease characterized by a small 6 kDa PrP fragment. *Acta Neuropathol.* **114,** 195–199 (2007).
10. Gambetti, P., Dong, Z., Yuan, J., Xiao, X., Zheng, M., Alshekhlee, A., Castellani, R., Cohen, M., Barria, M. A., Gonzalez-Romero, D., et al. A novel human disease with abnormal prion protein sensitive to protease. *Ann. Neurol.* **63,** 697–708 (2008).
11. Gambetti, P., Kong, Q., Zou, W., Parchi, P., and Chen, S. G. Sporadic and familial CJD classification and characterisation. *Br. Med. Bull.* **66,** 213–239 (2003).
12. Parchi, P., Strammiello, R., Notari, S., Giese, A., Langeveld, J. P., Ladogana, A., Zerr, I., Roncaroli, F., Cras, P., Ghetti, B., et al. Incidence and spectrum of sporadic Creutzfeldt-Jakob disease variants with mixed phenotype and co-occurrence of PrP^{Sc} types: An updated classification. *Acta Neuropathol.* **118,** 659–671 (2009).
13. Piccardo, P., Seiler, C., Dlouhy, S. R., Young, K., Farlow, M. R., Prelli, F., Frangione, B., Bugiani, O., Tagliavini, F., and Ghetti, B. Proteinase-K-resistant prion protein isoforms in Gerstmann-Straussler-Scheinker disease (Indiana kindred). *J. Neuropathol. Exp. Neurol.* **55,** 1157–1163 (1996).
14. Parchi, P., Chen, S. G., Brown, P., Zou, W., Capellari, S., Budka, H., Hainfellner, J., Reyes, P. F., Golden, G. T., Hauw, J. J., et al. Different patterns of truncated prion protein fragments correlate with distinct phenotypes in P102L Gerstmann-Straussler-Scheinker disease. *Proc. Natl. Acad. Sci. U. S. A.* **95,** 8322–8327 (1998).
15. Benestad, S. L., Arsac, J. N., Goldmann, W., and Noremark, M. Atypical/Nor98 scrapie properties of the agent, genetics, and epidemiology. *Vet. Res.* **39,** 19 (2008).

16. Gambetti, P., Puoti, G., Kong, Q., Zou, W., Tagliavini, F., and Parchi, P. Novel human prion disease affecting 3 prion codon 129 genotypes. the sporadic form of Gerstmann-Stäussler-Scheinker disease? [abstract]. *J. Neuropathol. Exp. Neurol.* **68**, 554 (2009).
17. Gambetti, P. A novel human prion disease affecting subjects with the three prion protein codon 129 genotypes. Could it be the sporadic form of Gerstmann-Sträussler? [abstract]. Prion2009 book of abstracts **45** (2009).
18. Goldfarb, L. G., Petersen, R. B., Tabaton, M., Brown, P., LeBlanc, A. C., Montagna, P., Cortelli, P., Julien, J., Vital, C., Pendelbury, W. W., et al. Fatal familial insomnia and familial Creutzfeldt-Jakob disease: Disease phenotype determined by a DNA polymorphism. *Science* **258**, 806–808 (1992).
19. Haik, S., Peoc'h, K., Brandel, J. P., Privat, N., Laplanche, J. L., Faucheux, B. A., and Hauw, J. J. Striking PrP^{Sc} heterogeneity in inherited prion diseases with the D178N mutation. *Ann. Neurol.* **56**, 909–910 (2004).
20. Cali, I., Castellani, R., Yuan, J., Al Shekhlee, A., Cohen, M. L., Xiao, X., Moleres, F.,J., Parchi, P., Zou, W. Q., and Gambetti, P. Classification of sporadic Creutzfeldt-Jakob disease revisited. *Brain* **129**, 2266–2277 (2006).
21. Hill, A. F., Joiner, S., Beck, J. A., Campbell, T. A., Dickinson, A., Poulter, M., Wadsworth, J. D., and Collinge, J. Distinct glycoform ratios of protease resistant prion protein associated with PRNP point mutations. *Brain***129**, 676–685 (2006).
22. Schaller, O., Fatzer, R., Stack, M., Clark, J., Cooley, W., Biffiger, K., Egli, S., Doherr, M., Vandevelde, M., Heim, D., et al. Validation of a western immunoblotting procedure for bovine PrP(Sc) detection and its use as a rapid surveillance method for the diagnosis of bovine spongiform encephalopathy (BSE). *Acta Neuropathol. (Berl)* **98**, 437–443 (1999).
23. Monari, L., Chen, S. G., Brown, P., Parchi, P., Petersen, R. B., Mikol, J., Gray, F., Cortelli, P., Montagna, P., Ghetti, B., et al. Fatal familial insomnia and familial Creutzfeldt-Jakob disease: Different prion proteins determined by a DNA polymorphism. *Proc. Natl. Acad. Sci. U. S. A.* **91**, 2839–2842 (1994).
24. IDEXX Laboratories. Instruction Manual for Bovine Spongiform Encephalopathy-Scrapie Antigen test kit.

10 Prions as Seeds and Recombinant Prion Proteins as Substrates

Giannantonio Panza, Lars Luers, Jan Stöhr, Luitgard Nagel-Steger, Jürgen Weiβ, Detlev Riesner, Dieter Willbold, and Eva Birkmann

CONTENTS

10.1 INTRODUCTION

The prion diseases like CreutzfeldtJakob disease in humans, scrapie in sheep or bovine spongiform encephalopathy are fatal neurodegenerative diseases, which can be of sporadic, genetic, or infectious origin. Prion diseases are transmissible between different species, however, with a variable species barrier. The key event of prion amplification is the conversion of the cellular isoform of the prion protein (PrP^C) into the pathogenic isoform (PrP^{Sc}). We developed a sodiumdodecylsulfate-based PrP conversion system that induces amyloid fibril formation from soluble α-helical structured recombinant PrP (recPrP). This approach was extended applying pre-purified PrP^{Sc} as seeds that accelerate fibrillization of recPrP. In the present study we investigated the interspecies coherence of prion disease. Therefore we used PrP^{Sc} from different species like Syrian hamster, cattle, mouse and sheep and seeded fibrillization of recPrP from the same or other species to mimic *in vitro* the natural species barrier. We could show that the *in vitro* system of seeded fibrillization is in accordance with what is known from the naturally occurring species barriers.

Prion diseases are fatal progressive neurodegenerative diseases of spontaneous, genetic, or infectious origin. The conversion of the host encoded prion protein (PrP^C) into the disease causing isoform PrP^{Sc} is the key molecular event in prion disease. The common hypothesis is, that the amplification of PrP^{Sc} is achieved by the conversion of the α-helical dominated cellular isoform PrP^C into β-sheet rich and insoluble PrP^{Sc} while PrP^{Sc} acts as template and catalyst for PrP^C conversion [1]. The pool of PrP^C is replenished by the cellular synthesis of PrP^C. Some mechanistic models have been suggested, including the heterodimer model [2], the cooperative model [3], and the model of seeded polymerization [4]. Most experimental data support the seeded polymerization. Prion diseases are not only transmissible within one species, but in some cases also between different species [5]. This so called "interspecies transmission" is of special interest since the origin of bovine spongiform encephalopathy is suspected to be caused by feeding meat and bone meal from Scrapie infected sheep to cattle, that is an interspecies transmission might have lead to the bovine spongiform encephalopathy (BSE)-epidemic in the UK [6]. In an experimental approach it was indeed shown that cattle are susceptible to infection by sheep Scrapie isolates [7]. On the other hand it is known that interspecies transmission can fail. For example,. Syrian hamster Scrapie is not transmissible to mice [8]. In that case the interspecies barrier would be too high for experimental observation. The molecular mechanism of prion disease transmissibility between different species is still not understood.

A molecular model for the infectious process of prions is the conversion of PrP^C to PrP^{Sc} induced by the invading PrP^{Sc}. As mentioned earlier, the widely accepted mechanistic model is that of PrP^{Sc} acting as seed for polymerization of PrP^C [9, 4, 10]. To study the mechanism of spontaneous and seed-depended fibrillization of recPrP different *in vitro* conversion assays were introduced, which lead to the formation of amyloid fibrils [11, 12, 13]. The *in vitro* conversion into amyloid led to the first generation of synthetic prions utilizing only murine recPrP [14], that is without PrP^{Sc} as seed, which would represent a model for the sporadic case of prion diseases.

In the present study we use the *sodium dodecyl sulfate* (SDS) PrP conversion system to simulate intra- and interspecies transmission *in vitro*. It is a minimal system in the sense, that only recPrP as substrate, buffer, and purified PrP^{Sc} as seed, but no cellular extract is needed. We extend the spontaneous and seed depended conversion, established for recombinant Syrian hamster PrP [11] and bovine recPrP [15], to recombinant ovine and murine PrP. It has been shown earlier that well balanced concentrations of sodium chloride and SDS have to be chosen for every substrate PrP in order to guarantee a suitable extent of partial denaturation, which is a prerequisite for fibrillization, spontaneous as well as seeded fibrillization.

To monitor amyloid formation of recPrP we used the amyloid specific marker Thioflavin T (ThT). We are able to mimic the species barrier *in vitro* in complete accordance with the species barrier as found with the *in vitro* infection of prions.

10.2 MATERIALS AND METHODS

10.2.1 Recombinant prion proteins

The recombinant prion protein (recPrP) was prepared and purified as described previously [28, 29]. The recPrP with the amino acid sequence of Syrian Hamster (90–231) (SHaPrP) PrP as well as recPrP with the amino acid sequence of cattle (25–241) (BovPrP) PrP was used in our studies before [15]. We adopted the purification protocols to full length recombinant ovine PrP (25–233) (OvPrP). Recombinant murine PrP (89–231) (MuPrP) was acquired from Allprion (Schlieren, Switzerland).

10.2.2 NaPTA precipitation of PrP^{Sc}

The PrP^{Sc} from brain tissue of different species was purified by NaPTA (Sodium phosphotungstate dibasic hydrate) precipitation [30, 31]. Additionally to Syrian hamster and bovine PrP^{Sc} ($SHaPrP^{Sc}$ and $BovPrP^{Sc}$) [11, 15] we adopt the purification protocol to ovine PrP^{Sc} ($OvPrP^{Sc}$) and to Murine PrP^{Sc} ($MuPrP^{Sc}$). The resulting pellet was resuspended in 10 mM NaPi by brief sonification (Sonificator, Labsonic U, Braun Dissel, Melsungen).

10.2.3 Circular dichroism spectroscopy

The circular dichroism (CD) spectra were recorded with a J-715 spectropolarimeter (Jasco, Easton, MD, USA) in a 0.1 cm quartz cuvette at room temperature. The scanning speed was 50 nm/min with resolution of 1 nm. For each sample 10 spectra were accumulated between 195 and 260 nm. The protein concentration was 150 ng/µl. Background spectra of buffer samples were subtracted from the respective protein spectra.

10.2.4 Analytical ultracentrifugation

The sedimentation–diffusion equilibrium experiments were performed, as described in Ref. [11], in a Beckman Optima XL-A analytical ultracentrifuge (Beckman Coulter) applying standard 12-mm double-sector cells at 20°C. The data were analysed by using the Global Fit procedure, which is implemented in the UltraScan II software package (Version 5.0 for UNIX) of B. Demeler (University of Texas Health Science Center, San Antonio, TX).

10.2.5 Electron microscopy, Negative Stain

A droplet of 5–10 μl containing the recPrP was placed on glow discharged grid and left to adsorb for 2 min. After adsorption to the grid surface the sample was washed briefly (50 μl of. 0.1 and 0.01 M NH_4 acetate) and stained with 2% ammoniummolybdate (50 μl). The samples were analysed with a Zeiss EM910 microscope at 80 kV.

10.2.6 Thioflavin T-assay

Fluorescence emission spectra of Thioflavin T (ThT) were measured at a concentration of 5 μM ThT and 10 ng/μl recPrP in 150 μl 10 mM NaPi pH 7.4. The emission spectra were recorded from 460 nm to 630 nm with a fixed excitation wavelength of 455 nm, average of λ_{em} 495 to 505 is shown for a time point. Fibrillization kinetics was followed in 96 well plates according to Stöhr *et al.* [11]. All measurements were performed in a Tecan saphire plate reader (Tecan Group, Maennedorf, Switzerland). The chosen regression line is polynomial fitted to original data points.

10.2.7 Spontaneous and Seeded Amyloid Formation of RecPrP

The spontaneous and seeded amyloid fibril formation of recPrP of the different species was monitored by ThT-assay as described. The buffer conditions especially the SDS-concentrations were adjusted according to the amyloid forming conditions of the spontaneous case for each species. The results are displayed as sum of the average fluorescence intensities of 495 to 505 nm in the saturation phase (24 h to 48 h in 30 min interval) of the curve. To determine the specificity of the seeding effect in interspecies transmission or interspecies barrier NaPTA-precipitated PrP^{Sc} from brain tissue of infected and non-infected animals was compared.

10.3 DISCUSSION

The conversion of the cellular prion protein (PrP^{C}) to PrP^{Sc} is the key event in prion infection [2]. Although, many studies with recPrP and with different conversion systems have been carried out the molecular mechanism is still not well understood [17 -19]. These studies are based mainly on the simulation of the structural properties of PrP^{Sc} like β-sheet content, PK-resistance or morphology of amyloid fibrils. However, these systems have not generated infectious PrP or only very low titers of infectivity were generated spontaneously [14]. Recently, the group of S. B. Prusiner and colleagues could show that subtle variations in the structure of *in vitro* generated fibrils give rise to a variety of infectious preparation with distinct strain properties [20]. In contrast to spontaneous fibrillization seed-dependent assays like protein misfolding cyclic amplification (PMCA) [21] and quaking-induced conversion (Quic) [22] were established and are in very good agreement with the infectious etiology of prion diseases. Both assays are carried out in cellular extracts from uninfected animals or cells, which cannot exclude the involvement of cellular compounds in the conversion reaction.

The group established a SDS-based conversion assay that works without cellular extracts, using solely purified compounds, like phosphate buffer, recPrP as substrate and prepurified PrP^{Sc} as seed [11]. The use of partially denaturing conditions in conversion of recombinant PrP into amyloid with physiochemical properties reminiscent of PrP^{Sc} seems to be a general concept in *in vitro* conversion studies of PrP and were studied in greater detail [18, 22]. Beside those classical protein denaturants other compounds have been identified (e.g. glycosaminclycans and oligonucleotides) to promote PrP conversion *in vitro* but their structural influence on PrP remains unknown [23, 24].

The PMCA and Quic were mainly developed for diagnostic purposes, system with the well-defined components was developed to describe quantitatively the prion propagation mechanism using hamster PrP. This SDS-based conversion assay was applied to additional species, in order to test if our conversion system can be used with prion proteins of different species and if the pre-amyloid state described with hamster PrP represent a general mechanism for amyloid formation in our in vitro conversion system. Furthermore, by combining seed and substrate from different species we have the opportunity to simulate the phenomena of species barriers for the first time on the level of a direct molecular interaction of prion seed and recPrP substrate without the influence of any other cellular component. In the presented study we were able to show that amyloid fibrils of recPrP can be formed within the SDS-based conversion system for all species investigated (cattle, sheep, mouse, and hamster). Only the SDS concentration had to be adapted. These results and the properties of the intermediate state will be discussed on a more detailed level later. The phenomenon of species barrier for prion transmission was successfully modeled as seed dependent *in vitro* fibrillization. Five interspecies transmissions have been observed (Figure 2). Not very surprisingly, the amyloid formation of interspecies transmission seems to be slower as compared to *intra* species transmission (Figure 2). Two interspecies transmissions were reported to have failed (SHaPrP^{Sc} to MuPrP(89–231) and BovPrP^{Sc} to SHaPrP(90–231)). Consequently, also a strict species barrier was found in the *in vitro* simulation (Figure 1). .For murine substrate, only one positive transmission (MuPrP^{Sc} to MuPrP(89–231)) and one negative (SHaPrP^{Sc} to MuPrP(89–231)) could be performed for reasons of shortage of mouse prions in our lab. As a main result of our work we found that the *in vitro* simulation of the species barrier is in complete agreement with the experimental data from *in vivo* transmission studies. Similar results on interspecies transmission and species barrier were reported by studies with PMCA [25] and by Cashman and co-workers [26], whereby both systems use cellular extracts. Because our conversion system does not include cellular extracts we conclude that the species barrier is encoded within the direct interaction of PrP^{Sc} and PrP^{C} [27]. The interspecies transmission as well as species barrier is well resembled in our system. Cellular factors might be beneficial for the conversion reaction, but only by enforcing the pre-existing interaction.

The interpretation of results on the molecular level shows the importance of partial denaturation of recPrP as substrate as described by Stöhr et al. [11]. The PrP in this intermediate or pre-amyloid state is soluble for weeks, suggesting that it is in a state of low free energy. Since, it is present in a monomer-dimer-equilibrium one can argue that the partially denaturated PrP is prone to intermolecular interactions possibly also with PrP^{Sc}. However, in conversion conditions PrP is in the state of lowest free energy if it is refolded for attachment to the fibrillar seed. We assume that this state is not present in measurable amounts in solution but only attached to the seed. The degree of denaturation in the intermediate state is critical, for different recPrP sequences, that is for different species, different SDS concentrations are needed. More than the optimal SDS concentration would lead to a more unfolded PrP^{C} whereas less SDS leads to a more refolded PrP^C state as compared to the optimal intermediate state. In both cases the conversion of PrP to fibrillar PrP would be too slow to be observable. For an interspecies transmission a partially unfolded state of the substrate PrP is required, but ΔG for substrate (PrP refolded in the complex) with the seed from the other species in the same way as for intraspecies transmission has to be sufficiently low, that is lower than in the intermediate state, to guarantee the transition. In summary, not the intermediate state of substrate PrP is critical for transition but its potency to refold into a conformation well adapted to the PrP^{Sc}-seed.

In future experiments, will extend experimental approach to other species like human CJD and cervid *chronic wasting disease* (CWD). Due to the high occurrence of CWD in Northern America, the combination of CWD-seed and human PrP as substrate would be of particular interest.

10.4 RESULTS

In previous studies, analysed spontaneous and PrP^{Sc}-seeded fibril formation of recombinant PrP (recPrP). Buffer conditions were established which consist of well selected SDS and NaCl concentrations, so that recPrP forms spontaneously amyloid fibrils within weeks, but the fibril formation was accelerated by seeding with PrP^{Sc} to hours or days. It is important to note, that the optimal buffer conditions had to be selected for the species of recPrP during spontaneous fibril formation, and these conditions were used for seeded fibril formation. It was discussed earlier that the buffer conditions allow a particular extent of partial denaturation of recPrP in a well-characterized pre-amyloid state. The spontaneous and seeded fibril formation of recPrP was analysed with bovine and hamster recPrP as substrate and the homologous NaPTA-precipitated PrP^{Sc} as seeds [11, 15]. In the present study, we established conditions for spontaneous and seeded fibril formation of recPrP from sheep (aminoacids 25–233) and mouse (aminoacids 89–231). The aim of this study was to combine recPrP-substrates and PrP^{Sc}-seeds of different species to investigate if the fibril formation *in vitro* does resemble the well-known phenomenon of species barrier for transmissibility. In system the species barrier phenomena are studied on the level of the molecular interaction of PrP and PrP^{Sc}.

10.4.1 Spontaneous Fibril Formation of Recombinant Ovine and Murine Prion Protein

To determine optimal buffer conditions in which ovine PrP (OvPrP(25–233)) forms amyloid fibrils, analysed OvPrP(25–233) fibril formation in 10 mM NaPi, 250 mM NaCl by varying the SDS-concentration from 0.01 to 0.05%. Fibril formation was followed in the Thioflavin T assay. Incubation was carried out at 37°C under constant agitation for 4 weeks and fibrils could be observed at 0.02% SDS after 3 to 4 weeks. The fibrillar character of OvPrP(25–233) was verified by electron microscopy. At higher or lower SDS-concentrations amyloid specific fluorescence increase could not be observed. Therefore, in all the following experiments with OvPrP(25–233), the buffer was NaPi pH 7.2, 250 mM NaCl and 0.02 percent SDS. As mentioned earlier amyloid formation depends upon of a specific pre-amyloid state of recPrP as long as it is soluble [11, 15]. Therefore, analyzed the pre-amyloid state of OvPrP(25–233).

The secondary structure, revealed with circular dichroism, showed in some contrast to formerly analysed initial-states a higher random-coil amount. The analytical ultra-centrifucation data revealed an equilibrium of monomeric (33%) and dimeric (67%) OvPrP(25–233) [11].

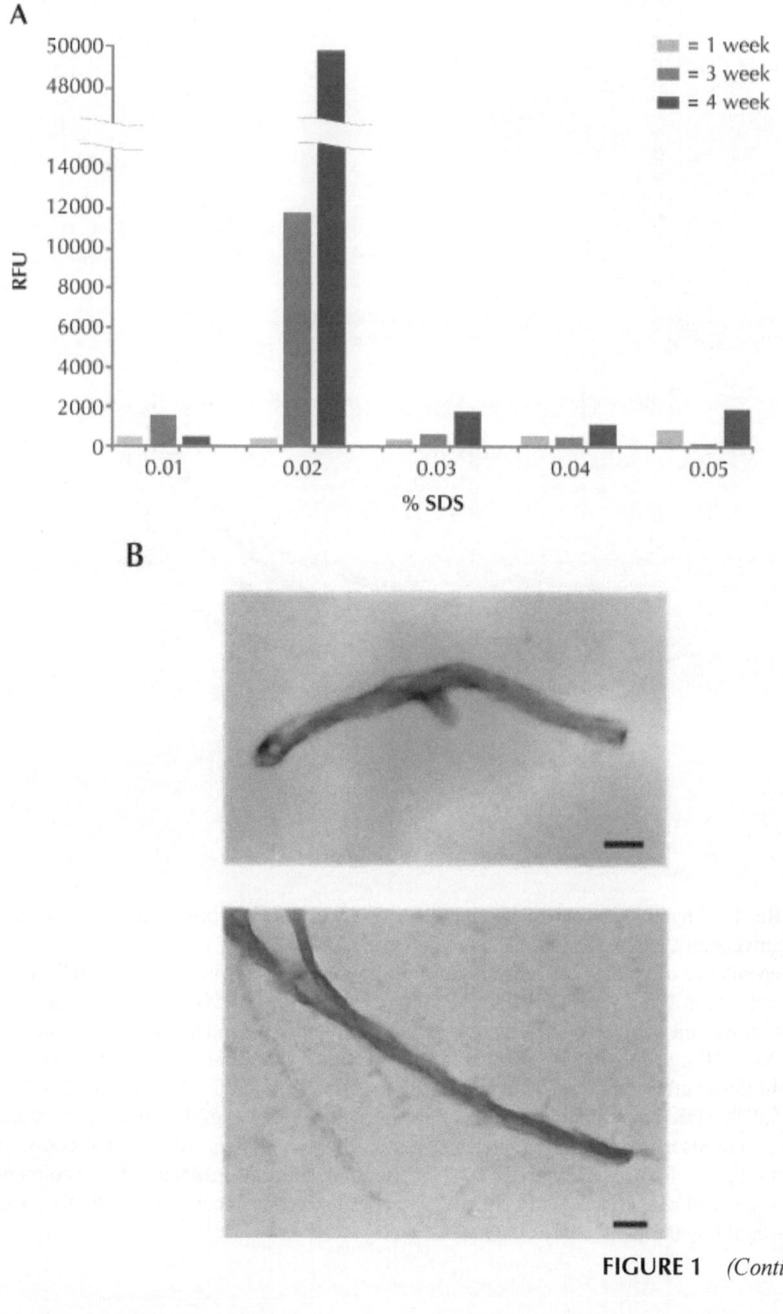

FIGURE 1 *(Continued)*

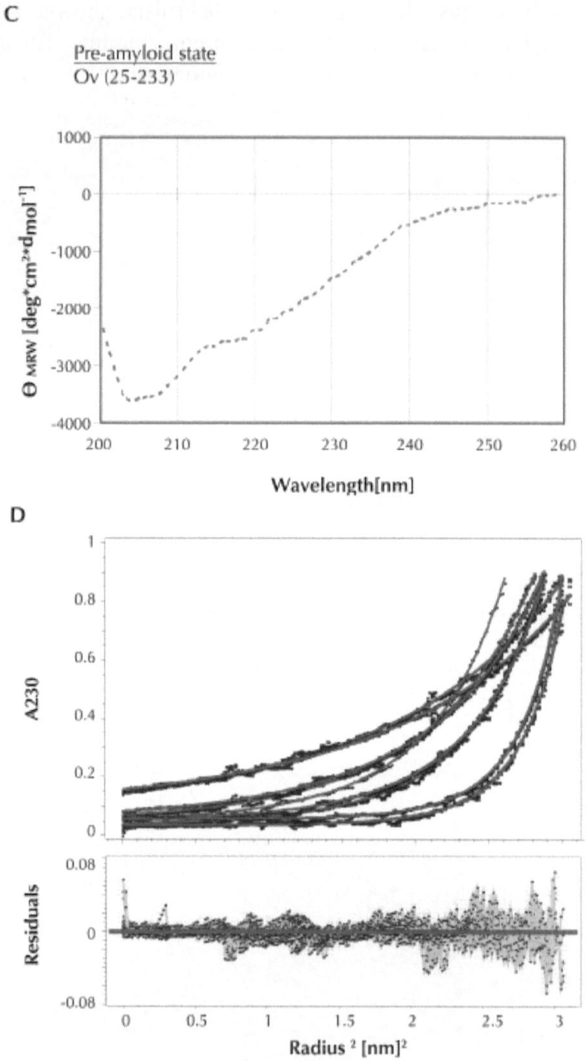

FIGURE 1 Characterization of spontaneous OvPrP(25–233) fibril formation within the *in vitro* conversion system.

(A) Dependence of OvPrP(25–233) amyloid formation on SDS-concentration. OvPrP(25–233) was incubated in 10 mM NaPi pH 7.4, 250 mM NaCl and 0.01–0.05% SDS. The amount of fibril formation was measured by ThT-fluorescence. Thioflavin T was added to a final concentration of 5 μM to 10 ng/μl OvPrP(25–233). (B) Electron micrographs show the typical structure of amyloid fibrils after 7 days of incubation of OvPrP(25–233) in 10 mM NaPi pH 7.4, 250 mM NaCl 0.02% SDS (bar = 20 nm). (C) Secondary structure analysis of the pre-amyloid state. CD-spectra were measured directly after adapting the SDS conditions, with a final concentration of 150 ng/μl OvPrP(25–233) in 10 mMNaPi pH 7.4 and 250 mMNaCl. (D) Sedimentation equilibrium centrifugation of OvPrP(25–233) after 7 days of incubation. *(left)* Experimental data overlaid by the fitted curves (*right*) residuals.

The buffer conditions for fibril formation of murine recPrP (MuPrP(89–231)) were analysed in the same way. It was found that a wider SDS-range was suitable for fibril formation, with an optimum at 0.04 percent SDS in NaPi pH 7.4, 250 mM.

10.4.2 Intraspecies PrP^{Sc}-seeded amyloid formation

In studies, it is showed that NaPTA-precipitated PrP^{Sc} from brain homogenate works very well as seed and could drastically accelerate fibril formation. As a control, NaPTA-precipitate from brain homogenate of uninfected animals did not show any acceleration effect. The presence of the N-terminal sequence (aminoacids 23–89) had no influence on the seeding effect. For the ovine system, NaPTA-precipitated PrP^{Sc} from brain tissue of Scrapie-infected sheeps ($OvPrP^{Sc}$) was used as seed and accelerated fibril formation was observed based on the increase of ThT-fluorescence within 10 to 20 hr. The approach was carried out with 40 ng/µl rec OvPrP(25–233) (1.8 µM) in conversion buffer and 0.02% SDS.

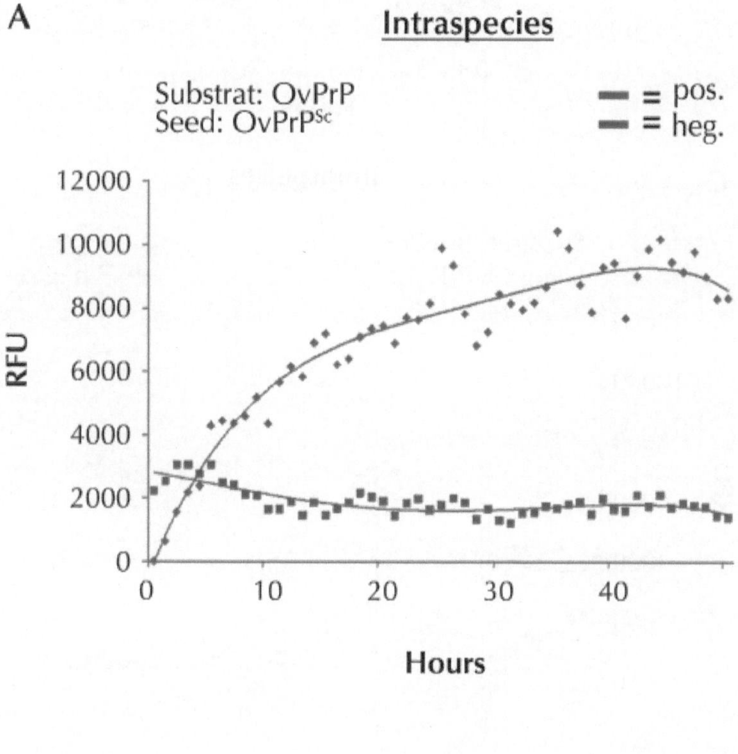

FIGURE 2 *(Continued)*

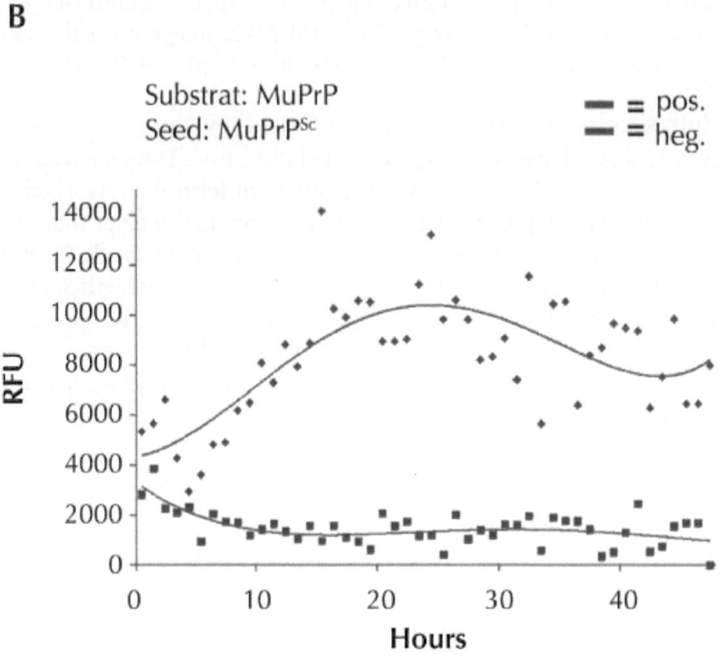

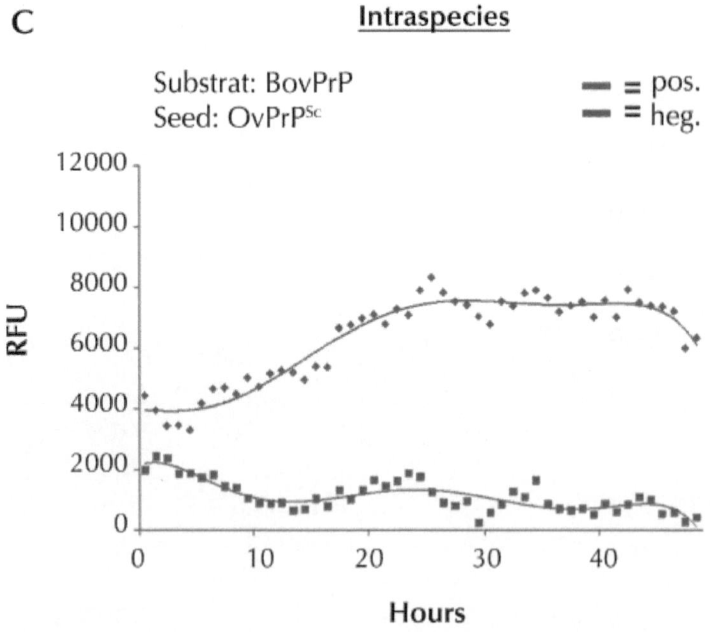

FIGURE 2 *(Continued)*

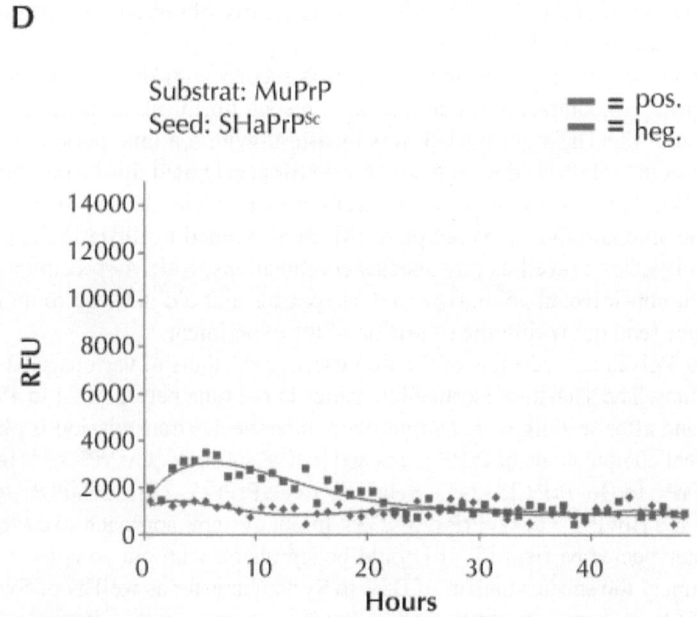

FIGURE 2 Kinetics of intra- and interspecies seeded recPrP amyloid formation. Amyloid formation of recPrP was monitored using ThT-fluorescence assay in 10 mM NaPi (pH 7.4) with 250 mM NaCl (at 37°C). *In vitro* intraspecies transmission. (A) Amyloid formation of 1.8 mM (40 ng/μl) OvPrP(25–233)-substrate seeded with OvPrP^{Sc} in 0.02% SDS (red line), (B) of 1.8 mM (30 ng/μl) MuPrP(89–231)-substrate seeded with MuPrP^{Sc} in 0.04% SDS (red line). *In vitro* interspecies transmission. (C) Amyloid formation of BovPrP(25–241)-substrate seeded with OvPrP^{Sc} in 0.02% SDS (red line). *In vitro* interspecies barrier. (D) In presence of SHaPrP^{Sc} no MuPrP(89–231) amyloid formation occurs (red line). NaPTA-precipitate of corresponding same treated brain tissue from healthy animals.

It is adopted the seed dependent fibril formation for MuPrP. NaPTA-precipitated PrP^{Sc} from mouse brain accelerated the process down to 10 to 20 hr while the spontaneous formation of amyloids takes at least 7 days. As a control NaPTA-precipitate of non-infected mouse brain tissue did not lead to any formation of amyloidogenic fibrils in the time range of the experiment.

10.4.3 *In Vitro* Species Barrier

The aim of this study was to analyse the species barrier of prion infection *in vitro*, that is with PrP^{Sc}-seeds and otherwise purified components only. From natural and experimental transmission data it is known that prion diseases are transmissible in some cases from one species to another and in other cases not. This phenomenon is commonly known as species barrier. PrP^{Sc} from brain homogenate of one species was taken as seeds and recPrP of another species as template. Transmission was simulated by fibrillization in our *in vitro* conversion assay with the exact concentrations of that particular species.

As an example of interspecies transmission of ovine Scrapie to cattle as recipient [7] is shown using OvPrP^{Sc} as seed and BovPrP(25–241) as substrate. Accelerated fibril formation of BovPrP by OvPrP^{Sc}-seed is clearly observed, very similar to the intraspecies seeded fibrillization in sheep.

In another experiment, murine recPrP was seeded with prions from Scrapie infected hamster. In contrast to the intraspecies seeded fibrillization of MuPrP(89–231) no increase of the ThT-fluorescence was measured within a time period of 48 hours. This shows that SHaPrP^{Sc} does not act as seed for accelerated fibril formation of MuPrP(89–231). This corresponds to the well-known species barrier between Syrian hamster Scrapie and mouse as recipient [8]. In all seeded fibrillization experiments, that is intraspecies as well as interspecies combinations, NaPTA- precipitate of brain tissue from non-infected animals served as controls and did not lead to an increased fluorescence read out within the timescale of the experiment.

Figure 3 gives an overview of the fluorescence readouts of various seed-substrate combinations. The ThT-fluorescence intensities in the time period of 24 to 48 hr of incubation and after seeding were summarized. Interspecies transmission is obvious for five different combinations of PrP^{Sc}-seeds and recPrP substrate. OvPrP^{Sc} to SHaPrP (90–231), OvPrP^{Sc} to BovPrP (25–241), SHaPrP^{Sc} to OvPrP(25–233), SHaPrP^{Sc} to BovPrP (25–241) and BovPrP^{Sc} to OvPrP (25–233). In our system, approach two well-known *in vivo* interspecies barriers [8, 16] could be simulated with our *in vitro*, conversion system namely the species barrier of BSE to Syrian hamster as well as of Syrian hamster scrapie to mouse. In summary, all of our *in vitro* intra- and interspecies transmission results resemble exactly the *in vivo* situation.

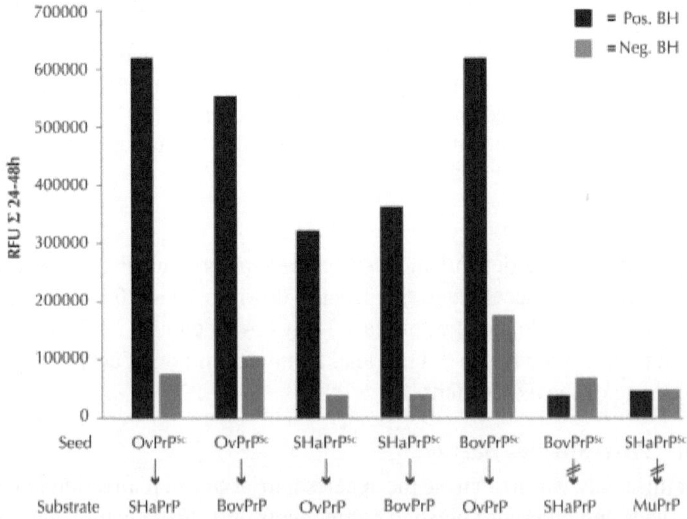

FIGURE 3 Comparison of different interspecies recPrP seeding. Amyloid formation in the seeding assay was monitored using Thioflavin T. The fluorescence signals were recorded every 30 min at 37°C at recPrP amyloid forming conditions. Fluorescence intensities were summarized from 24–48 h (saturation phase). (Left to right) interspecies transmission. OvPrP^{Sc} in SHaPrP(90–231), OvPrP^{Sc} in BovPrP(25–241), SHaPrP^{Sc} in OvPrP(25–233), SHaPrP^{Sc} in BovPrP(25–241), BovPrP^{Sc} in OvPrP(25–233). Interspecies barrier. BovPrP^{Sc} in SHaPrP(90–231), SHaPrP^{Sc} in MuPrP(89–231). As control parallel an approach was made with the corresponding intraspecies transmission.

KEYWORDS

- **Pathogenic isoform**
- **Thioflavin T**
- **Amyloid fibrils**
- **Syrian hamster**
- **Sodium dodecyl sulfate**

ACKNOWLEDGEMENTS

We would like to thank our collaborators for providing us brain tissue of prion infected as well as non-infected animals of different species. Stanley B. Prusiner (UCSF, San Francisco, USA) provided hamster brain, Martin Groschup (Friedrich-Loeffler-Institut, Institute for Novel and Emerging Infectious Diseases, Greifswald/Insel Riems, Germany) provided cattle brain, Carsten Korth (University of Duesseldorf Medical School, Department Neuropathology, Germany) provided mouse brain, Olivier Andréoletti (UMR INRA-ENVT.; Physiopathologie Infectieuse et Parasitaire des Ruminants, Ecole Nationale Vétérinaire, Toulouse, France) provided sheep brain. We are thankful for the guidance and help in cloning and expression of recombinant PrP by Tommy Agyenim and Bernd Esters.

AUTHOR CONTRIBUTORS

Conceived and designed the experiments: GP DR EB. Performed the experiments: GP LL JS LNS. Analyzed the data: GP LNS DR DW EB. Contributed reagents/materials/ analysis tools: JW. Wrote the paper: GP DR DW EB

REFERENCES

1. Watts, J. C., Balachandran, A., and Westaway, D. The expanding universe of prion diseases. *PLoS Pathog.* **2**, e26, (2006).
2. Cohen, F. E., Pan, K. M., Huang, Z., Baldwin, M., Fletterick, R. J., et al. Structural clues to prion replication. *Science* **264**, 530–531 (1994).
3. Eigen, M. Prionics or the kinetic basis of prion diseases. *Biophys. Chem.* **63**, A1–18 (1996).
4. Harper, J. D. and Lansbury, P. T. Jr. Models of amyloid seeding in Alzheimer's disease and scrapie. mechanistic truths and physiological consequences of the time-dependent solubility of amyloid proteins. *Annu. Rev. Biochem.* **66**, 385–407 (1997).
5. Beringue, V., Vilotte, J. L ., and Laude H. Prion agent diversity and species barrier. *Vet. Res.* **39**, 47 (2008).
6. Wells, G. A., Wilesmith, J. W., and McGill I. S. Bovine spongiform encephalopathy. A neuropathological perspective. *Brain Pathol.* **1**, 69–78 (1991).
7. Konold, T., Lee, Y. H., Stack, M. J., Horrocks, C., Green, R. B., et al. Different prion disease phenotypes result from inoculation of cattle with two temporally separated sources of sheep scrapie from Great Britain. *BMC Vet. Res.* **2**, 31 (2006).
8. Kimberlin, R. H. and Walker, C. A. Evidence that the transmission of one source of scrapie agent to hamsters involves separation of agent strains from a mixture. *J. Gen. Virol.* **39**, 487–496 (1978).
9. Laurent, M. Autocatalytic processes in cooperative mechanisms of prion diseases. *FEBS Lett.* **407**, 1–6 (1997).

10. Jarrett, J. T. and Lansbury, P. T. Jr. Seeding "one-dimensional crystallization" of amyloid—a pathogenic mechanism in Alzheimer's disease and scrapie? *Cell* **73**, 1055–1058 (1993).
11. Stohr, J., Weinmann, N., Wille, H., Kaimann, T., Nagel-Steger, L., et al. Mechanisms of prion protein assembly into amyloid. *Proc. Natl. Acad. Sci. U. S. A.* **105**, 2409–2414 (2008).
12. Vanik, D. L., Surewicz, K. A., and Surewicz, W. K. Molecular basis of barriers for interspecies transmissibility of mammalian prions. Mol. Cell. **14**, 139–145 (2004).
13. Colby, D. W., Zhang, Q., Wang, S., Groth, D., Legname, G., et al. Prion detection by an amyloid seeding assay. Proc. Natl. Acad. Sci. U. S. A. **104**, 20914–20919 (2007).
14. Legname, G., Baskakov, I. V., Nguyen H. O., Riesner D., Cohen F. E., et al. Synthetic mammalian prions. Science **305**, 673–676 (2004).
15. Panza, G., Stohr, J., Dumpitak, C., Papathanassiou, D., Weiss J. et al. Spontaneous and BSE-prion-seeded amyloid formation of full length recombinant bovine prion protein. Biochem. Biophys. Res. Commun. **373**, 493–497 (2008).
16. Raymond, G. J., Hope, J., Kocisko D. A., Priola S. A., Raymond L. D., et al. Molecular assessment of the potential transmissibilities of BSE and scrapie to humans. Nature **388**, 285–288 (1997).
17. Leffers, K. W., Wille, H., Stohr, J., Junger, E., Prusiner, S. B., et al. Assembly of natural and recombinant prion protein into fibrils. Biol. Chem. **386**, 569–580 (2005).
18. Baskakov, I. V., Legname, G., Baldwin, M. A., Prusiner, S. B., Cohen, F. E. Pathway complexity of prion protein assembly into amyloid. J. Biol. Chem. **277**, 21140–21148 (2002).
19. Bocharova, O. V., Breydo, L., Parfenov, A. S., Salnikov, V. V., and Baskakov, I. V. In vitro conversion of full-length mammalian prion protein produces amyloid form with physical properties of PrP(Sc). J. Mol. Biol. **346**, 645–659 (2005).
20. Colby, D. W., Wain, R., Baskakov, I. V., Legname, G., Palmer, C. G., et al. Protease-sensitive synthetic prions. *PLoS Pathog.* **6**(1), **e**1000736 (2010).
21. Saborio, G. P., Permanne, B., and Soto, C. Sensitive detection of pathological prion protein by cyclic amplification of protein misfolding. *Nature* **411**, 810–813 (2001).
22. Atarashi, R., Moore, R. A., Sim, V. L., Hughson, A. G., Dorward, D. W., et al. Ultrasensitive detection of scrapie prion protein using seeded conversion of recombinant prion protein. Nat. Methods **4**, 645–650 (2007).
23. Deleault, N. R., Harris, B. T., Rees, J. R., and Supattapone, S. Formation of native prions from minimal components in vitro. Proc. Natl. Acad. Sci. U. S. A. **104**(23), 9741–6 (2007).
24. Wong, C., Xiong, L. W., Horiuchi, M., Raymond, L., Wehrly, K., et al. Sulfated glycans and elevated temperature stimulate PrP(Sc)-dependent cell-free formation of protease-resistant prion protein. *EMBO J.***20**(3), 377–86 (2001).
25. Green, K. M., Castilla, J., Seward, T. S., Napier, D. L., Jewell, J. E., et al. Accelerated high fidelity prion amplification within and across prion species barriers. *PLoS Pathog.* **4**, **e**1000139 (2008).
26. Li, L., Coulthart, M. B., Balachandran, A., Chakrabartty, A., and Cashman N. R. Species barriers for chronic wasting disease by in vitro conversion of prion protein. *Biochem. Biophys. Res Commun.* **364**, 796–800 (2007).
27. Geoghegan, J. C., Miller, M. B., Kwak, A. H., Harris, B. T., and Supattapone, S. Trans-dominant inhibition of prion propagation in vitro is not mediated by an accessory cofactor. *PLoS Pathog.* **5**, **e**1000535 (2009).
28. Mehlhorn, I., Groth, D., Stockel, J., Moffat, B., Reilly, D., et al. High-level expression and characterization of a purified 142-residue polypeptide of the prion protein. *Biochemistry* **35**, 5528–5537 (1996).
29. Jansen, K., Schafer, O., Birkmann, E., Post, K., Serban, H. et al. Structural intermediates in the putative pathway from the cellular prion protein to the pathogenic form. *Biol. Chem.* **382**, 683–691 (2001).
30. Birkmann, E., Schafer, O., Weinmann, N., Dumpitak, C., Beekes, M., et al. Detection of prion particles in samples of BSE and scrapie by fluorescence correlation spectroscopy without proteinase K digestion. *Biol. Chem.***387**, 95–102 (2006).

31. Safar, J., Wille, H., Itri, V., Groth, D., Serban, H., et al. Eight prion strains have PrP(Sc) molecules with different conformations. *Nat. Med.* **4**, 1157–1165 (1998).

11 Quantitative Phosphoproteomic Analysis of Prion-infected Neuronal Cells

Wibke Wagner, Paul Ajuh, Johannes Löwer, and Silja Wessler

CONTENTS

11.1 INTRODUCTION

Prion diseases or transmissible spongiform encephalopathies (TSEs) are fatal diseases associated with the conversion of the cellular prion protein (PrP^{C}) to the abnormal prion protein (PrP^{Sc}). Since the molecular mechanisms in pathogenesis are widely unclear, we analyzed the global phosphoproteome and detected a differential pattern of tyrosine- and threonine phosphorylated proteins in PrP^{Sc}-replicating and pentosan polysulfate (PPS)-rescued N2a cells in two-dimensional gel electrophoresis. To quantify phosphorylated proteins, it was performed a SILAC (stable isotope labeling by amino acids in cell culture) analysis and identified 105 proteins, which showed a regulated phosphorylation upon PrP^{Sc} infection. Among those proteins, it was validated the dephosphorylation of stathmin and Cdc2 and the induced phosphorylation of cofilin in PrP^{Sc}-infected N2a cells in Western blot analyses. The analysis showed for the first time a differentially regulated phospho-proteome in PrP^{Sc} infection, which could contribute to the establishment of novel protein markers and to the development of novel therapeutic intervention strategies in targeting prion-associated disease.

11.2 FINDINGS

Transmissible spongiform encephalopathies (TSEs) are fatal neurodegenerative diseases occurring in many different host species including humans, which develop e.g. Creutzfeld Jacob disease (sCJD) [1]. The development of TSEs is associated with the self-propagating conversion of the normal host cellular prion protein (PrP^C) into the abnormal protease-resistant isoform (PrP^{Sc} or PrPres) in an autocatalytic manner [2]. The PrP^{Sc} plays a key role as an infectious agent in certain degenerative diseases of the central nervous system [3].

The cellular functions of PrP^C and PrP^{Sc} still remain enigmatic. The cellular prion protein can be variably glycosylated at two N-glycosylation sites and is C-terminally attached to the cell surface by a glycosyl phosphatidylinositol (GPI) anchor. The GPI-anchored proteins are found in lipid rafts, highly cholesterol- and glycolipid-enriched membrane domains associated with a large number of signaling molecules such as G-protein-coupled receptors and protein kinases suggesting that signaling transduction pathways might play role in TSEs [4]. Hence, previous publications described a functional role of PrP^C as a signaling molecule with major findings indicating that PrP^C interacts with and activates Src family kinases [5-7]. The increased levels of active Src kinases in scrapie-infected cells then led to the activation of downstream signal transduction pathways [8]. Recently, activation of the JAK-STAT signaling pathway in astrocytes of scrapie-infected brains was observed underlining that signal transduction pathways may play pivotal roles in prion pathogenesis [9]. Interestingly, it was demonstrated that inhibition of the non-receptor tyrosine kinase c-Abl strongly activates the lysosomal degradation of PrP^{Sc} [10]. These data indicate that specific interference with cellular signaling pathways could represent a novel strategy in treatment of TSEs.

A quantitative analysis performed the phosphoproteome to obtain a global insight into deregulated signal transduction pathways in scrapie-infected neuronal cells. Tyrosine- and threonine-phosphorylated proteins analyzed in the murine neuroblastoma cell line N2a58/22L, which were infected with the PrP^{Sc} strain 22L [11]. The N2a58/22L cells treated with pentosan polysulfate (PPS), a known inhibitor of 22L PrP^{Sc} replication in N2a cells [12], resulting in the PrP^{Sc}-rescued cell line N2a58# which served as an uninfected control. The successful rescue from PrP^{Sc} was demonstrated in the colony assay as reflected by the absence of proteinase K (PK)-resistant PrP^{Sc} in N2a58# cells after PPS treatment (Figure 1A). The PrP^{Sc} replication and the effect of PPS-treatment were further studied in an immunoblot. After PK digestion,

PrP^{Sc} replication was only observed in N2a58/22L cells (Figure 1B, lanes 2 and 4). Compared to 22L-infected N2a58/22L cells, PPS-treated N2a58# cells showed a different glycosylation profile as expected for PrP^{C} [13-15]. The glycosylation pattern of PrP^{C} in N2a58# cells displayed high amounts of di- and mono-glycosylated PrP^{C}, whereas in N2a58/22L cells predominantly mono- and non-glycosylated PrP^{Sc} was detected (Figure 1B, lanes 1 and 3). Altogether, PPS treatment of N2a58/22L cells successfully abolished PrP^{Sc} formation in N2a58# cells, which served as a non-infected control cell line in study.

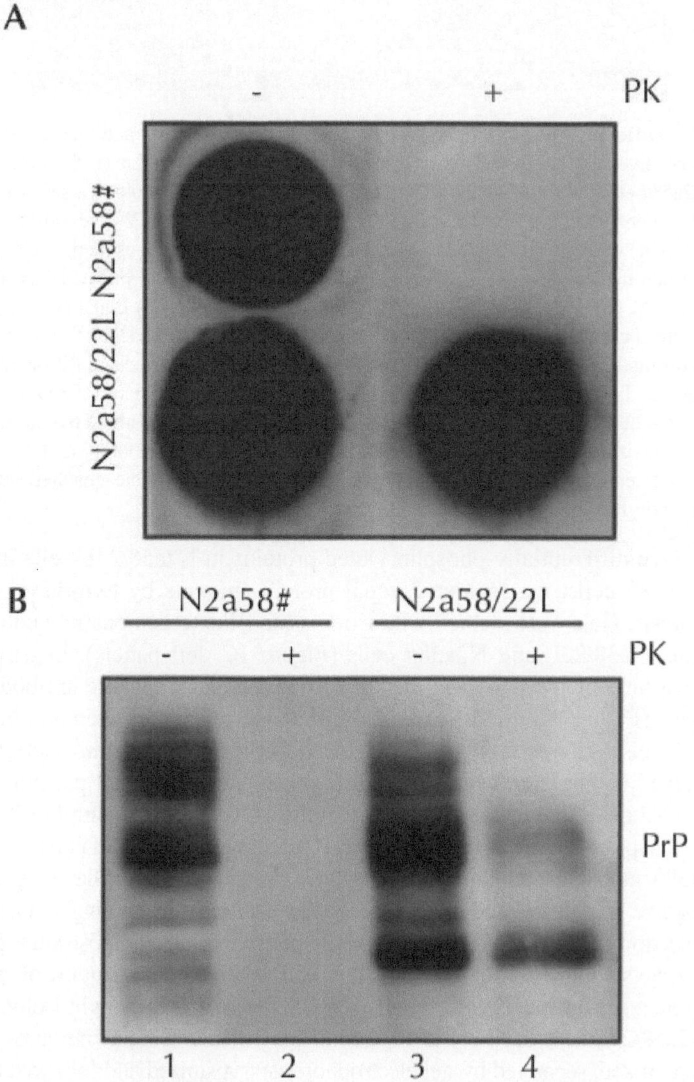

FIGURE 1 *(Continued)*

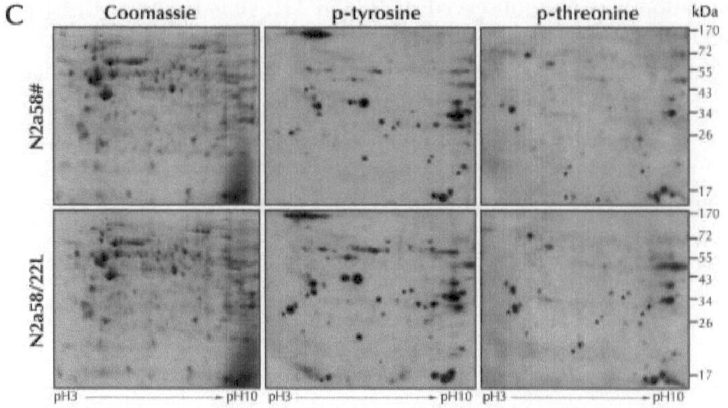

FIGURE 1 Differentially phosphorylated proteins in PrP[Sc]-positive and -negative N2a cells. (A) PrPres-positive N2a58/22L cells were treated with pentosan polysulfate (PPS) to obtain PrPres-negative N2a58# cells. The successful PPS treatment was validated in a colony assay. The cells were grown to confluence on cover slips and directly lysed on nitrocellulose. Where indicated 20 µg/ml proteinase K (PK) was added followed by the detection of PrP expression using the 6H4 monoclonal antibody. In non-treated cells (−), PrP was detected in both, cured and infected N2a cells. Upon digestion with PK (+), PrPres was only observed in N2a58/22L cells. (B) Equal amounts of protein lysates were incubated with 20 µg/ml PK or left untreated. PrP was detected with the 8H4 monoclonal antibody showing the typical migration pattern of PrP and PrPres in infected and PPS-treated N2a58# cells. In parallel, lysates were incubated with PK to visualize PK-resistant PrPres in N2a58/22L. (C) 150 µg of N2a58# or prion-infected N2a58/22L cell lysates were separated by two-dimensional gel electrophoresis followed either by Coomassie staining or immunoblotting for detection of tyrosine- and threonine-phosphorylated proteins. Black asterisks indicate changed intensities of protein phosphorylation.

To analyze differentially phosphorylated proteins in N2a58/22L cells in comparison to N2a58# cells, we separated equal protein amounts by two-dimensional gel electrophoresis. Gels were stained with Coomassie Blue to demonstrate equal protein amounts in N2a58/22L and N2a58# cells (Figure 1C, left panels). In parallel, gels were blotted onto membranes and incubated with phospho-specific antibodies to detect tyrosine- (Figure 1C, middle panels) or threonine-phosphorylated proteins (Figure 1C, right panels). Interestingly, considerable differences in phosphorylation patterns were observed (Figure 1C, asterisks), while other phosphorylated proteins were not changed in N2a58/22L and N2a58# cells (Figure 1C). These data imply differentially regulated phosphoproteins in response to 22L infection of neuronal cells.

Generally, global detection of phosphorylated proteins is still challenging, as antisera often recognize phosphorylated residues dependent on the surrounding sequence. For a general detection of proteins post-translationally phosphorylated at those sites, performed a SILAC analysis allowing the identification and relative quantification of differential phosphoprotein regulation. Therefore, N2a58# cells were grown in light isotope containing and N2a58/22L cells in heavy isotope containing medium. Equal amounts of protein lysates were mixed, separated by gel electrophoresis, trypsinized and followed by enrichment of phosphoproteins, which were then analyzed by mass spectrometry. The 109 different phosphoproteins identified, of which 105 were also quantified (Table 1 and 2). It was

TABLE 1 Proteins exhibiting increased phosphorylation in N2a58/22L cells.

No.	Uniprot	Protein names	Ratio[a]	Pept[b]	Sequence coverage [%]	PEP[c]	Biological process
1	P43276	Histone H1.5	0.46237	1	13.9	5.61E-16	Nucleosome assembly
2	P30681	High mobility group protein B2	0.48683	2	11	3.33E-02	Genome maintenance; differentiation
3	P11440	Cell division control protein 2 homolog	0.49428	5	7.7	3.03E-03	Cell cycles; protein phosphorylation
4	P97310	DNA replication licensing factor MCM2	0.54657	1	2.4	1.65E-05	Cell cycle; nucleosome assembly; transcription
5	P43275	Histone H1.1	0.56637	1	19.2	8.42E-04	Nucleosome assembly
6	P43274	Histone H1.4	0.58367	4	24.2	1.04E-14	Nucleosome assembly
7	Q9Z2X1-1	Heterogeneous nuclear ribonucleoprotein F	0.60054	2	6.5	5.45E-22	RNA processing
8	P70670	Nascent polypeptide-associated complex subunit alpha, muscle-specific form	0.60501	2	1.2	5.07E-11	Protein transport; transcription
9	P60843	Eukaryotic initiation factor 4A-I	0.65781	5	4.9	5.39E-03	Translation
10	P27659	60S ribosomal protein L3	0.65906	12	7.7	1.49E-03	Translation
11	P28656	Nucleosome assembly protein 1-like 1	0.70087	1	7.2	1.16E-05	Nucleosome assembly

TABLE 1 *(Continued)*

No.	Uniprot	Protein names	Ratio[a]	Pept[b]	Sequence coverage [%]	PEP[c]	Biological process
12	Q62167	ATP-dependent RNA helicase DDX3X	0.70816	3	3.3	3.12E-19	Putative helicase activity
13	P68040	Guanine nucleotide-binding protein subunit beta-2-like1	0.71371	1	7.9	5.15E-08	Unknown
14	P47911	60S ribosomal protein L6	0.71887	7	12	7.26E-07	Translation
15	Q61937	Nucleophosmin	0.72062	7	29.8	4.38E-07	Cell cycle; nuclear export
16	P15532	Nucleoside diphosphate kinase A	0.72108	2	17.8	7.93E-03	NTP biosynthesis; nervous system development
17	O70251	Elongation factor 1-beta	0.73137	1	24	2.20E-18	Translation
18	Q61656	Probable ATP-dependent RNA helicase DDX5	0.73832	4	3.7	1.41E-03	RNA processing; transcription
19	Q9ERK4	Exportin-2	0.74333	1	2.1	8.97E-05	Cell proliferation; protein transport
20	P09411	Phosphoglycerate kinase 1	0.74902	2	6.7	3.40E-03	Glycolysis; phosphorylation
21	P48962	ADP/ATP translocase 1	0.75149	6	18.8	3.87E-36	Transmembrane transport
22	Q9D8N0	Elongation factor 1-gamma	0.76191	3	7.8	1.45E-10	Translation

TABLE 1 *(Continued)*

No.	Uniprot	Protein names	Ratio[a]	Pept[b]	Sequence coverage [%]	PEP[c]	Biological process
23	P49312-2	Heterogeneous nuclear ribonucleoprotein A1	0.77303	5	12.1	7.20E-05	Alternative splicing; nuclear export/import
24	Q9CZM2	60S ribosomal protein L15	0.77591	9	10.3	2.04E-15	Translation
25	P97855	Ras GTPase-activating protein-binding protein 1	0.78021	2	7.3	1.11E-15	Protein transport
26	Q9EQU5-1	Protein SET	0.79357	7	7.9	5.77E-12	Nucleosome assembly
27	Q7TPV4	Myb-binding protein 1A	0.79496	2	1.5	2.39E-11	Cytoplasmic transport; transcription
28	P80318	T-complex protein 1 subunit gamma	0.79504	1	6.6	2.89E-05	Protein folding
29	P25444	40S ribosomal protein S2	0.79516	20	12.3	2.99E-04	Translation
30	P10126	Elongation factor 1-alpha 1	0.80203	5	18.6	1.06E-33	Translational elongation
31	P61979-2	Heterogeneous nuclear ribonucleoprotein K	0.80502	6	11.9	1.17E-15	RNA processing
32	P07901	Heat shock protein HSP 90-alpha	0.80637	2	32.7	1.38E-94	CD8 T-cell differentiation; chaperone activity
33	Q61598-1	Rab GDP dissociation inhibitor beta	0.81242	3	5.6	1.93E-05	Protein transport; regulation of GTPase activity

TABLE 1 *(Continued)*

No.	Uniprot	Protein names	Ratio[a]	Pept[b]	Sequence coverage [%]	PEP[c]	Biological process
34	P54775	26S protease regulatory subunit 6B	0.81795	2	7.2	1.59E-09	Blastocyst development; protein catabolism
35	Q20BD0	Heterogeneous nuclear ribonucleoprotein A/B	0.82349	3	19.3	3.26E-11	Nucleotide binding
36	P14206	40S ribosomal protein SA;Laminin receptor 1	0.8261	7	20.7	2.57E-17	Translation
37	P68134	Actin, alpha skeletal muscle	0.83457	10	27.3	4.38E-26	Cytoskeleton
38	P80314	T-complex protein 1 subunit beta	0.83579	2	11.6	1.05E-21	Protein folding
39	P50580	Proliferation-associated protein 2G4	0.84048	2	8.1	1.53E-03	rRNA processing; transcription; translation
40	P11983-1	T-complex protein 1 subunit alpha B	0.84687	4	8.8	5.73E-23	Protein folding
41	P35564	Calnexin	0.85134	1	6.3	2.08E-06	Protein folding
42	Q8BUP7	Putative uncharacterized protein;26S protease regulatory subunit 6A	0.85207	3	7.3	9.72E-03	Blastocyst development; protein catabolism
43	P63017	Heat shock cognate 71 kDa protein	0.85947	9	32.2	2.16E-96	Response to stress

TABLE 1 *(Continued)*

No.	Uniprot	Protein names	Ratio[a]	Pept[b]	Sequence coverage [%]	PEP[c]	Biological process
44	Q01768	Nucleoside diphosphate kinase B	0.86107	4	17.8	1.35E-04	NTP biosynthesis
45	P62082	40S ribosomal protein S7	0.86306	3	10.3	8.39E-03	Translation
46	P80315	T-complex protein 1 subunit delta	0.86423	1	6.9	3.81E-12	Protein folding
47	Q71LX8	Heat shock protein 84b	0.8654	4	32	6.13E-136	Protein folding; stress response
48	P58252	Elongation factor 2	0.86672	2	7.9	3.57E-25	Translation
49	P08249	Malate dehydrogenase, mitochondrial	0.86743	1	9.2	2.49E-35	Glycolysis
50	P70168	Importin subunit beta-1	0.87257	2	2.7	3.47E-13	Nuclear import
51	P51859	Hepatoma-derived growth factor	0.87395	2	12.7	2.95E-07	Transcription
52	P14152	Malate dehydrogenase, cytoplasmic	0.87522	1	7.8	1.09E-03	Glycolysis
53	P80313	T-complex protein 1 subunit eta	0.89251	1	8.6	1.22E-39	Protein folding
54	P62827	GTP-binding nuclear protein Ran	0.89794	3	22.7	2.27E-09	Cell cycle; nuclear import; signal transduction

TABLE 1 *(Continued)*

No.	Uniprot	Protein names	Ratio[a]	Pept[b]	Sequence coverage [%]	PEP[c]	Biological process
55	P20029	78 kDa glucose-regulated protein	0.89996	1	7	4.16E-04	Cerebellar Purkinje cell development/organization
56	P56480	ATP synthase subunit beta, mitochondrial	0.90399	1	18.5	1.85E-42	Proton transport; lipid metabolism
57	P17742	Peptidyl-prolyl cis-trans isomerase A	0.90719	10	20.4	4.97E-20	Neuron differentiation; protein folding
58	P20152	Vimentin	0.90814	10	26.2	2.22E-34	Cytoskeleton
59	P09103	Protein disulfide-isomerase	0.91019	2	4.7	8.84E-02	Redox homeostasis
60	P80317	T-complex protein 1 subunit zeta	0.91579	1	10.2	1.49E-02	Protein folding
61	Q569Z6	Thyroid hormone receptor-associated protein 3	0.93507	3	3.7	4.27E-04	Transcription
62	P09405	Nucleolin	0.93545	1	12.7	8.06E-34	Nucleotide binding
63	Q9D6F9	Tubulin beta-4 chain	0.93722	4	23.2	9.43E-32	Cytoskeleton
64	Q8C2Q7	Heterogeneous nuclear ribonucleoprotein H1	0.93795	3	5.7	2.30E-06	Nucleotide binding
65	Q8K019-1	Bcl-2-associated transcription factor 1	0.94132	4	5.3	2.22E-05	Transcription
66	P62908	40S ribosomal protein S3	0.94247	2	9.1	1.74E-06	Translation

TABLE 1 *(Continued)*

No.	Uniprot	Protein names	Ratio[a]	Pept[b]	Sequence coverage [%]	PEP[c]	Biological process
67	P99024	Tubulin beta-5 chain	0.9446	3	36.3	1.15E-54	Cytoskeleton
68	P15331-2	Peripherin	0.95087	3	17.2	3.41E-40	Cytoskeleton
69	P63038-1	60 kDa heat shock protein, mitochondrial	0.96089	6	13.8	9.17E-37	T cell activation; interferon production
70	P27773	Protein disulfide-isomerase A3	0.96765	1	7.9	1.26E-03	Redox homeostasis; apoptosis
71	P16858	Glyceraldehyde-3-phosphate dehydrogenase	0.96943	58	23.4	5.69E-66	Glycolysis
72	Q3TED3	Putative uncharacterized protein; ATP-citrate synthase	0.97174	2	4.3	4.63E-03	Acetyl-CoA biosynthesis
73	Q03265	ATP synthase subunit alpha, mitochondrial	0.98925	2	13.6	1.81E-22	Proton transport; lipid metabolism
74	Q9ERD7	Tubulin beta-3 chain	0.99071	2	27.3	5.05E-33	Cytoskeleton
75	P17751	Triosephosphate isomerase	0.9967	1	11.2	1.52E-04	Gluconeogenesis; glycolysis

[a]Ratio of N2a58/22L vs. N2a58# cells

[b]Number of identified peptides

[c]Posterior error probability (PEP) estimates the probability of wrong assignment of a spectrum to a peptide sequence

TABLE 2 Proteins exhibiting increased phosphorylation in N2a58/22L cells.

No.	Uniprot	Protein Names	Ratio[a]	Pept.[b]	Sequence coverage [%]	PEP[c]	Biological process
76	P80316	T-complex protein 1 subunit epsilon	1.017	1	6.1	8.63E-18	Protein folding
77	Q9CX22	Putative uncharacterized protein; Cofilin-1	1.0173	5	29.7	3.72E-41	Cytoskeleton; protein phosphorylation
78	P32067	Lupus La protein homolog	1.02	2	7.5	1.68E-02	RNA processing
79	A6ZI44	Fructose-bisphosphate aldolase	1.0328	8	12.2	2.75E-25	Glycolysis
80	P35700	Peroxiredoxin-1	1.0342	6	6.7	2.44E-10	Proliferation; redox homeostasis; stress response
81	P05202	Aspartate aminotransferase, mitochondrial	1.0344	1	12.3	1.04E-11	Aspartate biosynthesis; oxaloacetate metabolism
82	Q3TFD0	Serine hydroxymethyltransferase	1.0425	3	6.5	2.95E-03	Carbon metabolism
83	P63101	14-3-3 protein zeta/delta	1.0478	19	19.2	7.09E-27	Protein binding
84	Q9CZ30-1	Obg-like ATPase 1	1.0606	3	7.1	2.25E-16	ATP/GTP binding; hydrolase activity
85	P06745	Glucose-6-phosphate isomerase	1.0652	3	7.7	3.34E-15	Angiogenesis; gluconeogenesis; glycolysis
86	Q71H74	Collapsin response mediator protein 4A	1.0688	5	6.3	6.94E-12	Nervous system development
87	Q6P5F9	Exportin-1	1.0703	2	3.6	2.74E-20	Nuclear export; centrosome duplication

TABLE 2 *(Continued)*

No.	Uniprot	Protein Names	Ratio[a]	Pept.[b]	Sequence coverage [%]	PEP[c]	Biological process
88	Q01853	Transitional endoplasmic reticulum ATPase	1.0708	2	3.7	2.97E-05	Apoptosis; retrograde protein transport
89	Q3TCI7	L-lactate dehydrogenase	1.0746	4	18.3	1.43E-32	Glycolysis
90	P08113	Endoplasmin	1.08	1	2.6	1.05E-05	Protein folding
91	Q8VC46	Ubc protein;Ubiquitin	1.0848	17	30.7	2.18E-06	Protein binding
92	A0PJ96	Mtap1b protein	1.0871	2	3.2	2.72E-13	Cytoskeleton
93	Q61171	Peroxiredoxin-2	1.0934	2	9.1	4.16E-04	Signal transduction; redox homeostasis
94	Q9WVA4	Transgelin-2	1.106	2	11.8	9.37E-05	Muscle organ development
95	P52480-1	Pyruvate kinase isozymes M1/M2	1.1081	3	22.2	6.30E-39	Glycolysis
96	O08709	Peroxiredoxin-6	1.1204	5	13.8	1.10E-04	Redox homeostasis
97	P17182	Alpha-enolase	1.128	11	41.5	6.35E-26	Glycolysis
98	P54227	Stathmin	1.1422	4	18.1	1.94E-02	Cell cycle; cytoskeleton
99	Q922F4	Tubulin beta-6 chain	1.1427	1	19.2	1.01E-25	Cytoskeleton
100	Q60864	Stress-induced-phosphoprotein 1	1.1499	1	3.5	5.12E-05	Stress response
101	P52480-2	Pyruvate kinase isozymes M1/M2	1.2958	1	19.6	3.40E-33	Glycolysis
102	Q920E5	Farnesyl pyrophosphate synthetase	1.3534	1	8.2	8.44E-16	Cholesterol/isoprenoid biosynthesis

TABLE 2 *(Continued)*

No.	Uniprot	Protein Names	Ratio[a]	Pept.[b]	Sequence coverage [%]	PEP[c]	Biological process
103	P38647	Stress-70 protein, mitochondrial	1.357	2	4.6	6.57E-08	Nuclear export; protein folding
104	P14824	Annexin A6	1.6252	4	7.9	2.47E-08	Ca^{2+} transport; muscle contraction
105	Q61753	D-3-phosphoglycerate dehydrogenase	1.7927	3	8.1	6.28E-15	Cell cycle; neural development; serine biosynthesis

[a]Ratio of N2a58/22L vs. N2a58# cells
[b]Number of identified peptides
[c]Posterior error probability (PEP) estimates the probability of wrong assignment of a spectrum to a peptide sequence

Among quantified phosphoproteins, we then considered specific phosphosites in selected target proteins, such as Cdc2, stathmin, and cofilin as analyzed by mass-spectrometry (Table 3). An increase of cofilin^{S3} phosphorylation in N2a58/22L cells was suggested by a ratio 1.63, while the amount of the two tyrosine phosphorylation sites (Y15, Y160) in Cdc2 were decreased upon 22L infection. Stathmin phosphopeptides containing serine 38 were increased, whereas the amount of stathmin phosphopeptides harboring serine 25 in N2a58/22L cells was significantly lower (Table 3).

that observed 75 proteins with a ratio of identified peptides in N2a58/22L versus N2a58# cells ranging from 0.46 to 0.99 (Table 1). Conversely, 30 phosphoproteins showed a ratio between 1.01 and 1.79 (Table 2). The proteins defined as exhibiting a ratio <0.70 as dephosphorylated proteins and proteins with ratios between 0.70 and 1.40 as proteins, whose phosphorylation was not altered in 22L-infected N2a58/22L cells. Ratios >1.40 were considered as proteins whose phosphorylation increased upon Scrapie infection.

TABLE 3 Identified phosphorylation sites.

Protein names	Ratio (total)[a]	Phosphosite	Ratio (specific phospho-site)[b]
Cdc2	0.49428	Y15	0.43086
		Y160	0.64359
Stathmin	1.1422	S25	1.2155
		S38	0.45601
Cofilin	1.0173	S3	1.6328

[a]Ratio of phosphorylation N2a58/22L vs. N2a58# cells
[b]Ratio of phospho sites in N2a58/22L and N2a58# cells

To validate the results obtained in the SILAC phosphoproteomic analysis we performed Western blots for cofilin 1, Cdc2, and stathmin using antibodies for the detection of specific phosphosites. As predicted by the SILAC analysis, cofilin 1 phosphorylation was significantly induced in Scrapie-infected N2a58/22L cells compared to PPS-treated N2a58# cells (Figure 2, left panels). The cofilin represents a potent regulator of the actin filaments, which is controlled by phosphorylation of serine 3 mediated through the LIM-kinase 1 (LIMK-1) *in vitro* and *in vivo* [16]. These data support previous studies indicating a direct interaction of PrP^{Sc} with cofilin [17]. Together with findings that phosphorylation of cofilin is induced in PrP^{Sc}-infected neuronal cells; the results indicate a significant role for the protein in neurodegeneration processes. The stathmin acts as an important regulatory protein of microtubule dynamics, which can be directly targeted by Cdc2 [18]. The analysis showed that $stathmin^{S38}$ phosphorylation was decreased (Figure 2, middle panels), which correlates with the inactivation of Cdc2 in N2a58/22L cells (Figure 2, right panels) implying that there is a functional interaction. The Cdc2 is a crucial kinase in starting M phase events during the cell cycle progression and regulates important mitotic structure changes, including nuclear envelope breakdown and spindle assembly [19]. The dephosphorylation of $stathmin^{S38}$ led to an inhibition of cells at G2/M phase, lack of spindle assembly, and growth inhibition [20, 21]. Together with the finding that the prion gene is transcriptionally activated in the G1 phase in confluent and terminally differentiated cells [22], it assume that control of the cell cycle might be important in prion diseases.

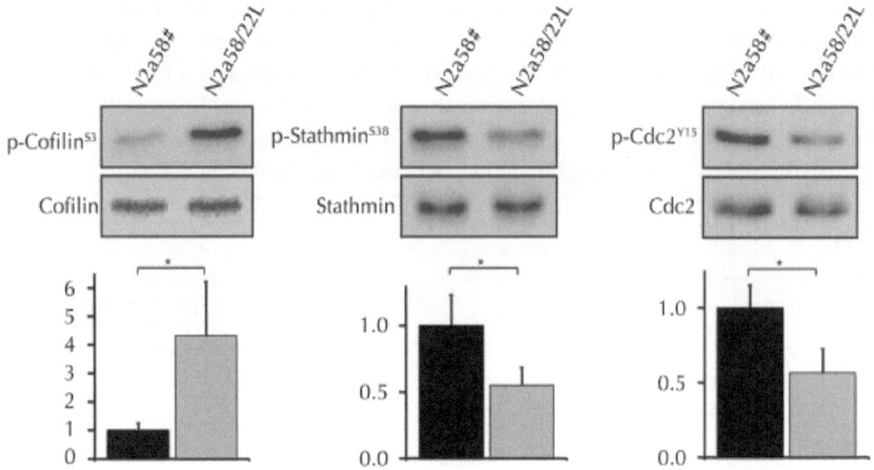

FIGURE 2 Specific regulation of cofilin, Cdc2, and stathmin phosphorylation in scrapie-infected neuronal cells. Cell lysates of N2a58# and 22L-infected N2a58/22L cells were analyzed by Western blot using phospho-specific antibodies to detect p-cofilin^{S3}, p-cdc2^{Y15}, and p-stathmin^{S38} (left panels). As loading controls, equal amounts of cofilin, Cdc2, and stathmin were shown. Quantification of intensities of phosphorylation signal was performed by normalizing the corresponding loading control (* p <0.05) (right panels).

Aberrant signal transduction pathways are implicated in many diseases. However, perturbations in phosphorylation-based signaling networks are typically studied in a hypothesis-driven approach. In this study, it was performed that the first global analysis of the phosphoproteome of scrapie-infected neuronal cells, since the knowledge of PrP-dependent deregulation of the signalling network is poor. The SILAC provides a powerful and accurate technique for relative proteome-wide quantification by mass-spectrometry. Its versatility has been demonstrated by a wide range of applications, especially for intracellular signal transduction pathways [23–25]. Since, it was applied, the SILAC for the quantitative detection of the phosphoproteome in scrapie-infected neuroblastoma cells, it is found 105 different phosphoproteins. Among identified proteins, validated the regulated phosphorylation of cofilin, stathmin and Cdc2 indicating that the identification of phosphoproteins in scrapie-infected neuronal cells by SILAC is reliable. The future work is necessary to determine whether the identified novel phosphoproteins are involved in prion diseases and if they probably represent sensitive and specific biomarkers for diagnosis or therapeutic intervention strategies.

11.3 METHODS

11.3.1 Cell Culture

The N2a58/22L cells have been described [11] and were kindly provided by Prof. Schätzl (LMU, Munich). The cells were cultured in DMEM containing 10% FCS and 4 mM L-glutamine at 37°C. Cells were treated with 5 μg/ml pentosan polysulfate (Cartrophen Vet, A. Albrecht GmbH + Co. KG, Germany) for two passages, resulting in a stable rescued cell line for more than 15 passages (N2a58# cells). Cell lysates were prepared by scraping cells in lysis buffer containing 150 mM NaCl, 0.5% Triton X-100, 0.5% DOC, 50 mM Tris pH 7.5, 1 mM Na-vanadate, 1 mM Na-molybdate, 20 mM NaF, 10 mM NaPP, 20 mM β-glycerophosphat, 1× protease inhibitor cocktail (Roche, Mannheim, Germany). For digestion with proteinase K (PK) 80 μg protein were treated with 20 μg/ml PK for 30 min at 37°C. PK digestion was stopped by addition of laemmli sample buffer and protein denaturation at 95°C for 7 min.

11.3.2 Colony Assay

The colony assay was performed as previously described with minor modifications [26]. In brief, cells were grown on glass cover slips to confluence using a 24 well plate. The cell layer was soaked in lysis buffer (150 mM NaCl, 0.5% Triton X-100, 0.5% DOC, 50 mM Tris pH 7.5) on a nitrocellulose membrane. After drying for 30 min at room temperature, the membrane was incubated in lysis buffer containing 5 μg/ml proteinase K (PK) for 90 min at 37°C, rinsed twice with water, and incubated in 2 mM PMSF for 10 min. The membrane was shaken in 3 M guanidinium thiocyanate, 10 mM Tris-HCl (pH 8.0) for 10 min, followed by rinsing five times with water. Five percent nonfat dry milk in TBS-T was used for blocking for 1 hr at room temperature. PrP was detected using an anti-PrP antibody 6H4 (Prionics) and a HRP-conjugated sheep anti-mouse antibody (GE Healthcare).

11.3.3 SDS-PAGE and Western Blot

Proteins were separated by 12% SDS-PAGE and transferred to polyvinylidene difluoride membranes (PVDF, Millipore) by semidry blotting. PrP was detected using the PrP-specific mouse mAb 8H4 (Alicon AG). For validation of phosphorylated proteins anti-phospho-stathmin (Ser38) (#3426, Cell Signaling Technology), anti-phospho-cdc2 (Tyr15) (#4539, Cell Signaling Technology), and anti-phospho-cofilin (Ser3) antibodies (#3313, Cell Signaling Technology) were used. Antibodies recognizing stathmin (#3352), cdc2 (#9112) and cofilin (#3312) were also obtained from Cell Signaling Technology.

11.3.4 Two-dimensional Gel Electrophoresis

For 2D electrophoresis 150 μg protein of cell lysates were purified by trichloroacetic acid precipitation and re-suspended in DeStreak Rehydration Solution (Amersham Biosciences) containing 0.5% Bio-Lyte pH3-10 (Bio-Rad Laboratories GmbH, München). The isoelectric focusing was run on IPG strips with a non-linear pH range of 3–10 and a length of 7 cm (Bio-Rad) using the ZOOM® IPGRunner™ system from Invitrogen. After focussing strips were equilibrated in 50 mM Tris, 1 mM Urea, 30% Glycerin, 2% SDS, 1% DTT for 25 min and in 50 mM Tris, 1 mM Urea, 30% Glycerin, 2% SDS, 5% Iodacetamid for 25 min. Strips were then separated in 10% SDS-PAGE gels in the second dimension and analyzed by Coomassie staining or immunoblotting using an anti-phospho-tyrosine (sc-7020, Santa Cruz) or an anti-phospho-threonine antibody (#9381, Cell Signaling Technology).

11.3.5 SILAC Phosphoproteomics Analysis

The SILAC ready-to-use cell culture media and dialyzed FBS were obtained from Dundee Cell Products Ltd, UK. While N2a58# cells were cultured in control SILAC DMEM media containing unlabelled arginine and lysine amino acids (R0K0), N2a58/22L cells were cultured in ready-to-use SILAC DMEM medium containing ^{13}C labeled arginine and lysine amino acids (R6K6) for seven cell division cycles. After preparation of cell lysates and measurement of protein concentration, lysates of N2a58# and N2a58/22L cells were mixed in a ratio 1:1. Each sample was reduced in SDS PAGE loading buffer containing 10 mM DTT and alkylated in 50 mM iodoacetamide prior to separation by one-dimensional SDS-PAGE (4–12% Bis-Tris Novex mini-gel, Invitrogen) and visualization by colloidal Coomassie staining (Novex, Invitrogen). The entire protein gel lane was excised and cut into 10 gel slices each. Every gel slice was subjected to in-gel digestion with trypsin [27]. The resulting tryptic peptides were extracted by 1% formic acid, acetonitrile, lyophilized in a speedvac (Helena Biosciences).

11.3.6 Phosphopeptide Enrichment

The lyophilized peptides above were resuspended in 5% acetic acid (binding buffer) and phosphopeptide enrichment was carried out using immobilized metal ion affinity chromatography (IMAC). Immobilized gallium in the Pierce Ga-IDA Phosphopeptide Enrichment Kit was used to enrich for phosphopeptides prior to MS/MS analysis according to the manufacturer's instructions (Thermo Scientific).

11.3.7 LC-MS/MS

Trypsin digested peptides were separated using an Ultimate U3000 (Dionex Corporation) nanoflow LC-system consisting of a solvent degasser, micro and nanoflow pumps, flow control module, UV detector and a thermostated autosampler. The 10 µl of sample (a total of 2 µg) was loaded with a constant flow of 20 µl/min onto a PepMap C18 trap column (0.3 mm id × 5 mm, Dionex Corporation). After trap enrichment peptides were eluted off onto a PepMap C18 nano column (75 µm × 15 cm, Dionex Corporation) with a linear gradient of 5–35% solvent B (90% acetonitrile with 0.1% formic acid) over 65 min with a constant flow of 300 nl/min. The HPLC system was coupled to a LTQ Orbitrap XL (Thermo Fisher Scientific Inc) via a nano ES ion source (Proxeon Biosystems). The spray voltage was set to 1.2 kV and the temperature of the heated capillary was set to 200°C. Full scan MS survey spectra (m/z 335–1,800) in profile mode were acquired in the Orbitrap with a resolution of 60,000 after accumulation of 500,000 ions. The five most intense peptide ions from the preview scan in the Orbitrap were fragmented by collision-induced dissociation (normalised collision energy 35%, activation Q 0.250 and activation time 30 m) in the LTQ after the accumulation of 10,000 ions. Maximal filling times were 1,000 ms for the full scans and 150 ms for the MS/MS scans. Precursor ion charge state screening was enabled and all unassigned charge states as well as singly charged species were rejected. The dynamic exclusion list was restricted to a maximum of 500 entries with a maximum retention period of 90 s and a relative mass window of 10 ppm. The lock mass option was enabled for survey scans to improve mass accuracy [28]. The data were acquired using the Xcalibur software.

11.3.8 Quantification and Bioinformatic Analysis

The quantification was performed with MaxQuant version 1.0.7.4 [29], and was based on two-dimensional centroid of the isotope clusters within each SILAC pair. To minimize the effect of outliers, protein ratios were calculated as the median of all SILAC pair ratios that belonged to peptides contained in the protein. The percentage variability of the quantitation was defined as the standard deviation of the natural logarithm of all ratios used for obtaining the protein ratio multiplied by a constant factor 100.

The generation of peak list, SILAC- and extracted ion current-based quantitation, calculated posterior error probability, and false discovery rate based on search engine results, peptide to protein group assembly, and data filtration and presentation was carried out using MaxQuant. The derived peak list was searched with the Mascot search engine (version 2.1.04; Matrix Science, London, UK) against a concatenated database combining 80,412 proteins from International Protein Index (IPI) human protein database version 3.6 (forward database), and the reversed sequences of all proteins (reverse database). Alternatively, database searches were done using Mascot (Matrix Science) as the database search engine and the results saved as a peptide summary before quantification using MSQuant http://msquant.sourceforge.net/website. Parameters allowed included up to three missed cleavages and three labeled amino acids (arginine and lysine). Initial mass deviation of precursor ion and fragment ions were up to 7 ppm and 0.5 Da, respectively. The minimum required peptide length was set to six amino acids. To pass statistical evaluation, posterior error probability (PEP) for peptide identification (MS/MS spectra) should be below or equal to 0.1. The required false positive rate (FPR) was set to 5% at the peptide level. False positive rates or PEP for peptides were calculated by recording the Mascot score and peptide sequence length-dependent histograms of forward and reverse hits separately and then using Bayes' theorem in deriving the probability of a false identification for a given top scoring peptide. At the protein level, the false discovery rate (FDR) was calculated as the product of the PEP of a protein's peptides where only peptides with distinct sequences were taken into account. If a group of identified peptide sequences belong to multiple proteins and these proteins cannot be distinguished, with any unique peptide reported, these proteins are reported as a protein group in MaxQuant. Proteins were quantified if at least one MaxQuant-quantifiable SILAC pair was present. Identification was set to a false discovery rate of 1% with a minimum of two quantifiable peptides. The set value for FPR/PEP at the peptide level ensures that the worst identified peptide has a probability of 0.05 of being false; and proteins are sorted by the product of the false positive rates of their peptides where only peptides with distinct sequences are recognized. During the search, proteins are successively included starting with the best-identified ones until a false discovery rate of 1% is reached; an estimation based on the fraction of reverse protein hits.

Enzyme specificity was set to trypsin allowing for cleavage of N-terminal to proline and between aspartic acid and proline. The carbamidomethylation of cysteine was searched as a fixed modification, whereas *N*-acetyl protein, oxidation of methionine and phosphorylation of serine, threonine and tyrosine were searched as variable modifications.

COMPETING INTERESTS

The authors declare that they have no competing interests.

AUTHORS' CONTRIBUTIONS

WW carried out the experimental work, drafted and wrote the manuscript. PA performed and interpreted the SILAC analysis. JL participated in the design of the study. SW conceived of the study, and participated in its design and coordination and wrote the manuscript. All authors read and approved the final manuscript.

ACKNOWLEDGMENTS

We thank Prof. Schätzl from the LMU in Munich for providing N2a58/22L cells.

KEYWORDS

- **Cellular prion protein**
- **Colony assay**
- **Creutzfeld Jacob disease**
- **False discovery rate**
- **Pentosan polysulfate**
- **Prion diseases**
- **Stathmin phosphopeptides**
- **Transmissible spongiform encephalopathies**

REFERENCES

1. Prusiner, S. B. Prions. *Proc Natl Acad Sci USA* **95,** 13363–13383 (1998).
2. Prusiner S. B. Novel proteinaceous infectious particles cause scrapie. *Science* **216,** 136–144 (1982).
3. Aguzzi, A. and Polymenidou M. Mammalian prion biology: one century of evolving concepts. *Cell* **116,** 313–327 (2004).
4. Taylor, D. R. and Hooper N. M. The prion protein and lipid rafts. *Mol Membr Biol* **23,** 89–99 (2006).
5. Mattei, V., Garofalo T., Misasi R., Circella A., Manganelli V., Lucania G., Pavan A., and Sorice M. Prion protein is a component of the multimolecular signaling complex involved in T cell activation. *FEBS Lett* **560,** 14–18 (2004).
6. Hugel, B., Martinez, M. C., Kunzelmann, C., Blattler, T., Aguzzi, A., and Freyssinet, J. M. Modulation of signal transduction through the cellular prion protein is linked to its incorporation in lipid rafts. *Cell Mol Life Sci* **61,** 2998–3007 (2004).
7. Mouillet-Richard, S., Ermonval, M., Chebassier, C., Laplanche, J. L., Lehmann, S., Launay, J. M., and Kellermann, O.. Signal transduction through prion protein. *Science* **289,** 1925–1928 (2000).
8. Gyllberg, H., Lofgren K., Lindegren H., and Bedecs K. Increased Src kinase level results in increased protein tyrosine phosphorylation in scrapie-infected neuronal cell lines. *FEBS Lett* **580,** 2603–2608 (2006).

9. Na Y. J., Jin J. K., Kim J. I., Choi E. K., Carp R. I., and Kim Y. S. JAK-STAT signaling pathway mediates astrogliosis in brains of scrapie-infected mice. *J Neurochem* **103,** 637–649 (2007).
10. Ertmer, A., Gilch S., Yun S. W., Flechsig E., Klebl B., Stein-Gerlach M., Klein M. A., and Schatzl H. M. The tyrosine kinase inhibitor STI571 induces cellular clearance of PrP^{Sc} in prion-infected cells. *J Biol Chem* **279,** 41918–41927 (2004).
11. Nishida, N., Harris D. A., Vilette D., Laude H., Frobert Y., Grassi J., Casanova D., Milhavet O., and Lehmann S. Successful transmission of three mouse-adapted scrapie strains to murine neuroblastoma cell lines overexpressing wild-type mouse prion protein. *J Virol* **74,** 320–325 (2000).
12. Kocisko, D. A., Engel A. L., Harbuck K., Arnold K. M., Olsen E. A., Raymond L. D., Vilette D., and Caughey B. Comparison of protease-resistant prion protein inhibitors in cell cultures infected with two strains of mouse and sheep scrapie. *Neurosci Lett* **388,** 106–111 (2005).
13. Kascsak, R. J., Rubenstein R., Merz P. A., Carp R. I., Wisniewski H. M., and Diringer H. Biochemical differences among scrapie-associated fibrils support the biological diversity of scrapie agents. *J Gen Virol* **66**(Pt 8), 1715–1722 (1985).
14. Collinge, J., Sidle K. C., Meads J., Ironside J., and Hill A. F. Molecular analysis of prion strain variation and the aetiology of 'new variant' CJD. *Nature* **383,** 685–690 (1996).
15. Somerville, R. A., Chong A., Mulqueen O. U., Birkett C. R., Wood S. C., and Hope J. Biochemical typing of scrapie strains. *Nature* **386,** 564 (1997).
16. Yang, N., Higuchi O., Ohashi K., Nagata K., Wada A., Kangawa K., Nishida E., and Mizuno K. Cofilin phosphorylation by LIM-kinase 1 and its role in Rac-mediated actin reorganization. *Nature* **393,** 809–812 (1998).
17. Giorgi, A., Di Francesco L., Principe S., Mignogna G., Sennels L., Mancone C., Alonzi T., Sbriccoli M., De Pascalis A., Rappsilber J., et al. Proteomic profiling of PrP27-30-enriched preparations extracted from the brain of hamsters with experimental scrapie. *Proteomics* **9,** 3802–3814 (2009).
18. Beretta, L., Dobransky T., and Sobel A. Multiple phosphorylation of stathmin. Identification of four sites phosphorylated in intact cells and in vitro by cyclic AMP-dependent protein kinase and p34cdc2. *J Biol Chem* **268,** 20076–20084 (1993).
19. Satyanarayana, A. and Kaldis P. Mammalian cell-cycle regulation: several Cdks, numerous cyclins and diverse compensatory mechanisms. *Oncogene* **28,** 2925–2939 (2009).
20. Brattsand, G., Marklund U., Nylander K., Roos G., and Gullberg M. Cell-cycle-regulated phosphorylation of oncoprotein 18 on Ser16, Ser25 and Ser38. *Eur J Biochem* **220,** 359–368 (1994).
21. Marklund, U., Osterman O., Melander H., Bergh A., and Gullberg M. The phenotype of a "Cdc2 kinase target site-deficient" mutant of oncoprotein 18 reveals a role of this protein in cell cycle control. *J Biol Chem* **269,** 30626–30635 (1994).
22. Gougoumas, D. D., Vizirianakis I. S., and Tsiftsoglou A. S. Transcriptional activation of prion protein gene in growth-arrested and differentiated mouse erythroleukemia and human neoplastic cells. *Exp Cell Res* **264,** 408–417 (2001).
23. Zhang, G. and Neubert T. A. Use of stable isotope labeling by amino acids in cell culture (SILAC) for phosphotyrosine protein identification and quantitation. *Methods Mol Biol* **527,** 79–92 (2009).
24. Amanchy, R., Kalume D. E., Iwahori A., Zhong J., and Pandey A. Phosphoproteome analysis of HeLa cells using stable isotope labeling with amino acids in cell culture (SILAC). *J Proteome Res* **4,** 1661–1671 (2005).
25. Kruger, M., Kratchmarova I., Blagoev B., Tseng Y. H., Kahn C. R., and Mann M. Dissection of the insulin signaling pathway via quantitative phosphoproteomics. *Proc Natl Acad Sci USA* **105,** 2451–2456 (2008).
26. Klohn, P. C., Stoltze L., Flechsig E., Enari M., and Weissmann C. A quantitative, highly sensitive cell-based infectivity assay for mouse scrapie prions. *Proc Natl Acad Sci USA* **100,** 11666–11671 (2003). (PubMed Central Full Text)
27. Shevchenko, A., Wilm M., Vorm O., and Mann M. Mass spectrometric sequencing of proteins silver-stained polyacrylamide gels. *Anal Chem* **68,** 850–858 (1996).

28. Olsen, J. V., de Godoy L. M., Li G., Macek B., Mortensen P., Pesch R., Makarov A., Lange O., Horning S., and Mann M. Parts per million mass accuracy on an Orbitrap mass spectrometer via lock mass injection into a C-trap. *Mol Cell Proteomics* **4,** 2010–2021 (2005).
29. Cox, J. and Mann M. MaxQuant enables high peptide identification rates, individualized p.p.b.-range mass accuracies and proteome-wide protein quantification. *Nat Biotechnol* **26,** 1367–1372 (2008).

Authors' Notes

CHAPTER 1

Authors' Contributions

VMG designed and performed experiments, data analysis and wrote the article. GWZ supervised the research project and edited the manuscript. The authors read and approved the final manuscript.

Acknowledgements

This work was supported by an operating grant to GWZ from the PrioNet Canada. GWZ is a Scientist of the Alberta Heritage Foundation for Medical Research (AHFMR) and a Canada Research Chair in Molecular Neurobiology. VMG is supported by an AHFMR Fellowship and by a Fellowship from the Hotchkiss Brain Institute (HBI). We thank Dr. Clint Doering for genotyping and breeding paradigms, Dr. Frank R. Jirik for providing the outbred PrP null mouse line, and Stephan Bonfield for the blind analysis for experiment 1. *Prion Protein's Protection Against Pain was originally published as "Cellular Prion Protein Protects from Inflammatory and Neuropathic Pain" in Molecular Pain 2011, 7:59.* Used with permission

CHAPTER 2

Authors' Contributions

SC, RY, and GT performed the RNA purifications and labeling and microarray hybridizations, RY, SLG, and GT performed the QPCR experiments. MV, BP, and CP bred and obtained the transgenic mice and BP.; GT and RY collected the tissue samples used. FB, LST, SB, MLMM and JPR supervised the microarray experiment and performed statistical analysis of the results. VB, FLP, HL, and JLV designed and supervised the overall experiment and prepared the manuscript. All authors read and approved the final manuscript.

Acknowledgments

We are most grateful to John Collinge and colleagues for providing the Tg37 and NFH-Cre transgenic mice and for agreeing to include the data obtained with them in this article and to S. Prusiner for providing the FVB/N *Prnp*$^{-/-}$ mice. RY is a post-doctorant supported by the French Ministry of Research and the ANR-09-BLAN-OO15-01. *Prion Protein-Encoding Genes was originally published as "Brain Transcriptional Stability upon Prion Protein-Encoding Gene Invalidation in Zygotic or Adult Mouse" in BMC Genomics 2010, 11:448.* Used with permission.

CHAPTER 3

A provisional patent application was submitted by the institution to which authors are affiliated.

Acknowldegments

This work was supported by by NIH grant NS045585 to IVB.; Baltimore Research and Education Foundation and the Beatriu de Pinos Fellowship to NGM with the support of the Commission for Universities and Research of the Department of Innovation, Universities and Enterprise of the Government of Catalonia. The funders had no role in study design, data collection and analysis, decision to publish, or preparation of the manuscript. *Protein Misfolding Cyclic Amplification was originally published as "Highly Efficient Protein Misfolding Cyclic Amplification" in PLoS Pathology 2011, 7(2):e1001277.* Used with permission.

Ethics Statement

This study was carried out in strict accordance with the recommendations in the Guide for the Care and Use of Laboratory Animals of the National Institutes of Health. The protocol was approved by the Institutional Animal Care and Use Committee of the University of Maryland, Baltimore (Assurance Number A32000-01, Permit Number. 0309001).

CHAPTER 4

The authors thank Dr. Quingzhong Kong for providing PrP-knockout mice. This work was supported in part by National Institutes of Health grants NS44158 and AG14359. *The Amplification of the Scrapie Isoform of Prion Protein was originally published as "The Role of Glycophospatidylinositol Anchor in the Amplification of the Scrapie Isoform of Prion Protein in Vitro" in FEBS Letters 2009, 19:583(22).* Used with permission.

CHAPTER 5

Acknowledgements

We thank Dazhi Tang for mass spectrometry analyses, Drs. Anthony Lau and Cleo Salisbury for helpful discussions and comments on the manuscript, Grace Ching for her assistance, and Drs. Philip Minor (National Institute for Biological Standards and Control, Potters Bar, United Kingdom) and Adriano Aguzzi (Institute of Pathology at the University Hospital of Zürich) for generously providing samples and for critical reading of the manuscript. *The Octarepeat Region of the Prion Protein was originally published as "The Octarepeat Region of the Prion Protein Is Conformationally Altered in PrP^{Sc}" in PLoS One 2010, 5(2):e9316.* Used with permission.

Competing Interests

The authors of this work were or are employed by Novartis Vaccines & Diagnostics, Inc. This work is also part of a pending patent application, for which there is no direct financial benefit to the authors. This does not alter the authors adherence to all the PLoS ONE policies on sharing data and materials.

Funding

This work was funded by Novartis Vaccines & Diagnostics, Inc. The authors are or were employed by the funder and, as such, the funders played a role in the study design, data collection and analysis, decision to publish, and preparation of the manuscript.

CHAPTER 6

Authors' Contributions

AR and AB conceived the study and together with JPML were responsible for study design and coordination. All PMCA assays were performed by JP and all peptide-arrays were performed at Pepscan presto B.V. by DPT. Supporting experiments and data-analysis were performed by AR, AR, JPML, and AB were responsible for data interpretation. The experiments for the study were facilitated by grants under the supervision of AB, JPML, and FGvZ. AR drafted the manuscript and AB and JPML critically read the manuscript before submission.

Acknowledgements

We thank Drs. Jaques Grassi (SPI.; CEA) and Martin Groschup (FLI, Reims, GE) for generously supplying the monoclonal antibodies Sha31 and L42. This work was supported by grant 903-51-177 from the Dutch Organization for Scientific Research (NWO), by a grant from the Dutch Ministry of Agriculture, Nature Management and Fisheries (LNV) and by EU NeuroPrion project STOPPRIONs FOOD-CT-2004-506579. *Prion Protein Self-Peptides was originally published as "Prion Protein Self-Peptides Modulate Prion Interactions and Conversion" in BMC Biochemistry 2009, 10:29.* Used with permission.

CHAPTER 7

Competing Interests

The authors declare that they have no competing interests.

Authors' Contributions

WMW carried out the PET blot studies, participated in immunohistochemistry, Western blot, tissue acquisition and the design of the study and drafted the manuscript. SLB participated in immunohistochemistry, Western blot, tissue acquisition and design of the study and co-edited the manuscript. AW, WEW, BeB and BjB participated in tissue acquisition and diagnosing the cases. WJSS conceived the study, participated in its design and coordination and co-edited the manuscript. All authors read and approved the final manuscript.

Acknowledgements

We would like to thank Tatjana Pfander, Nadine Rupprecht and Kerstin Brekerbohm for their skilful technical assistance. The work was supported by the VolkswagenStiftung (grants ZN 1294 and ZN 2168 to W.J.S.S). *PrP^{Sc} Spreading Patterns and Prion Types was originally published as "PrP^{Sc} Spreading Patterns in the Brain of Sheep Linked to Different Prion Types" in Veterinarian Research 2011, 42:32.* Used with permission.

CHAPTER 8

Competing Interests

The authors declare that they have no competing interests.

Authors' Contributions

Conception and design of experiments. TY and HO. Conduction of experiments. HO.; YI, MI, KM, and YM. Intracerebral inoculation of H-type BSE isolate and collection of samples from H-type BSE-infected cattle. YI, HO, MI, KM, YS, and KK. Manuscript draft preparation and data analysis. HO, YI, MI, and YM. Participation in scientific discussion of the results. SC. Study supervision. SM. All the authors have read and approved the final manuscript.

Acknowledgements

We thank Dr Yuichi Tagawa (National Institute of Animal Health) for providing the anti-prion mAb T1 and Dr Motoshi Horiuchi (Graduate School of Veterinary Medicine, Hokkaido University) for providing mAbs 44B1 and 43C5. Expert technical assistance was provided by Mutsumi Sakurai, Miyo Kakizaki, Junko Endo, Noriko Amagai, Tomoko Murata, Naomi Furuya, Naoko Tabeta, Nobuko Kato, and the animal caretakers. This work was supported by grants from the BSE and other Prion Disease Control Projects of the Ministry of Agriculture, Forestry and Fisheries of Japan and from the Canadian Food Inspection Agency. *Experimental H-Type Bovine Spongiform Encephalopathy was originally published as "Experimental H-Type Bovine Spongiform Encephalopathy Characterized by Plaques and Glial- and Stellate-Type Prion Protein Deposits" in Veterinarian Research 2011, 42:79.* Used with permission

CHAPTER 9

Consent

Written informed consent was obtained from the next of kin of the patient for publication of this case report. A copy of the written consent form is available for review from the editor-in-chief of this journal.

Competing Interests

The authors declare that they have no competing interests.

Authors' Contributions

ABRM carried out the molecular analyses, and drafted the manuscript. ABRM.; JMG and RAJ designed the molecular analyses. JJZ helped to draft the neurological and neuropathological sections of the manuscript. JMA provided neurological examination data and helped to draft the manuscript. MMP carried out sequencing analysis. BAP and IFA performed neuropathological studies and IFA drafted the neuropathological section. MJB diagnosed the patient. RAJ coordinated all the information and completed the writing of the final manuscript. All authors read and approved the final manuscript.

Acknowledgements

This work received a Support to Health Research grant from the Planning and Arranging Director of the Department of Health of the Basque Government (grant #2006111037). We thank the DNA Bank of the University of the Basque Country for technical support. We also thank Dr. Natalia Elguezabal and Dr. Ana Hurtado for English revision. *A Novel Form of Human Disease was originally published as "A Novel Form of Human Disease with a Protease-Sensitive Prion Protein and Heterozygosity Methionine/Valine at Codon 129: Case Report" in BMC Neurology 2010, 10:99.* Used with permission

CHAPTER 10

Acknowledgements

We would like to thank our collaborators for providing us brain tissue of prion infected as well as non-infected animals of different species. Stanley B. Prusiner (UCSF, San Francisco, USA) provided hamster brain, Martin Groschup (Friedrich-Loeffler-Institut, Institute for Novel and Emerging Infectious Diseases, Greifswald/Insel Riems, Germany) provided cattle brain, Carsten Korth (University of Duesseldorf Medical School, Department Neuropathology, Germany) provided mouse brain, Olivier Andréoletti (UMR INRA-ENVT.; Physiopathologie Infectieuse et Parasitaire des Ruminants, Ecole Nationale Vétérinaire, Toulouse, France) provided sheep brain. We are thankful for the guidance and help in cloning and expression of recombinant PrP by Tommy Agyenim and Bernd Esters. *Prions as Seeds and Recombinant Prion Proteins as Substrates was originally published as "Molecular Interactions between Prions as Seeds and Recombinant Prion Proteins as Substrates Resemble the Biological Interspecies Barrier in Vitro" in PLoS One 2010, 5(12): e14283.* Used with permission.

Author Contributions

Conceived and designed the experiments: GP DR EB. Performed the experiments: GP LL JS LNS. Analyzed the data: GP LNS DR DW EB. Contributed reagents/materials/ analysis tools: JW. Wrote the paper: GP DR DW EB.

CHAPTER 11

Competing Interests

The authors declare that they have no competing interests.

Authors' Contributions

WW carried out the experimental work, drafted and wrote the manuscript. PA performed and interpreted the SILAC analysis. JL participated in the design of the study. SW conceived of the study, and participated in its design and coordination and wrote the manuscript. All authors read and approved the final manuscript.

Acknowledgements

We thank Prof. Schätzl from the LMU in Munich for providing N2a58/22L cells. *Quantitative Phosphoproteomic Analysis of Prion-Infected Neuronal Cells was originally published as "Quantitative Phosphoproteomic Analysis of Prion-Infected Neuronal Cells" in Cell Communication and Signaling 2010, 8:28.* Used with permission.

Index